AF839323

# Environmental History

Volume 13

**Series Editor**

Mauro Agnoletti, Florence, Italy

More information about this series at https://link.springer.com/bookseries/10168

Józef Hernik · Maria Walczycka ·
Edward Sankowski · Betty J. Harris
Editors

# Cultural Heritage—Possibilities for Land-Centered Societal Development

Springer

*Editors*
Józef Hernik
Department of Land Management
and Landscape Architecture
University of Agriculture in Krakow
Krakow, Poland

Edward Sankowski
Department of Philosophy
University of Oklahoma
Norman, OK, USA

Maria Walczycka
Department of Animal Product Processing
University of Agriculture in Krakow
Krakow, Poland

Betty J. Harris
Department of Anthropology
University of Oklahoma
Norman, OK, USA

Polish National Agency for Academic Exchange
This book has been prepared as part of a project (no. PPI/APM/2018/1/00010/U/001), "Cultural heritage of small homelands" funded by the Polish National Agency for Academic Exchange, as part of the program for International Academic Partnerships.

ISSN 2211-9019 ISSN 2211-9027 (electronic)
Environmental History
ISBN 978-3-030-58091-9 ISBN 978-3-030-58092-6 (eBook)
https://doi.org/10.1007/978-3-030-58092-6

© Springer Nature Switzerland AG 2022
This work is subject to copyright. All rights are reserved by the Publisher, whether the whole or part of the material is concerned, specifically the rights of translation, reprinting, reuse of illustrations, recitation, broadcasting, reproduction on microfilms or in any other physical way, and transmission or information storage and retrieval, electronic adaptation, computer software, or by similar or dissimilar methodology now known or hereafter developed.
The use of general descriptive names, registered names, trademarks, service marks, etc. in this publication does not imply, even in the absence of a specific statement, that such names are exempt from the relevant protective laws and regulations and therefore free for general use.
The publisher, the authors and the editors are safe to assume that the advice and information in this book are believed to be true and accurate at the date of publication. Neither the publisher nor the authors or the editors give a warranty, expressed or implied, with respect to the material contained herein or for any errors or omissions that may have been made. The publisher remains neutral with regard to jurisdictional claims in published maps and institutional affiliations.

This Springer imprint is published by the registered company Springer Nature Switzerland AG
The registered company address is: Gewerbestrasse 11, 6330 Cham, Switzerland

# Preface

In the transition from a more domestically agriculturally focused to a less domestically agriculturally focused society, in many places around the world, the preservation of more local, agricultural, cultural heritage becomes more of an issue, and the development of a desirable balance between (or integration of) two types of land use (agricultural/rural, and what we can call "multifarious-productive"/urban) becomes more of a question for social-psychological adjustment and social policy. Both types of land use are conceptually complex. Agricultural land use, as examined in our book, targets production mainly at the satisfaction of human needs or desires to eat and drink. Multifarious-productive land use, as the name suggests, concerns satisfaction of many types of needs or desires, e.g., manufacturing physical non-edible, non-drinkable objects (autos, computers, industrial materials such as steel, constructing much urban architecture, etc.) for use; providing human services beyond offering food or drink; generating intangible goods beyond the gustatory, etc. Even the language of "consumption" (in English) tries to express a hoped for re-connection of the two types of social orders with their respective dominant patterns of land use. One can speculate that diminished direct awareness of agricultural processes and associated land use can be only at best partially satisfactorily replaced psychologically by more indirect or episodic experiences (agro-tourism, increasing emphasis on manufactured, even very skillful enhancements of food and drinkable liquids, etc.). Still, such sublimations, including well-communicated education about features of culture's agricultural origins and history, urban farms or gardens, etc., may encourage more people's appreciation of land as a source of basic sustenance. Without the sublimations and with the diminishment of the societal meaning of rural lives, as some commentators have suggested, as well as notable rural poverty and less drastic rural economic challenges around the world, there is a threat that the arrival of a near totally dominant multifarious-productive/urban existence will dilute or swamp a vivid experience of pre-urbanized land as a source of bodily sustenance, an experience arguably necessary to adequately motivate widespread appreciation and care for much of the current environment and the society's future.

A societal effort to connect its agricultural land-use past and present (one central aspect of cultural heritage) with a continuing viable future, when it focuses especially on local and sub-national regional land-centered phenomena, is also a goal-directed effort to resist and redirect trends toward cultural homogeneity at the national, multinational, and global levels. Such homogenization and uniformity are threatened by some features of globalization and national government and large-scale corporate powers. Protecting, preserving, and furthering uniqueness and variety in local and sub-national regional cultural heritage includes the agricultural/rural aspects of cultural heritage but extends to broader issues about land-use planning for a future in which people appreciate varieties of cultural heritage ("their own" and that of "others"). By doing this, a society arguably addresses current-generation and future-generation needs and distributive-justice concerns, thus addressing sustainable development in important parts by an emphasis on certain land-use practices. Resistance to excessive uniformity in cultural heritage helps improve versions of sustainable development that might otherwise unfortunately interpret globalization as necessarily in service to cultural heritage uniformity.

These remarks may help explain why our book is a type of research that aims to further variety in cultural heritage while also acknowledging the potential value in national sovereignty and positive aspects of multi-national and global co-operation. Local uniqueness may sometimes positively interact with national and global opportunities. This has political, economic, and cultural dimensions. In our book, we begin with basics about not only more local food and drink but also research examples of further issues about urbanization, distribution of land, nation-states, global trends, and so on.

Norman, USA — Prof. Edward Sankowski
Kraków, Poland — Prof. Józef Hernik

# Contents

# Contributors

**Mauro Agnoletti** Laboratory for Landscape and Cultural Heritage, Department of Agriculture, Food, Environment and Forestry (DAGRI), University of Florence, Florence, Italy

**Lucia Benešová** Faculty of Biotechnology and Food Sciences, Department of Food Hygiene and Safety, Slovak University of Agriculture, Nitra, Slovakia

**Emilia Bernaś** Department of Plant Products and Nutrition Hygiene, Faculty of Food Technology, University of Agriculture in Krakow, Krakow, Poland

**Wiktor Berski** Department of Carbohydrate Technology and Cereal Processing, Faculty of Food Technology, University of Agriculture in Krakow, Krakow, Poland

**Katarzyna Cegielska** Department of Land Management and Landscape Architecture, Faculty of Environmental Engineering and Land Surveying, University of Agriculture in Krakow, Krakow, Poland

**Gaiying Chen** Department of Landscape Architecture, Beijing University of Agriculture, Beijing, China

**Bohdan Cherkes** Department of Land Management and Landscape Architecture, Faculty of Environmental Engineering and Land Surveying, University of Agriculture in Krakow, Krakow, Poland;
Institute for Architecture and Design, Lviv Polytechnic National University, Lviv, Ukraine

**Barbara Czesak** Department of Land Management and Landscape Architecture, Faculty of Environmental Engineering and Land Surveying, University of Agriculture in Krakow, Krakow, Poland

**Jacek Domagała** Department of Animal Product Processing, Faculty of Food Technology, University of Agriculture in Krakow, Krakow, Poland

**Zuzana Drdolová** Faculty of Biotechnology and Food Sciences, Department of Food Hygiene and Safety, Slovak University of Agriculture, Nitra, Slovakia

**Martina Fikselová** Faculty of Biotechnology and Food Sciences Department of Food Hygiene and Safety, Slovak University of Agriculture, Nitra, Slovakia

**Agnieszka Filipiak-Florkiewicz** Department of Plant Products and Nutrition Hygiene, Faculty of Food Technology, University of Agriculture in Krakow, Krakow, Poland

**Adam Florkiewicz** Department of Food Analysis and Quality Assessment, Faculty of Food Technology, University of Agriculture in Krakow, Krakow, Poland

**Ján Gažo** Faculty of Agrobiology and Food Resources, Department of Genetics and Plant Breeding, Slovak University of Agriculture, Nitra, Slovakia

**Piotr Gębczyński** Department of Plant Products and Nutrition Hygiene, Faculty of Food Technology, University of Agriculture in Krakow, Krakow, Poland

**Jozef Golian** Faculty of Biotechnology and Food Sciences, Department of Food Hygiene and Safety, Slovak University of Agriculture, Nitra, Slovakia

**Betty J. Harris** Department of Anthropology, College of Arts and Sciences, University of Oklahoma, Norman, OK, USA

**Józef Hernik** Department of Land Management and Landscape Architecture, Faculty of Environmental Engineering and Land Surveying, University of Agriculture in Krakow, Krakow, Poland

**Lucia Hodulová** Faculty of Veterinary Hygiene and Ecology, Department of Plant Origin Food Science, University of Veterinary Sciences Brno, Brno, Czech Republic

**Josef Kameník** Faculty of Veterinary Hygiene and Ecology, Department of Animal Origin Food and Gastronomic Sciences, University of Veterinary Sciences Brno, Brno, Czech Republic;
Faculty of Veterinary Hygiene and Ecology, Department of Gastronomy, University of Veterinary Sciences, Brno, Czech Republic

**Radovan Kasarda** Faculty of Agrobiology and Food Resources, Department of Animal Genetics and Breeding Biology, Slovak University, Nitra, Slovakia

**Martin Král** Faculty of Veterinary Hygiene and Ecology, Department of Plant Origin Food Science, University of Veterinary Sciences Brno, Brno, Czech Republic

**Karol Król** Department of Land Management and Landscape Architecture, Faculty of Environmental Engineering and Land Surveying, University of Agriculture in Krakow, Krakow, Poland

**Anita Kukulska-Kozieł** Department of Land Management and Landscape Architecture, Faculty of Environmental Engineering and Land Surveying, University of Agriculture in Krakow, Krakow, Poland

**Marcin Łukasiewicz** Department of Engineering and Machinery for Food Industry, Faculty of Food Technology, University of Agriculture in Krakow, Krakow, Poland

**Slavomír Marcinčák** Department of Food Hygiene, Technology and Safety, University of Veterinary Medicine and Pharmacy in Košice, Košice, Slovakia

**Andrea Mendelová** Faculty of Biotechnology and Food Sciences Department of Technology and Quality of Plant Products, Slovak University of Agriculture, Nitra, Slovakia

**Wladysław Migdał** Department of Animal Product Processing, Faculty of Food Technology, University of Agriculture in Krakow, Krakow, Poland

**Łukasz Migdał** Department of Animal Genetics, Breeding and Ethology, University of Agriculture in Krakow, Krakow, Poland

**Nina Moravčíková** Faculty of Agrobiology and Food Resources, Department of Animal Genetics and Breeding Biology, Slovak University, Nitra, Slovakia

**Dorota Najgebauer-Lejko** Department of Animal Product Processing, Faculty of Food Technology, University of Agriculture in Krakow, Krakow, Poland

**Maria Nawieśniak-Caesar** Department of Land Management and Landscape Architecture, Faculty of Environmental Engineering and Land Surveying, University of Agriculture in Krakow, Krakow, Poland

**Tomasz Noszczyk** Department of Land Management and Landscape Architecture, Faculty of Environmental Engineering and Land Surveying, University of Agriculture in Krakow, Krakow, Poland

**Barbara Olczak** Department of Land Management and Landscape Architecture, Faculty of Environmental Engineering and Land Surveying, University of Agriculture in Krakow, Krakow, Poland

**Maria Pazdan** Department of Land Management and Landscape Architecture, Faculty of Environmental Engineering and Land Surveying, University of Agriculture in Krakow, Krakow, Poland

**Matej Pospiech** Faculty of Veterinary Hygiene and Ecology, Department of Plant Origin Food Science, University of Veterinary Sciences Brno, Brno, Czech Republic

**Barbara Prus** Department of Land Management and Landscape Architecture, Faculty of Environmental Engineering and Land Surveying, University of Agriculture in Krakow, Krakow, Poland

**Renata Różycka-Czas** Department of Land Management and Landscape Architecture, Faculty of Environmental Engineering and Land Surveying, University of Agriculture in Krakow, Krakow, Poland

**Edward Sankowski** Department of Philosophy, College of Arts and Sciences, University of Oklahoma, Norman, OK, USA

**Antonio Santoro** Laboratory for Landscape and Cultural Heritage, Department of Agriculture, Food, Environment and Forestry (DAGRI), University of Florence, Florence, Italy

**Boris Semjon** Department of Food Hygiene, Technology and Safety, University of Veterinary Medicine and Pharmacy in Košice, Košice, Slovakia

**Jacek Słupski** Department of Plant Products and Nutrition Hygiene, Faculty of Food Technology, University of Agriculture in Krakow, Krakow, Poland

**Wojciech Sroka** Department of Economics and Food Economy, University of Agriculture in Krakow, Krakow, Poland

**Marta Szylar** Department of Land Management and Landscape Architecture, Faculty of Environmental Engineering and Land Surveying, University of Agriculture in Krakow, Krakow, Poland

**Sylwester Tabor** Department of Production Engineering, Logistics and Applied Computer Science, University of Agriculture in Krakow, Krakow, Poland

**Kinga Topolska** Department of Plant Products and Nutrition Hygiene, Faculty of Food Technology, University of Agriculture in Krakow, Krakow, Poland

**Peter Turek** Department of Food Hygiene, Technology and Safety, University of Veterinary Medicine and Pharmacy in Košice, Košice, Slovakia

**Maria Walczycka** Department of Animal Product Processing, Faculty of Food Technology, University of Agriculture in Krakow, Krakow, Poland

**Magdalena Wilkosz-Mamcarczyk** Department of Land Management and Landscape Architecture, Faculty of Environmental Engineering and Land Surveying, University of Agriculture in Krakow, Krakow, Poland

**Gabriela Zięć** Department of Carbohydrate Technology and Cereal Processing, Faculty of Food Technology, University of Agriculture in Krakow, Krakow, Poland

**František Zigo** Department of Animal Nutrition and Husbandry, University of Veterinary Medicine and Pharmacy in Košice, Košice, Slovakia

# Chapter 1
# The Need to Preserve Cultural Heritage

**Karol Król, Józef Hernik, Barbara Prus, and Marta Szylar**

**Abstract** Is it possible to link unique architectural heritage with a culinary tradition? Can biodiversity and landscapes be deemed historic cultural heritage? Can they determine the cultural identity and uniqueness of a particular place? Finally, can cultural heritage play an important role in binding local communities together and resolving conflicts between them? These and many other questions have become the focus of numerous activities undertaken worldwide, which have adopted formal frameworks and are being, or have been, implemented in the form of various initiatives including scientific projects. The chapter reviews and describes selected initiatives related to the preservation and promotion of cultural heritage.

**Keywords** Cultural heritage · Identity · Tradition · Heritage conservation · Regional development

## 1.1 Introduction

The issue of cultural heritage has in recent years been experiencing something of a revival. The notion of a "historic monument" highlighted so far is more and more frequently replaced by a broader notion of "heritage", with an accompanying increase in the awareness of the protection of cultural landscape (which is also referred to as architectural landscape). The notion of cultural heritage does not only include tangible but also intangible forms, which for several years have been anchored in the European Union legislation (Ahmad 2006). Increasing public awareness of the need for and significance of the protection of each form of cultural heritage is an

K. Król (✉) · J. Hernik · B. Prus · M. Szylar
Department of Land Management and Landscape Architecture, Faculty of Environmental Engineering and Land Surveying, University of Agriculture in Krakow, Krakow, Poland
e-mail: karol.krol@urk.edu.pl

J. Hernik
e-mail: rmhernik@cyf-kr.edu.pl

© Springer Nature Switzerland AG 2022 
J. Hernik et al. (eds.), *Cultural Heritage—Possibilities for Land-Centered Societal Development*, Environmental History 13,
https://doi.org/10.1007/978-3-030-58092-6_1

opportunity for the development of new initiatives for monitoring, protection and preservation of valuable assets inherent in the identities of countries and nations (Prus et al. 2020).

Going Christmas carolling, whisperers of Podlasie, the skill of bulrush weaving, mazurka and oberek dances, the martial-arts dance of capoeira, horse breeding in Janów Podlaski, nursery rhymes and local dishes, e.g., bokkoms, samoosas and sosaties, morogo and pap, avocado mousse, and Coronation chicken and many others—all of these are legacy of local and regional communities. It creates an ethnic and national identity as well as cultural heritage, which undergo transformations in time and space. This legacy demonstrates the remarkable diversity and multifacetedness of the phenomenon of tradition (Brzezińska 2013).

Heritage, in the broad sense, comprises tangible assets that include the natural and cultural environment, landscapes, historical sites, buildings and monuments (Król et al. 2019). The term "tangible cultural heritage" usually refers to physical objects, e.g. buildings, paintings, books, artefacts and historic monuments. Cultural heritage is not an anonymous production process; it is created by people, groups of people or a specific person being a community member (Skowroński 2012). Heritage also includes intangible and legal assets that are manifested in (1) traditions and oral lore; (2) performing arts; (3) customs, rituals and ceremonies; (4) knowledge and practices; and (5) craft-related skills (Brzezińska 2013). Intangible heritage is comprised of the language, dance, songs, religions, festivals and dishes as well as traditions, practices and customs that often form part of the culture passed down from generation to generation (Ramli et al. 2016). The impressions of experiencing cultural heritage are described as "*the unspoiled, pristine, genuine, untouched and traditional*" (Handler 1986, p. 2), and regarded as something "*exceptional in its actuality, and valuable*" (Trilling 1972, p. 93).

Cultural heritage is a group of resources inherited from the past which people identify, independently of ownership, as a reflection and expression of their constantly evolving values, beliefs, knowledge and traditions. It includes all aspects of the environment resulting from the interaction between people and places through time (Faro Convention 2005). Some people believe that heritage comprises the entire legacy left by ancestors: everything that the passing generations preserve for their posterity. Others, however, claim that heritage is only the legacy that the next generations want to continue, the one that has been accepted and has a chance to be passed on to future generations (Nieroba et al. 2009).

For individuals, cultural heritage has an individual and personal significance. For nations as well as religious and ethnic minorities, it is a common treasure. It is also a determinant of mores and social development, and a factor of identity preservation. Cultural heritage is a source of strength and a sense of self-worth that develops the awareness of cultural roots. For many regions that have failed to establish scientific or industrial centres, cultural heritage may be the highest of values which arises from the multitude of cultures drawing on from religious, ethnic and national traditions (Szkopińska 2013; Król et al. 2019).

Until now, heritage used to have a special status; it was regarded as a specific element of the past which has emerged in the present and, therefore, requires different treatment. Heritage was an asset to be preserved and protected. The community, while taking care of heritage, strived to save it. Nowadays, cultural heritage belongs to everyone, and access to it is one of the fundamental human rights (Purchla 2010). Heritage is therefore a part of everyday life, and plays a significant role in it. It has acquired a utilitarian dimension, ceased to be a value "in itself" and has become a value that serves the community (Owsianowska and Banaszkiewicz 2015).

Nowadays, cultural heritage is, on the one hand, the object of protection, and on the other, it has potential, including an economic one that should be exploited for the future development (Król and Prus 2018). The transition from the passive and static thinking about heritage perceived in terms of the *sacrum* to its protection under the conditions of spontaneous processes of public space privatisation and commercialisation currently requires fundamental changes in the management system. At present, a historic monument is not only a *sacrum* but also a *commoditum* that increasingly becomes an object of a market game in which *inter alia* investors and local authorities participate. At the same time, it should be borne in mind that heritage is a resource with unique characteristics. Its management should be based on social economy and not on overexploitation of principles (Purchla 2010). The aim of the study is to review and describe selected initiatives related to the preservation and promotion of cultural heritage.

### *1.1.1 Cultural Heritage as a Driving Force Behind the Development of Regions*

Conservation of cultural heritage sites, regeneration of historical cities and protection of natural heritage are measures of economic and social development. Investments in the protection of cultural heritage and its integration with the sustainable development of local communities can generate income and stimulate the development of entrepreneurship. The economic values of cultural goods are determined by the value usually attributed to them by people. Proper management of cultural heritage can enhance these values and make them easier to use while effectively conserving the resources. For the economy, cultural heritage is an actual "product"; it generates profit and turnover as tourists want to be its consumers (Nypan 2012).

By undertaking measures aimed at assessing the state of preservation of cultural heritage resources and its protection, one can search for the strengths and weaknesses as well as the opportunities and risks. A detailed analysis of an object enables an in-depth diagnosis of the state of its preservation and the definition of emerging problems. An example here can be a register maintained in Poland that supports the protection and monitoring of objects of historical value on a national scale (Report 2017). Digitisation of cultural heritage sites enables easier and diversified access to resources, which can stimulate the development of cultural tourism of regional,

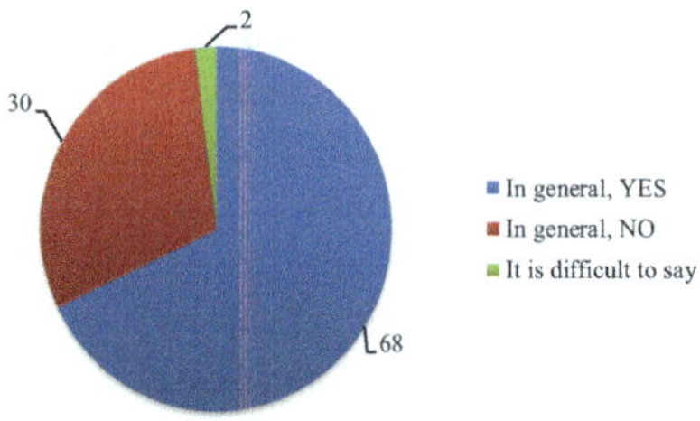


(a)

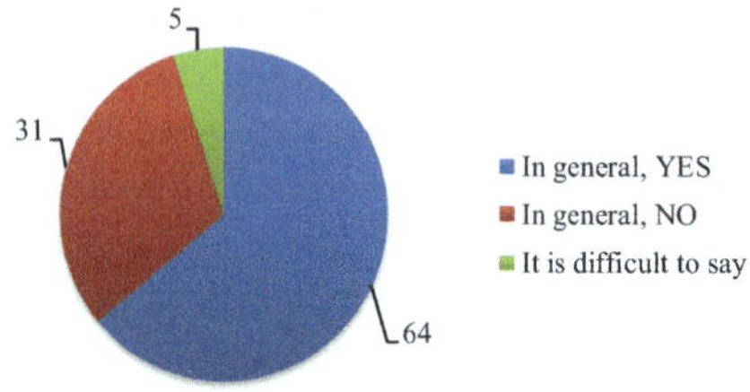


(b)

**Fig. 1.1** Respondents who expressed an interest in broadening their knowledge of Europe's cultural heritage (in percentage terms). *Source* own study based on the report "Europeans and cultural heritage" (Eurobarometer 2017)

national and international significance. The important role associated with undertaking protection measures results *inter alia* from the actual loss of historical sites over time due to their being uncared for and unused, and because of the impact of natural and anthropogenic determinants. Other (intangible) cultural values are also subject to loss, which is related to generational renewal.

Inhabitants of the European Union countries appreciate the significance of Europe's cultural heritage. Eurobarometer 466 (Eurobarometer 2017) published in December 2017 is the first cross-sectional public survey carried out in the field of cultural heritage throughout the European Union. The issues addressed in the report include an interest in cultural heritage, participation, barriers, awareness and beliefs, values and expectations related to the promotion, funding, commercialisation and protection of the related assets (Głowacki 2018). The survey showed that most Poles would like to know more about the cultural heritage of Europe (Fig. 1a) and Poland (Fig. 1b).

Policies of most European Union countries are aimed at protecting, managing, improving access to and informing the public about historic monuments. There is an apparent striving for stimulation in order to maintain historic monuments in good shape through financial support for their owners as well as legal and advisory assistance. In accordance with the principle of sustainable development, cultural heritage is taken into account in spatial planning at both the central and local levels. What can be observed is the increased local authorities' attention to the image of localities, including local cultural heritage. The restoration of historic monuments is usually preceded by archeological and historical research. The conservation and protection of cultural heritage are not perceived as an obstacle but as a driving force behind the development of regions that enhances their attractiveness and render the landscape unique.

In the United Kingdom, the view is that heritage protection and historic monument renovation and conservation are an effective solution to social and economic problems of urban areas. Research conducted by the English Heritage Trust has shown that expenditures on the protection of heritage are an investment catalyst, as each 10.000 pounds invested in heritage protection yields approximately 48.000 pounds of investments from the private sector and other sources. This is why the United Kingdom has placed an emphasis on both the proper development and use of objects of historical value and the economic activity of historical areas. A similar approach can be observed in France where the performance of tasks related to heritage protection is, to a large extent, subject to the principle of competitiveness. This results in a large proportion of private businesses engaging in work and research on historic monuments and sites.

In Lithuania, over 80% of cultural heritage sites and buildings are owned by the State Treasury or local government. Due to such an ownership structure, it is the state that is mainly responsible for the protection and condition of cultural heritage in this country. The situation is quite different in Norway where 99% of objects of historical value are privately owned. Therefore, the preservation of cultural heritage is largely determined by the proper management of historic monuments by their owners. In this case, the protection of heritage is based on the cooperation between the public and private sectors. The role of the state is to develop a logical and effective legal framework, and to support owners of historical objects with grants. By contrast, in Germany there are major differences in the functioning of the historic monument protection system in different federal states, which is due to the division of competencies in the field of cultural heritage protection (Skaldawski et al. 2011).

## 1.2 Selected Cultural Heritage Preservation Initiatives

There is a need to protect and sustainably use heritage as well as natural and cultural resources, which are subject to increasing environmental and economic pressure, and to conflicts of use. The heritage protection system is not only shaped by government agencies dedicated to the conservation of historic monuments, funding methods or regulations; it is also shaped by political, economic and cultural determinants as well as social life, education and the media. An increasingly important role in the protection and management of cultural heritage is being played by initiatives taking the form of international, institutionally funded projects. These are as diverse as cultural heritage objects and the relationships existing between heritage and local communities can be.

### *1.2.1 Central Europe Programme (2014–2020)*

CENTRAL EUROPE projects (Interreg CENTRAL EUROPE, Natural and cultural resources) support the preservation and use of cultural heritage to foster knowledge, strengthen regional identity and increase the region's attractiveness. They protect and reconstruct historic heritage like "Listen to the voice of villages" or "Thetris". Projects such as Danube Limes increase the accessibility and visibility of cultural sites or set up culture routes, like CrossCulTour and Forte Cultura (Breznik 2014).

These projects address the improved protection and management of cultural heritage sites as well as the valorisation of cultural heritage, e.g. through cultural routes or by better use and management of sites. The CENTRAL EUROPE Programme is based on the conviction that transnational cooperation can enhance the potential of both the public and private sectors involved in the protection and sustainable use of cultural heritage by supporting an integrated approach. The development and implementation of a strategy and policy for the evaluation of cultural heritage and the exploitation of its potential may affect socio-economic development.

The CENTRAL EUROPE Programme which was implemented during the years 2007–2013 created many opportunities for the cooperation between public authorities, institutions and privately owned companies from nine Central European countries, namely Austria, the Czech Republic, Germany, Hungary, Italy, Poland, Slovakia, Slovenia and Ukraine. By co-financing 124 projects, the CENTRAL EUROPE Programme has helped to improve local and regional innovation, to increase accessibility, to preserve the environment and to enhance the competitiveness and attractiveness of regions within central Europe. In the programming period 2014–2020, the CENTRAL EUROPE Programme is continuing to support regional cooperation among central European countries. Croatia is the latest country to have joined the programme.

### *1.2.2 Citizens' Catalogue of Historic Monuments*

The "Open Monuments" (Polish: *Otwarte Zabytki*) project is run by the Digital Centre Project: Poland (Polish: *Centrum Cyfrowe Projekt*: *Polska*). The aim of the project is to involve local communities in activities targeted at protecting objects of historical value and facilitating citizens' access to knowledge of Polish historical objects. The "Open Monuments" project is also addressed to educators, activity organisers and enthusiasts in various parts of the country, who work towards the popularisation of cultural heritage resources (Janus et al. 2014). "Openness" is a key factor that enables the promotion and use of cultural heritage resources. Data or contents are open if there are no limitations for anyone to use, reuse and disseminate them (Werner et al. 2014). An open monument is one which has been included in the "Open Monuments" website (otwartezabytki.pl) and described by the website users as part of the so-called Digital Community Work. All content is made available on the website under the license of "Creative Commons Uznanie Autorstwa 3.0 Polska". This allows them to be used freely, as "open" means available to all legally. The "Open Monuments" website also serves as a platform for knowledge exchange and a means of sharing information and resources with other Internet users.

The "Open Monuments" project was launched in 2012 with the so-called Digital Community Work, i.e., a citizens' campaign for the verification of data originating from the Register of Historic Monuments and made available by the National Heritage Board of Poland. Over 7.000 people have participated in the campaign, and information on historical objects has been edited over 10.000 times. As part of the project, an open Citizens' Catalogue of Historic Monuments based on the Register of Historic Monuments has been established. The entire catalogue, namely the parts originating from the Register and those added by users, is available for further use as a common asset.

### *1.2.3 Eat Your Way Through Edinburgh*

Searching for culinary traditions has in recent years become one of the major research fields of food studies. Cuisine, having gained the status of culinary culture, is no longer regarded exclusively in terms of gastronomy. Cuisine, in particular its national varieties, has begun to be regarded as a cultural heritage. It has also been incorporated into numerous cultural processes that influence the shaping of the national community. The phenomenon of a culinary return to local roots and the search for national culinary traditions are an effect of globalisation processes. The growing domination of selected national cuisines and their universal character have contributed to the heyday of local cuisines (Matras 2016).

We eat to live, and some of us live to eat. However, eating is an activity with a multifaceted dimension. Consuming food is associated with flavours and aromas, unique recipes, products and dishes that are frequently found only in selected locations or regions. Culinary tourism, i.e. the pursuit of regional products and dishes is becoming increasingly popular. Enthusiasts for traditional products and gourmets are able to travel long distances in search of new experiences. By tasting new dishes, a tourist learns about the behind-the-scenes secrets of the country they are visiting, and discovers it via unique flavours and aromas (Mason and Paggiaro 2012).

Consuming meals has become a cultural activity. It often takes the form of a sign of communication with people of different cultures. Due to the rituals associated with dish preparation, the sequence of adding particular products, and the specific mystique that surrounds cuisine, tasting food can become a real feast for all the senses. The history of the development of local cuisine flavours enables the construction of various historical narratives, both personal and political.

So far, the work associated with cultural heritage has usually focused on heritage sites and objects; until recently, food was not regarded as heritage (Vecco 2010). Meanwhile, food, cuisine, dishes, flavours and aromas are an interesting way of discovering new places. Family stories and the history of locations and events can be presented from the angle of culinary recipes passed down from generation to generation. Biographies of prominent figures and historical events can be presented in the light of culinary experiences. An example here can be the book entitled *Hunger for Freedom: The Story of Food in the Life of Nelson* which describes the story of Nelson Mandela, his family and their fight for freedom from the angle of shared meals (Trapido and Goode 2008).

For hundreds of years, inhabitants of the city of Edinburgh have been buying food products at street markets. This tradition continues to this day at the Edinburgh Farmers' Market. Street trading has left its mark in many names of places and streets that can still be seen in the Edinburgh Old Town in locations like the fish market and Fleshmarket Close. At the Grassmarket, cattle used to be purchased and sold, and stalls were set out there with fruit and vegetables originating from the areas in the vicinity of Tron Kirk. Tourists' attention is drawn to historic buildings in the Old Town where oyster shells embedded in the walls by masons can still be seen (Edinburgh World Heritage 2019).

Edinburgh World Heritage is an independent charity with the aim of ensuring the city's World Heritage status is a dynamic force that benefits everyone. Edinburgh World Heritage was created in 1999 through a merger of the Edinburgh Old Town Renewal Trust and the Edinburgh New Town Conservation Committee. The Edinburgh Food Heritage Trail offers a way to experience the city's World Heritage Site, revealing some of the hidden links between the city's unique built heritage and its food traditions. The trail allows one to discover nooks and crannies of both the Old and New Town, to learn about its history, to see how the architecture of buildings has been changing over centuries, to get to know culinary traditions, and to try dishes in some of its historical locations. The trail has been supported by the VisitScotland Year of Food and Drink Growth Fund. Edinburgh's World Heritage Site has over 1,600 listed buildings, so there are many opportunities to eat in stunning, historic surroundings. Eating within the World Heritage Site can be a great opportunity to soak in some of the city's historic sights. Because of the extraordinary topography of the site, one can also find many places with a fine view across the city's rooftops and skyline (Edinburgh World Heritage 2019).

### *1.2.4 HeAT—Heritage and Threat (2015–2018)*

Today's world contains a host of phenomena and situations that constitute threats to objects, sites and practices deemed "heritage". Yet there is still a dearth of systematic information about this broad palette of threats, a dearth that constitutes a gap in our general knowledge and an obstacle to the purposeful activity of governments and institutions at times of crisis evaluation and intervention or post-crisis reconciliation (HeAT 2019). The coordinator of the HeAT—Heritage and Threat—project was the University of Copenhagen (Denmark), representing a total of 10 institutions from all over the world, including from China (Fudan University, Shanghai) and from Poland (The Central and Eastern Europe Research Centre, OBEŚW), were involved in project activities. The research conducted in Copenhagen focused on establishing the typology of threats to cultural heritage in Syria and Northern Iraq in the context of military conflicts (Kinzel et al. 2018). The Romanian part of the project included a theoretical approach to the notion of cultural heritage associated with memorial sites. The research focused on various changes that occurred during the communist period and in the post-communist era. The research conducted in Poland was devoted to the impact of the Second World War on the preservation and perception of cultural heritage in Poland. From another perspective, research carried out in Italy was aimed at assessing the long-term effects of the construction of dams and artificial lakes on heritage, and the threats they pose to archeological sites and cultural landscapes.

### 1.2.5 Cultural Heritage and Diversity: STEPS Project

Contemporary societies are increasingly culturally diverse, in particular in major urban and metropolitan centres. This can lead to misunderstandings and conflicts that frequently arise from prejudices and stereotypes. Meanwhile, cultural activities can provide an open space for meetings between members of various cultures, and be a place for dialogue on complex issues concerning cultural differences and conflicts that arise from them. From this standpoint, cultural heritage plays an important role in ensuring the cohesion of communities and solving conflicts between them.

Cultural heritage of a city is an expression of the city's identity. Unlike national identities, urban identities can include all city dwellers across state borders or citizen boundaries. An intercultural approach to cultural heritage allows the city to open up to all communities while increasing trust and mutual recognition, and finally contribute to the cohesion of communities.

"Building specialisation strategies on local participation and heritage resources" (STEPS) is a Council of Europe/European Union project aimed at the creation and strengthening of community cohesion and promoting trust, dialogue and mutual understanding in various communities through cultural heritage. For the pilot project, two cities were selected, namely Lisbon (Portugal) and Rijeka (Croatia), due to their rich cultural heritage and the diversity of communities. The STEPS project is based on the paradigm of intercultural integration that includes respect for individuals (who are entitled to both freedom and duties), coexistence and cultural reciprocity, and the willingness to accept cultural hybridisation as a factor of change and development. The key operational elements of this paradigm include (1) the separation of powers not only between people of different origin but also between public institutions and civil society; (2) supporting cultural diversity in public institutions and public spaces, and (3) the development of cultural competencies (STEPS 2019).

### 1.2.6 Jewish Cultural Heritage Project (2013–2017)

The principal objective of the Jewish Cultural Heritage programme was to recover and disseminate the legacy of Polish Jews through education. The wide range of educational activities proposed as part of the project was a response to the need for saving the Jewish cultural heritage in Poland from oblivion. Indeed, this history is turbulent and multithreaded; it suffices to mention the example of the city of Łódź which was established through the cooperation of many nations, with the main role in its formation being played by Poles, Germans and Jews, even though the industrial Łódź was also co-created by representatives of other nations. However, the most significant nationalities that built the industrial Łódź as merchants, bankers and factory owners were the Jews of Łódź (Cudny et al. 2011).

The Jewish Cultural Heritage project was a response to society's demand for both the regaining of knowledge about Polish Jews and their culture, and passing it on to younger generations. The aim of the project was to support an open and tolerant society, and to combat intolerance, anti-Semitism, exclusion and negative stereotyping. The main beneficiaries of the project were children and young people including the disadvantaged and the disabled. The programme was based on the conviction that the exposure to the rich and dramatic history of Polish Jews offers more than just historical information, as it instills respect for people of various ethnic groups and religious communities, strengthens the determination to fight against xenophobia and prepares young people to live in a diverse society. The project coordinator was the Museum of the History of Polish Jews.

During the 32 months of project implementation, from September 2013–April 2016 (with the project end date in April 2017), 3.000 educational activities have been organised, with 435.000 participants in 240 localities across Poland and Norway, and there have been 4.600.000 Internet user sessions and 500.000 views of educational films. The Jewish Cultural Heritage project was awarded the 2017 European Union Prize for Cultural Heritage by the European Commission and Europe Nostra (Jewish Cultural Heritage 2019).

### 1.2.7 Cultural Routes

Launched by the Council of Europe in 1987, the Cultural Routes demonstrate, by means of a journey through space and time, how the heritage of different countries and cultures of Europe contributes to a shared and living cultural heritage. Through the Cultural Routes programme, the Council of Europe has developed a model of intra-national culture and tourism management, and enabled a synergy between national, regional and local authorities and a wide range of socio-economic associations and entities (Cultural Routes 2019). The central goal of the Cultural Routes programme of the Council of Europe is to contribute towards the promotion of European identity and citizenship through knowledge and awareness of the common heritage of Europe and the development of cultural links and dialogue within Europe as well as with other countries and regions. It seeks to form a common cultural space through the development of cultural routes in order to encourage sensibility towards heritage, education, the creation of networks, high-quality international sustainable tourism and other related activities (Historic Gardens 2019).

The Cultural Routes programme is aimed at making Europeans aware of their cultural identity and European citizenship, preserving and protecting heritage as a source of social, economic and cultural development, assigning particular significance to locations associated with cultural tourism while observing the principles of sustainable development, giving priority to cultural tourism as a form of leisure and promoting the broadly understood culture of Europe. The Council of Europe's Cultural Routes is an invitation to discover the rich and diverse heritage of Europe. More than 30 Cultural Routes guarantee a wealth of recreational and educational activities, and can contribute to sustainable development. Cultural Routes include a number of themes, from architecture and landscape to religious influences, and from gastronomy and intangible heritage to the major figures of European art, music and literature.

Moreover, a detailed procedure for planning routes was developed, which covers three areas, namely (1) people; (2) migrations, and (3) the spread of major European trends in philosophy, art, religion, science, technology and trade. At the same time, the notion of "European Cultural Route" that is supposed to run through more than one country or region, with a common root, was formulated. It can also be a route set out on the basis of the existing geographical trail due to particular assets of the natural environment. The route should be based on numerous and varied cultural attractions, with sites of particular historical significance that are representative of European culture as a whole.

### *1.2.8 National Register of Historical Rural Landscapes*

A cultural landscape is defined as a landscape transformed by humans as a result of the development of civilisation. Cultural landscapes represent a combination of creations of nature and products of human labour. They illustrate the evolution of human relationships with the place of existence under the influence of physical limitations and the opportunities to use the natural environment. Human activities, carried out with various intensities, can lead to a certain kind of harmony or disharmony of the space.

Cultural heritage is an element of rural identity (Chigbu 2012). One of the major problems of the landscape architecture related to the development of rural areas is the preservation of their cultural identity and the specificity of the landscape. In Italy, there is a diversity of unique landscapes shaped by humans over the centuries, which represent different civilisations. These landscapes are unique heritage and an expression of Italy's cultural identity.

Historical rural landscapes have become a subject of research carried out under the supervision of Prof. Mauro Agnoletti (Laboratory for Landscape and Cultural Heritage, CultLab, University of Florence), sponsored by the Ministry of Agriculture, Food and Forestry Policies, in cooperation with 14 Italian universities and selected international research institutions. The study involved 123 landscapes located throughout the regions of Italy. In the course of the work under study, descriptions of landscapes were collected with account taken of their historical, environmental and economic value, along with the typical products and critical issues that posed a threat to their integrity. At the same time, guidelines for landscape evaluations were proposed (National Register 2019).

Numerous good practices of synergy between the protection and conservation of historic monuments and the commercial, tourism and promotional functions indicate the possibilities for a wider perception of cultural assets and cultural heritage (Król 2019). This is an opportunity for preventive measures aimed *inter alia* at the protection of vanishing traditions, old professions, traditional construction techniques and technologies, etc. which positively affect the structure and authenticity of cultural heritage elements.

## 1.3 Summary

Cultural heritage is an important part of a community's prosperity. National governments and European institutions increasingly recognise the value of cultural heritage and its importance to sustainable development and the need for broad public participation in discussions about. Preservation of cultural heritage has an effect on the identity of local communities, builds cultural values and emphasises the unique character of a place (Tweed and Sutherland 2007). Cultural heritage encompasses resources from the past in a variety of aspects and forms. These include monuments, traditions, sites, transmitted knowledge and expressions of human creativity, as well as collections conserved and managed by museums, libraries and archives. Meanwhile, cultural heritage is more than historic monuments left behind from the past. It is all the things, places and practices that define who we are as individuals, as communities, as regions. The sustainable use of cultural heritage is essential to ensure that Central Europe will remain a popular destination to visit and to live in (Breznik 2014).

Projects supporting the preservation of cultural heritage are mostly international and interdisciplinary in nature. They focus on preserving cultural heritage while using it in regional development. They are used to enhance the competitiveness of regions while taking account of their local cultural resources. They present cultural heritage as an asset of a particular place that determines its unique character. Such projects are implemented at the social (supporting multiculturalism, social dialogue), economic (tourism, traditional products) and cultural levels (preservation and promotion of local traditions, promotion of a particular culture and integration of multicultural communities). They make local cultural heritage globally accessible, help to rediscover it and prevent it from being regarded only in terms of conservation and preservation. Cultural heritage has both a chance to be rediscovered and the potential to play an important role as an element (factor) of local development policy and an element of social life. Furthermore, it can be a stimulus for an increase in the sense of civic duty manifested by actions aimed at cultivating traditions of local and regional culture.

## References

Ahmad Y (2006) The scope and definitions of heritage: from tangible to intangible. Int J Herit Stud 12(3):292–300. https://doi.org/10.1080/13527250600604639

Breznik C (2014). Project stories from the CENTRAL EUROPE programme. In: Cultural heritage and creative resources. CENTRAL EUROPE Programme, Vienna. Austria, https://goo.gl/BJKcxF

Brzezińska AW (2013) Reifikacja dziedzictwa kulturowego w świetle Konwencji UNESCO z 2013 roku. Nauka 1:109–128

Chigbu UE (2012) Village renewal as an instrument of rural development: evidence from Weyarn Germany. Commun Dev 43(2):209–224. https://doi.org/10.1080/15575330.2011.575231

Faro Convention (2005). Convention on the value of cultural heritage for society. Culture and cultural heritage. Council of Europe. https://goo.gl/noMkRW

Cudny W, Kubiak E, Rouba R (2011) Dziedzictwo kulturowe łódzkich Żydów i jego rola w rozwoju współczesnej Łodzi i regionu łódzkiego. Przegląd Nauk Hist 10(1):117–136

Edinburgh World Heritage (2019). Food heritage. Eat your way through Edinburgh. https://ewh.org.uk/trails/food-heritage/

Eurobarometer (2017). Europejczycy i dziedzictwo kulturowe. Special Eurobarometer 466. September—October 2017. European Commission

Historic Gardens (2019) The European cultural route project. European Network of Historic Gardens. http://europeanhistoricgardens.eu/

Głowacki W (2018) Eurobarometr 466. Nowości Badawcze NCK 1:40–41

Handler R (1986) Authenticity. Anthropol Today 2(1):2–4

HeAT (2019). HeAT—heritage and threat. Department of Cross-Cultural and Regional Studies. University of Copenhagen. https://goo.gl/Y5ixmb

Janus A, Haratyk P, Majdecka E, Radziwiłko K, Sawko K, Sielicka K, Śliwowski K, Werner K (2014) Otwarte Zabytki. Przewodnik dla nauczycieli, bibliotekarzy i animatorów kultury (A guide for teachers, librarians and cultural event organisers). Centrum Cyfrowe, Warszawa

Jewish Cultural Heritage (2019) Jewish cultural heritage. Polish-Norwegian bilateral activities in 2017. https://www.polin.pl/en/heritage

Kinzel M, Thuesen MB, Thuesen I (Eds.) (2018) Conflict & culture. Understanding threats to heritage. Copenhagen Forlaget Orbis

Król K (2019) Forgotten agritourism: abandoned websites in the promotion of rural tourism in Poland. J Hosp Tour Technol 10(3):461–472. https://doi.org/10.1108/JHTT-09-2018-0092

Król K, Prus B (2018) Application of interactive charts in the evaluation of socio-economic development of regions; the case of Poland. Acta Sci Pol. Form Circum 17(3):141–151. https://doi.org/10.15576/ASP.FC/2018.17.3.141

Król K, Kao R, Hernik J (2019) The scarecrow as an indicator of changes in the cultural heritage of rural Poland. Sustainability 11(23):6857. https://doi.org/10.3390/su11236857

Mason MC, Paggiaro A (2012) Investigating the role of festivalscape in culinary tourism: the case of food and wine events. Tour Manag 33(6):1329–1336. https://doi.org/10.1016/j.tourman.2011.12.016

Matras A (2016). Performowanie polskiego smaku: kuchnia narodowa jako niematerialne dziedzictwo kulturowe. Nowa Biblioteka Usługi Technol Inform Media 3(22):105–114

Nieroba E, Czerny A, Szczepański MS (2009) Między nostalgią a nadzieją. Dziedzictwo kulturowe jako dyskursywny obszar rzeczywistości społecznej. In: Red E, Nieroba A, Czerny MS (eds) Szczepański Między nostalgią a nadzieją. Dziedzictwo kulturowe w ujęciu interdyscyplinarnym, Opole, 17–36

Nypan T (2012) Ekonomiczny potencjał dziedzictwa—nie tylko turystyka. Biznesowa przygoda Li-Pena i jaką rolę odegrało dziedzictwo kulturowe. Ochrona Zabytków, 65/3–4(258–259):175–178

Owsianowska S, Banaszkiewicz M (2015) Trudne dziedzictwo a turystyka O Dysonansie Dziedzictwa Kulturowego. Turystyka Kulturowa 11:6–24

Prus B, Król K, Gawroński K, Sankowski E, Hernik J (2020) From classic (Analogue) to digital forms of cultural heritage protection in Poland. In: Kremers H (ed) Digital cultural heritage. Springer, Cham, (255–278). https://doi.org/10.1007/978-3-030-15200-0_17
Purchla J (2010) W stronę systemu ochrony dziedzictwa kulturowego w Polsce. Zarządzanie Publ 2(12):69–82
Ramli AM, Zahari MSM, Suhaimi MZ, Talib SA (2016) Determinants of food heritage towards food identity. Environ Behav Proc J 1(1):207–216. https://doi.org/10.21834/e-bpj.v1i1.217
National Register (2019) National register of historical rural landscapes. http://landscapeunifi.it/en/national-register-of-historical-rural-landscapes
Report (2017). Raport o stanie zachowania zabytków nieruchomych w Polsce. Zabytki wpisane do rejestru zabytków (księgi rejestru A i C). National Heritage Board of Poland, Warszawa
Cultural Routes (2019) Explore all cultural routes. Council of Europe. https://www.coe.int/en/web/cultural-routes
Skaldawski B, Chabiera A, Lisiecki A (2011) System ochrony zabytków w wybranych krajach europejskich. Kurier Konserwatorski 11:5–33
Skowroński H (2012) Dziedzictwo kulturowe ojczyzny jako kategoria aksjologiczna dla osoby w nauczaniu kard Stefana Wyszyńskiego. Studia Prymasowskie 6:35–50
STEPS (2019) Cultural heritage and diversity: STEPS project. Council of Europe. https://www.coe.int/en/web/interculturalcities/cultural-heritage-and-diversity
Szkopińska A (2013) Obecna nieobecność. Ewangelicy na Sejneńszczyźnie. Dziedzictwo kulturowe jako źródło historii. The Philos Soc Yearbook Civitas Hominibus 8:77–85
Trapido A, Goode R (2008) Hunger for freedom: the story of food in the life of Nelson Mandela. Jacana Media, Auckland Park
Trilling L (1972) Sincerity and authenticity. Oxford University, London
Tweed C, Sutherland M (2007) Built cultural heritage and sustainable urban development. Landsc Urban Plan 83(1):62–69. https://doi.org/10.1016/j.landurbplan.2007.05.008
Vecco M (2010) A definition of cultural heritage: from the tangible to the intangible. J Cult Herit 11(3):321–324. https://doi.org/10.1016/j.culher.2010.01.006
Werner K, Sawko K, Janus A, Majdecka E, Rybicka K, Lis R, Niezgódka P, Śliwowski K (2014) Otwarte Zabytki Digitalizacja. Przewodnik. Centrum Cyfrowe Projekt: Polska, Warszawa

# Chapter 2
# The Italian National Register of Historical Rural Landscapes

**Mauro Agnoletti and Antonio Santoro**

**Abstract** In recent years, the role assigned to rural landscapes has gained importance in Italy, both at scientific and political levels. Some political decisions in the field of agriculture and planning have recognized the multifunctional role of traditional rural landscapes. It is widely recognized by the scientific community that these landscapes can be of fundamental importance for the economy of many rural areas, for their connections with tourism, for high-quality productions, for the conservation of agrobiodiversity and for reducing hydrogeological risk. In Italy, one of the main changes concerning the Italian rural landscape is the Decree n. 17070 of 2012 by the Ministry of Agriculture Food and Forestry Policies about the institution of the "National Observatory of Rural Landscape, Agricultural Practices and Traditional Knowledge". Among the tasks of the National Observatory of Rural Landscape can be found the surveying of landscape, of agricultural practices and of traditional knowledge considered to be of particular value, and the promotion of research activities for studying the values associated with the rural landscape, its preservation, its management and planning and even advancing the goal of bio-cultural diversity. It must also develop general principles and guidelines for the protection and enhancement of the rural landscape with particular reference to action taken under the Common Agricultural Policy. In addition to the landscape, the decree is aimed at the preservation and enhancement of "agricultural practices and traditional knowledge", defined as "complex systems based on ingenious and diversified techniques, on local knowledge expressed by rural civilization, which have made a major contribution to the construction and maintenance of traditional landscapes". This decree has also established the "National Register of Rural Landscape, Agricultural Practices and Traditional Knowledge". Through this Register, the Ministry identifies and catalogs "the traditional rural landscapes or landscapes of historical interest present within the national territory and the connected traditional practices and knowledge, defining

M. Agnoletti (✉) · A. Santoro
Laboratory for Landscape and Cultural Heritage, Department of Agriculture, Food, Environment and Forestry (DAGRI), University of Florence, Via San Bonaventura 13, 50145 Florence, Italy
e-mail: mauro.agnoletti@unifi.it

A. Santoro
e-mail: antonio.santoro@unifi.it

© Springer Nature Switzerland AG 2022

J. Hernik et al. (eds.), *Cultural Heritage—Possibilities for Land-Centered Societal Development*, Environmental History 13,
https://doi.org/10.1007/978-3-030-58092-6_2

their significance, integrity and vulnerability, taking account both of the opinion of scholars and of the values ascribed to these landscapes, practices and knowledge by the concerned communities, subjects and populations". There are currently 27 landscapes and 3 traditional practices inscribed in the Register. The Register is also the first step to access international programs, such as the Globally Important Agriculture Heritage Systems (GIAHS) program developed by the FAO or the UNESCO World Heritage List.

**Keywords** Cultural landscape · Rural development · Landscape monitoring · Historical landscape

## 2.1 Introduction

Italy still boasts a rich heritage of rural landscapes built up over thousands of years: landscapes that, while continuing to evolve, still retain evident testimonies of their historical origin and maintain an active role in society and the economy. These landscapes are indissolubly tied to traditional practices handed down from one generation of farmers, shepherds and woodsmen to the next, complex sets of ingenious and diversified techniques that have contributed in a fundamental way to the construction and conservation of our historical, cultural and natural heritage. These techniques were a means to continuously adapt to difficult environmental conditions to provide multiple goods and services, thereby improving people's standard of living as well as giving rise to landscapes of great beauty. Landscape heritage and the related traditional knowledge are fundamental resources that need to be safeguarded. The speed and extension of the technological, cultural and economic changes that have taken place over the last few decades are threatening landscapes and the rural societies associated with them. Multiple pressures are constraining farmer innovation and this often leads to unsustainable practices, resource depletion, productivity decline and excessive specialization, making the preservation of landscapes an economic, cultural and environmental resource in serious jeopardy. The result is not only an interruption in the transmission of the traditional knowledge required for local landscape maintenance but also socioeconomic destabilization of rural areas and a loss of competitiveness of agriculture.

## 2.2 The National Register

The research for the development of the National Register of Historical Rural Landscapes was meant as a testimony, not only of the importance of the Italian landscape as one of the most representative historical expressions of the country's cultural identity, due to the prevalent role of rural civilization in its history, but also of the universal value of the Italian rural landscape in the cultural heritage of humanity

(Agnoletti 2012). This is a value that seems to have been often forgotten today. The research intended to lay a foundation for the identification, conservation and dynamic management of historical landscape systems and traditional practices, in the face of economic and cultural globalization, climate changes and inappropriate policies, favors the creation of a national register of historical landscapes. Actually, the term "historical" in itself is not significant semantically. All areas that have been anthropized for a few decades can be legitimately said to have a historical footprint. But the landscapes of Italy, as we well know, reach back far beyond this minimal threshold. What distinguishes the complexity of the historical character of the Italian peninsula's landscape—even compared to other European landscapes that were anthropized in ancient times—is the multiplicity and stratification of the footprints left by so many distinct civilizations on our countryside. We only have to think of the changes determined by land reclaiming works carried out by Greek settlers, Etruscans, Romans and Arabs. In the course of time, these civilizations provided such an incomparably vast contribution to our agriculture, in the form of new plant species, cultivation techniques, plantation and land delimitation methods, water collection and use, and buildings and land works that the historical character of our landscape acquired a special value compared to that of other European countries. We should also not forget that, just as a landscape merges in an original synthesis the beauty of a site or plantation with the historical character of its use and manipulation for economic purposes, the buildings scattered in our countryside, immersed in the most diverse habitats, are at once documents of past agrarian civilizations and artistically valuable constructions, aesthetically prestigious works, admirable for their magnificence and the genius of their builders.

Nowadays, we are witnessing increasing interest in the subject at the European level, as stated by the European Landscape Convention,[1] signed in Florence in 2000, which addresses the deep changes in course in modern society. As Roberto Gambino explains, the need to preserve the identity and meaning of places expressed by the current "demand for landscape" reflects a deeper malaise that certainly has to do with globalization processes and their effects: on the one hand, homologation and modernization, on the other, imbalances and inequalities that need to be addressed (Gambino 1994). In this perspective, the introduction of landscape in the national rural policies reflects a change in the conception of the role of this resource, as well as that of rural territory in general. The role of landscape and its perception has indeed changed over time. Today it is no longer an elite aesthetic and cultural construct, isolated from its socioeconomic context; it has become, instead, an essential element in the definition of an adequate development model for the national rural context.

---

[1] The research has received the patronage of the Council of Europe for its contribution to the implementation of the European Landscape Convention. Article 6.C.1 of the convention requires identification and assessment, which states that each party undertakes.

– a. i to identify its own landscapes throughout its territory;

ii to analyze their characteristics and the forces and pressures transforming;

iii to take note of changes;

– b. to assess the landscapes thus identified, taking into account the particular values assigned to them by the interested parties and the population concerned.

The prevalence of aesthetic considerations in past conceptions of landscape, as well as their more recent superimposition on the concept of "nature", has led to an emphasis on deterioration caused by urban dynamics, or criteria for the assessment of landscape quality based on its ecological characteristics, reductively understood as its flora and fauna, or as a series of natural habitats. All this has pushed in the background both the strong human print on our country's landscape and the fact that, while urban expansion certainly played a role in this, the transformation of the rural landscape was largely endogenous, something that few have remarked. While it is evident, as Emilio Sereni explained (Sereni 1961), that the agrarian landscape is "the form that man, in the course and for the ends of his agricultural productive activities, impresses on the natural landscape", it is equally evident that not all agricultures produce good landscapes. Unfortunately, ordinary conservation legislation based on protected area systems or landscape restrictions is ineffective as a means to preserve the rural landscape. It is this realization that persuaded all of the scholars who contributed to our catalog of the need to draw it up, that it is finally time for the issue to be addressed by agricultural policies. Conserving the quality of a rural landscape, which by its own nature is always evolving, can only be done by setting up a socioeconomic system capable of supporting and reproducing it; hence the decisive importance of strategies and actions undertaken in the framework of agricultural policies. The new guidelines for rural development policies associating them with local development are a major step forward in this direction. The objective is to make the most of all the resources of rural areas, emphasizing the local dimension, the new role of farmers and the involvement of new actors in the social and geographical space designated today as "rural" (Ploeg 2006). Important landmarks for the rise of this new vision of rural policies in Europe were the *Rural White Paper* published by the English government in 2000 and the *National Agenda for a Living Countryside* produced by Holland (2004)—a country where the preservation of the rural landscape is entrusted to the Ministry of Agriculture. Both documents indicate landscape conservation and restoration as a priority in national rural policies (Moreira et al. 2006).

In the local dimension of Italian rural policies, the landscape dimension plays a paradigmatic role, as it corresponds to the transition from individual business projects to projects at territorial scale, for which a landscape-oriented approach is undoubtedly more suitable, because of the peculiar characteristics of our country, than an industrial or environmental one, even in a development perspective. Indeed, today the notion that conservation is an obstacle to development in any form has given way to the realization that conservation is the new face of innovation in contemporary society. An authentic innovation is one that adds to a store of values slowly accumulated over the ages. Conversely, there can be no authentic conservation without the production of new values. In this perspective, the restoration and promotion of actions implemented in Italy by the National Rural Development Plan (2007–2013) have already introduced instruments by which the Italian regions can begin to modify the orientation of Rural Development Plans to address landscape issues, although at this initial stage the new landscape orientations of regional agricultural policies, especially in regions with vast and valuable landscape heritage, do not appear very effective (Fig. 2.1).

**Fig. 2.1** 1:250,000 map of the Italian territory resulting from an interpretation of Corine Level 4 data. The map highlights the polarization of the rural landscape, which today appears divided between forest areas (in green), prevalently located in mountain areas, and agricultural areas (in beige). Although the adopted scale overemphasizes the phenomenon, socioeconomic dynamics have indeed undermined the historical integration among woods, pastures and agriculture, reducing the complexity of Italy's landscape mosaic and biodiversity by favoring, instead, simplification and structural homogeneity (Agnoletti 2010)

## 2.3 The Investigations

Our research is not meant as an exhaustive overview of Italy's landscape heritage. Rather, it is intended to contribute to the development of a methodology for the identification and classification of landscapes of historical interest, and, at the same time, to provide a preliminary sample of the substance and state of the country's landscape heritage. This will hopefully be the first step in the drawing up of a truly comprehensive inventory of the Italian rural landscape, on the desirability of which there appears to be a wide consensus today among both scholars and agricultural policy makers. We decided not to focus on the strictly environmental features of Italian rural landscapes—climate, geomorphology and vegetation—since these have been examined in depth in the existing literature. We strove, instead, to take a more detailed look at the structure and organization of rural landscapes. Thus, we did not focus on ecological and naturalistic aspects, nor aesthetic ones, although these are also mentioned in the individual area descriptions. Rather, we adopted as our landmark Emilio Sereni's pioneering work (1961), which examined the "forms" impressed by humankind on the natural substrate, but left open the question of their characterization and conservation at a national scale. Our purpose was to carry forward Sereni's work by combining traditional historiographies of agriculture, forestry and, more in general, the landscape with approaches highlighting the material elements of landscape structure, as found in important studies by European scholars, especially English ones such as Oliver Rackham (1986), and also in some remarkable investigations conducted in Italy by workgroups led by Moreno (1988) on the agropastoral sector and Pietro Piussi on forests (1990).

Our project's board of advisors gathered scholars with competences in the domains of history, geography, agrarian and forest science and architecture. Coordinators were nominated for one or more regions, each of whom selected collaborators to conduct investigations at a local scale. About 80 researchers from 14 universities thus contributed to the catalog, as well as some professional studios and independent researchers. An international committee of experts was formed to assess the work. Some foreign institutions were also involved in the project, including the Committee for Cultural Heritage and Landscape of the European Council, the European Society of Environmental History (ESEH) and the International Union of Forest Research Organizations (IUFRO).

One of the methodological problems we had to deal with in the initial stage of our research was the definition of its spatial and chronological scale. As regards the chronological scale, no limits were set. The origins of the landscapes under investigation were traced as far back as available sources allowed. As regards the spatial scale, we decided to analyze areas with extensions between 300 and 5000 ha, large enough, that is, to include management units such as the typical Italian sharecropping farm or the *latifundium*, and to encompass spatial relationships between land uses, in consideration of the importance of the spatial scale in UNESCO parameters for world heritage sites. In the area descriptions, we decided to indicate only the geographical coordinates of the center of each area, while a complete GIS analysis have been carried out during the second stage of the research. The main reason for this was the difficulty, which we will discuss further on, of accurately determining the geographical boundaries of areas with non-contiguous cultivated zones.

Each area was illustrated in a separate descriptive text. Although the area descriptions were based on a common template, due to the many different competences of the scholars involved in the research, there were differences in individual sections of each description. The collected information was hence homogenized to make published descriptions of equal length and make sure they contained the same kind of data, also for the purpose of making them more easily comparable. It is important to specify that the photographs are meant as an accompaniment to the text, but are not themselves the object of the catalog. They are merely meant as a support to the descriptions, not having been taken with the highlighting of aesthetic parameters in mind. This reflects the general approach followed in this work, which is to highlight mostly the historical character of landscapes in connection with aspects such as aesthetic quality, typical products, tourism and biodiversity.

By the end of the first 12 months of the project, 123 areas had been singled out. The number of areas per region varies from a minimum of 2 to a maximum of 8, which were the limits we set for local researchers in their choice of representative areas. We tried to reduce the effect of differences in the relative abundance of historical landscapes between one region and the other by carefully employing selective criteria.

## 2.4 The Major Transformations of the Rural and Forest Landscapes of Italy from Its Unification to the Present Day

For the reader to fully understand not only the situation "photographed" by our catalog but also the urgency of such an investigation, we need to briefly go over the evolution of the Italian rural landscape since the country's unification, not so much in terms of socioeconomic changes, but rather as regards land use, which gives a measure of the dramatic changes that occurred in this period. It is undoubtedly a limited time frame, considering the remote historical origins of the Italian landscape. However, as environmental historians have shown, this is the period when the abundance and intensity of changes at the global level occurred with a speed that had no precedent in the history of human civilization, and Italy is no exception (McNeill 2001). At least until the second postwar period, much of the country's rural landscape was still strongly influenced by traditional agro-silvo-pastoral models developed during the previous century, and sometimes going all the way back to the Etruscan period and Greek civilization. The following decades, however, witnessed deep transformations. Due to demographic growth and the expansion of agriculture into mountain areas, the rural landscape attained the peak of its development in the decades between the late nineteenth and early twentieth centuries. The resulting landscape was one of great complexity, enhanced by the stratification of the prints left by so many civilizations on the land, and the country's complex orography and climatic variability. In the second postwar period, however, we observe a gradual simplification and homogenization of the rural landscape that can be analyzed in terms of its effects on its two main components: woods and crops (Agnoletti 2010).

### 2.4.1 The Evolution of Agricultural Surfaces

The image of rural Italy at the time of the country's unification is one of great complexity. Adaptation to different and difficult local conditions, as well as differences in economic and social structures, had diversified the national territory over the centuries. Except in a few limited areas in the country, the history of Italian agriculture had been one of continuous and laborious adaptation to a difficult natural environment, mostly made up of mountains and high hills, originally covered with impenetrable forests and extensive marshes, to create favorable conditions for agriculture. The result was an extraordinary landscape whose value has been recognized by the Western culture at least since the sixteenth century. At the end of that century, Michel De Montaigne, going through the Garfagnana in Tuscany, observed in amazement that the land was cultivated and terraced from the foot of the mountains to their summit, appearing to him as a garden (Trechmann 1929). Those who followed in his wake echoed his admiration, from Grand Tour travelers of the eighteenth and nineteenth centuries—who were impressed not just by Italy's monuments but also by its rural and forest landscapes—down to present-day tourists.

Morphological differences, farming systems, settlement patterns and local styles of rural buildings placed their distinctive stamp on the landscape of rural areas. The main agricultural systems, such as those revolving around local types of the farmhouse—the Lombard *cascine*, sharecroppers' farms and farmhouses, the farmhouses of the grain-growing latifundia of Maremma, the Roman *casali*, or the *masserie* of southern Italy—are the most visible manifestation of a much more complex reality. In spite of Italy's great variability, however, there were some common traits, such as the extension of arable land with a prevalence of cereal cultivation. Italy's vast "bread lands" (*terre da pane*) reflected a strong orientation of agricultural production toward self-consumption and maintained their prevalence in the agricultural landscape until the 1960s, even in mountain areas. Another unequivocal sign of the importance of production for self-consumption was the multiplicity of crops and mixed cultivation, as well as the presence of extensive terracing providing horizontal surfaces to allow crops to be sown in acclivitous areas, an enlightening example of ingenious adaptation to difficult environmental conditions to solve the food problem. In this context of low-intensity farming (Moreira et al. 2005), agriculture in the post-unity period appears as the country's main economic motor, and displays strong continuity going back several centuries (Fig. 2.2).

From the twentieth century onward, the percentage of the population employed in agriculture, which used to comprise almost the total working force, slowly began to decline under the impulse of great socioeconomic changes. Today, the sector employs only 4% of the working population and its share of the GNP is equal to ca. 3%. These changes, however, occurred with different speeds and intensities in different parts of the country. The trend established itself much earlier in the industrial regions of the Northwest, where between the two wars workers employed in agriculture were already down to 35% of the total working population. In the rest of Italy, the tipping of the scales between the primary and the secondary sectors only occurred

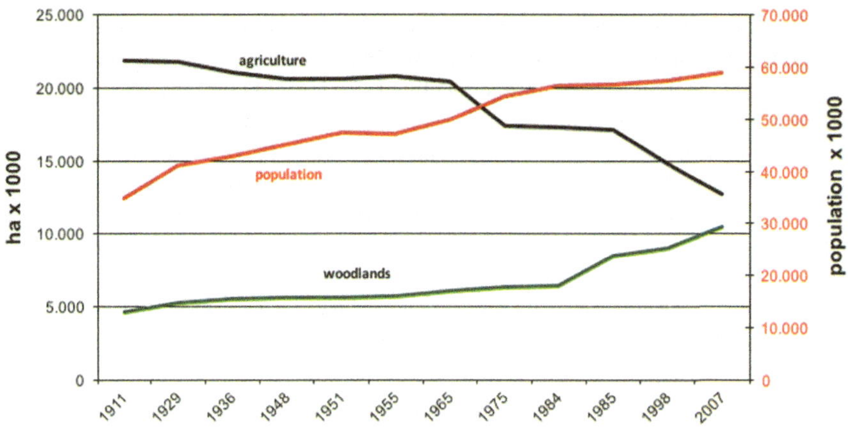


**Fig. 2.2** Evolution of agricultural, wooded and unproductive surfaces and of the Italian population from 1861 to 2007. One can observe the strong reduction of agricultural surfaces and the increase of woodland. The increase in woodland is due to the abandonment of farmed land and pastures

on the morrow of the Second World War. Accordingly, landscape transformations of different areas of the country followed different timelines. As shown by the graph on the evolution of agricultural, forest and unproductive surfaces, the importance of cropland makes it a dominant element in the Italian landscape, down to the present day. Nevertheless, today it has lost millions of hectares to the expansion of woods and unproductive surfaces, a category that also includes urban areas. In their turn, agricultural surfaces have undergone internal transformations that have changed the landscape fabric.

From the second postwar period onward, available data show a sudden decrease of agricultural surfaces, the symptom of a transformation reflecting the dominant role of socioeconomic factors in the Italian landscape. The increase of unproductive and urbanized surfaces, on the one hand, and the increase of forests, on the other, are just different facets of the same problem, namely the abandonment of agriculture. They are the result of an epochal transformation of our landscape that took place in just one hundred years, and which has gone largely unnoticed.

Within agricultural surfaces, the most significant reduction was of arable land, followed by that of meadows and pastures. The decline of grain-growing has special significance and symbolic relevance in a country that fought a "battle for grain" in the 1920s. The decrease of grain field surface was only partially compensated by increases in productivity, so that today Italy imports most of its grain. Various factors intervened in bringing about this deep mutation of the rural landscape. Among these, especially worthy of mention are demographic evolution, the spread of important technological innovations such as chemical fertilizers and pesticides and mechanization, which ended up favoring rather than limiting the exodus from the countryside. The employment of mechanical farming equipment, which considerably reduced labor requirements, along with the country's industrialization contributed to the abandonment of many cultivated surfaces, beginning with marginal mountains and high-hill areas. This evolution went hand in hand with a change in the structure of farming businesses, whose number declined sharply, although the average surface per farm has not changed much, a distinctively Italian trait that contrasts with the trend in countries like Spain or France. The low-to-middle size typical of sharecroppers' holdings and family-run farms has given way to a growing gap between large and small farms. An increasing trend to use externally hired rather than resident labor is breaking the bond between farmers and their holdings. From the 1970s onward, changes made to the Common Agricultural Policy (CAP) to limit surpluses favored the spread of non-food crops such as soy, colza and sunflower; vast industrial monocultures that have accentuated the simplification of the agricultural mosaic and are now facing a crisis.

In the second half of the twentieth century, along with the reduction of cultivated surfaces, there were radical changes in crops, livestock and the activities of the agricultural sector. One of the most significant phenomena was the internal transformation of agricultural surfaces, with a trend toward specialized cultivations. This transition applied to all the typical sectors of agricultural and food production. New cultivation techniques were introduced to increase productivity and product quality: a quality, however, in which the landscape and its specific environmental contents played no role. Wine-, olive-, vegetable- and citrus-growing, as well as livestock and dairy farming, have all been impacted by these new trends, which have led to an intensification of production that is often incompatible with landscape quality. Slopewise planting has replaced terraces (Romero Díaz et al. 2007). Tree rows, mixed cultivations and widely spaced cultivations have made way to intensive specialized cultivations with reduced labor costs.

In the years of postwar reconstruction, Italian agriculture adopted a development model aimed at maximizing production to meet internal food demand and compete on foreign markets. At first, the policies of the European Union had the same objective. However, today this "battle for production" has been lost. The sector has proved unable either to meet the national food demand or to compete on international markets in terms of quantity. Over recent decades, the fate of both the grain and livestock businesses has depended on the changing moods of CAP funding rather than on the free market. The livestock industry, in particular, has become almost entirely independent of meadow and pasture resources, once abundant in the Italian landscape and much reduced today. In the context of this "imperfect" market, influenced by the orientations of the CAP and external global phenomena, the need and opportunity have arisen to associate product quality with landscape quality, to take advantage of an added value that the competition cannot reproduce, and, at the same time, implement low-intensity agricultural models more compatible with environmental quality and revive extensive livestock farming methods.

As mentioned above, urban expansion partially accounts for the increase of unproductive surfaces in our country. Urban growth is often branded as the main enemy of the rural landscape, something on which there is usually a broad agreement among the public, farmers and environmentalists. While it is true that the permanence of agriculture acts as a barrier against urban expansion, it is equally true that the most significant changes in the rural sector are due to abandonment, on the one hand, and endogenous changes that are not as obvious, but much more in-depth and enduring, on the other. Urban surface, according to the most up-to-date European mapping system (Corine Land Cover 2000), does not exceed 5% of the total surface of Italy. It is true, however, that scattered urbanization eludes Corine. The Italian Ministry of Agriculture, Food and Forest Policies hence resolved to establish a new category of the rural area labeled *poli urbani*, including areas still classified as rural, but with high settlement densities. Table 2.1 details surface extensions for the five first-level CLC classes in 2000 and 1990. As one can see, agricultural areas are not only the prevalent category in terms of the total surface but also the category that changed most significantly, with a 1434 km$^2$ decline. In relative terms, instead, the class that evolved the most from 1990 to 2000 is that of artificial surfaces, with a 6% increase.

**Table 2.1** Land cover changes in Italy from 1990 to 2000 as recorded by the Corine satellite system, promoted by the European Environmental Agency

| Land cover, CLC Level 2 | 2000 (km$^2$) | 1990 (km$^2$) | 2000–1990 (km$^2$) | (2000–1990)/1990 (%) |
|---|---|---|---|---|
| Residential urban areas | 10,819.60 | 10,315.70 | 503.9 | 4.88 |
| Industrial and commercial areas, and infrastructure | 2,631.90 | 2,377.90 | 254 | 10.68 |
| Mineral extraction, construction and dump sites; artificial and abandoned areas | 565.1 | 514.7 | 50.4 | 9.79 |
| Artificial non-agricultural vegetated areas | 299.6 | 281.1 | 18.4 | 6.56 |
| Arable land | 83,121.90 | 83,760.60 | −638.7 | −0.76 |
| Permanent crops | 21,780.00 | 21,871.20 | −91.2 | −0.42 |
| Permanent pastures | 4,475.30 | 4,552.20 | −76.9 | −1.69 |
| Heterogeneous agricultural areas | 47,075.60 | 47,702.90 | −627.3 | −1.31 |
| Forests | 79,025.60 | 78,190.40 | 835.2 | 1.07 |
| Areas with shrub and/or herbaceous vegetation | 36,685.90 | 36,969.50 | −283.6 | −0.77 |
| Open spaces with little or no vegetation | 11,112.30 | 11,065.00 | 47.2 | 0.43 |
| Inland wetlands | 159 | 158.5 | 0.6 | 0.36 |
| Coastal wetlands | 531.8 | 532.3 | −0.4 | −0.08 |
| Inland waters | 2,186.20 | 2,175.10 | 11.1 | 0.51 |
| Marine areas | 945.5 | 947.9 | −2.4 | −0.261 |

Extending the analysis to the second level of Corine, the land use class that expanded the most in absolute terms is that of wooded areas (by over 800 km$^2$). Interestingly, over 900 km$^2$ of shrublands and herbaceous areas evolved into woods. Within the class of artificial areas, although urban areas for residential purposes have expanded the most in absolute terms (over 500 km$^2$), in percentage terms the largest expansion was that of industrial, commercial and infrastructure areas (10.68%). This bears witness to the strong impulse to urbanization over the last years, whose visual impact on the general public is higher than that of changes in agriculture, since these can only be perceived by a trained eye, capable of interpreting changes in the rural landscape mosaic. In other words, while the great majority of the public can perceive the higher aesthetic quality of a Tuscan farmhouse compared to a suburban condominium, not all can appreciate the difference between a mixed cultivation area and an industrial monoculture area. This is why the solution of Italy's "rural landscape question" depends on the degree of cultural maturity of its society and on its understanding of landscape evolution.

### 2.4.2 The Evolution of Wooded Surfaces

The Italian forest landscape can be historically interpreted as the result of changes brought about by human beings to the natural vegetation, following a well-defined historical sequence of culturally determined landscapes. The beauty of Italian forest landscapes was celebrated by Grand Tour travelers as much as that of the country's rural landscapes. Stendhal and Shelley were impressed by the splendid, dense chestnut groves extending down the slopes of the mountains around the Como Lake almost to its banks. Edward Lear describes with admiration groups of huge oaks, as well as the incredibly diverse landscapes he encountered during a journey to Calabria in 1847, which he contrasts with the "forests dense as carpets" and "monotonous expanses of greenery" found in other countries (Lear 1964). Like its agricultural landscape, the wooded landscape of Italy today appears simpler and more homogeneous than in the past. Its diversity is presently mainly a matter of specific composition rather than spatial arrangement. This is partially a result of the presently clear-cut separation between the woods and agriculture, after many centuries of integration. The natural substrate of the Italian forest landscape was modified long before the Roman period, but the general public is largely unaware of our forests' historically determined character. This is partly due to the scientific trends of recent years, which have seen a prevalence of environmental approaches in the study and management of forests, constantly looking for "natural areas" to be protected: a quest that fails to take adequate account, however, of centuries of human influence. The truth is that the actions of human beings in historical and protohistorical times constantly modified the ecosystem. Identifying truly "natural" landscapes in Italy is thus not an easy task (Moreno 1988). The last few decades have witnessed a trend in forest studies to relegate the historical reality of wooded landscapes to the background in favor of a naturalistic interpretation. This of course has affected management policies and led to conflicts with farmers and livestock breeders. Significantly, our catalog highlights many cases of woods that are losing their historical characteristics due not only to the abandonment of traditional practices but also to management policies aimed at transforming them into more natural formations.

The statistical data available show that in the period between the unification of Italy and the years immediately preceding World War I there was a significant reduction of Italian forests, mainly due to the expansion of agricultural land and pastures as a consequence of increasing demographic pressure in mountain areas. One of the interesting elements highlighted by the graph in Fig. 2.1 is the relationship between forest surface and demographic trends. As we can see, from the unification of the country to ca. 1910, demographic growth went hand in hand with a shrinking of the wooded surface. This is a typical landscape trend in developing countries, where the woods give way to pastures and fields to meet the urgent food demands of a growing population. In spite of some not negligible problems in the data-recording criteria, it seems certain that from the 1920s onward there was a stable reversal in this trend, with a more than twofold increase of forest surface, although accurate statistics are not available (Agnoletti 2005). Thus, in this period the ratio between population and woods extension changed, since the latter continued to expand independently of demographic growth, an indication that Italian society's food supply no longer depended on the availability of cultivable land. The 1920s thus marked the end of the last phase of surface reduction in the history of Italian woods, which had seen several expansion and reduction cycles from the Roman period onwards. The new expansion was the result of the gradual abandonment of mountain and high-hill areas, a trend that is already apparent during the Fascist period and became unstoppable in the second postwar period. The secondary, post-cultural forestation process affected all of the country's regions, especially those where the abandonment of agriculture and animal husbandry was more intense, even extending to lower altitudes. This led to a gradual reduction of the pre-existing landscape mosaic, a strong and often uncontrolled increase of wild fauna and a strong decrease of cultivated land. Today, Italian agricultural products are grown on much smaller surfaces, thanks to yield increases. Above all, however, the country imports them massively from abroad, a model it shares with Europe, North America and other industrialized countries, including some in Asia. All these countries have been experiencing for years a gradual growth of forest surface, a concomitant shrinking of agricultural surface and growing recourse to external resources.

Along with the reduction of agricultural surface, to which it is indissolubly tied, reforestation is one of the most important phenomena to affect the Italian rural landscape in the last century. The expansion of the woods from 10% of the national territory in 1920 to the present 34% has changed the face of whole regions. This statistic, however, also partially reflects changes in the notion of "woods". The forest inventory of 2007 regards as "forest formations" populations of trees or shrubs meeting all three of the following requirements: a surface larger than 5000 $m^2$, a foliage cover percentage higher than 10% and an area width higher than 20 m.[2] The inventory

[2] Ministry of Agricultural, Food and Forest Policies, Inventory of Forests and Carbon Reserves, http://www.sian.it/inventarioforestale/jsp/home.jsp.

includes the following categories: woods and other wooded areas; prairies, pastures and uncultivated land; sparsely vegetated areas; lumber farms, isolated groves and linear formations (tree rows). The land classified as "woods" accounts for 83.7% of the total forest surface, "other wooded areas" for 16.3%. According to this new classification, the forest surface of Italy is about 10,528,000 ha. Clearly, however, the above criteria also gather under the heading "woods" shrub and areas that are actually pastures or wooded pastures with trees or shrubs. These would require distinctive management approaches to adequately preserve their role in the landscape.

Among landscape changes induced by forestation, the almost threefold increase of woods in Sicily and Emilia Romagna is especially remarkable. The Italian territorial districts with the higher percentage of land surface classified as "woods" are Alto Adige, Trentino, Friuli Venezia Giulia, Liguria, Tuscany, Umbria, Abruzzo, Calabria and Sardinia. The most densely wooded regions are Liguria and Trentino, with a respective cover percentage of 62.6 and 60.5%, while the less wooded regions are Puglia (7.5%) and Sicily (10%). "Other wooded areas" are constituted by 58.0% of shrubland, with a large component of Mediterranean maquis and shrubland. If we consider the sum of all the surfaces classified as "woods" in the inventory, however, the most wooded region in Italy is Sardinia, because here "other wooded areas", that is sparsely treed areas and areas with shrub vegetation, mainly used for grazing, are the most extensive in Italy. The "woodland" of this region thus abounds with features classified as "low woods", "low-density woods" and "shrubs", making it very distinctive among Italian landscapes. This is a very interesting example of the unsuitability of the traditional concept of "woods" to a situation where wooded or treed pastures, maquis and pollarded groves—a vegetation perfectly adapted to the needs of the local economy—dominate the landscape, rather than woods intended as continuous and clearly bounded cover. Typically, this kind of vegetation is seen as a deterioration of "natural" vegetation, intended as tall woods, and is hence frequently steered to evolve in that direction.

Forestation is advancing in Italy at a rate of ca. 70,000 ha per year, which is also indicative of the rate at which agricultural surfaces are being abandoned. The advance of woods contributes to reducing the landscape diversity of complex rural landscape mosaics, at such a rate that in Tuscany about 70% of this diversity has been lost since the nineteenth century (Agnoletti 2007). This diversity, as indicated by studies of the Tuscan landscape monitoring system on some mountain areas in the region, arose from a great variety of land uses that have given way to a homogenization and banalization of the landscape (Agnoletti 2002). It is true, although not always, that the expansion of woods can increase biodiversity as a result of the increase in the number of tree species. Concomitantly, however, there is a decrease in herbaceous species associated with meadows and pastures, and in animal species populating cultivated habitats, as well as a reduction of diversity at the landscape scale. Farina (1993) provides significant testimony about this trend. His research indicates that the replacing of olive groves with woods has determined a reduction of avifaunal diversity. From a silvicultural and landscape perspective, it would be much more desirable to have less woods, but better managed ones, with a higher level of spatial diversity. Furthermore, reforestation occurred on dry stone terraces due to the crisis of traditional agriculture which was identified as the main cause of failure during heavy rainfall events causing landslides in the Cinque Terre area (Italy) in 2011 (Brandolini et al. 2018) (Fig. 2.3).

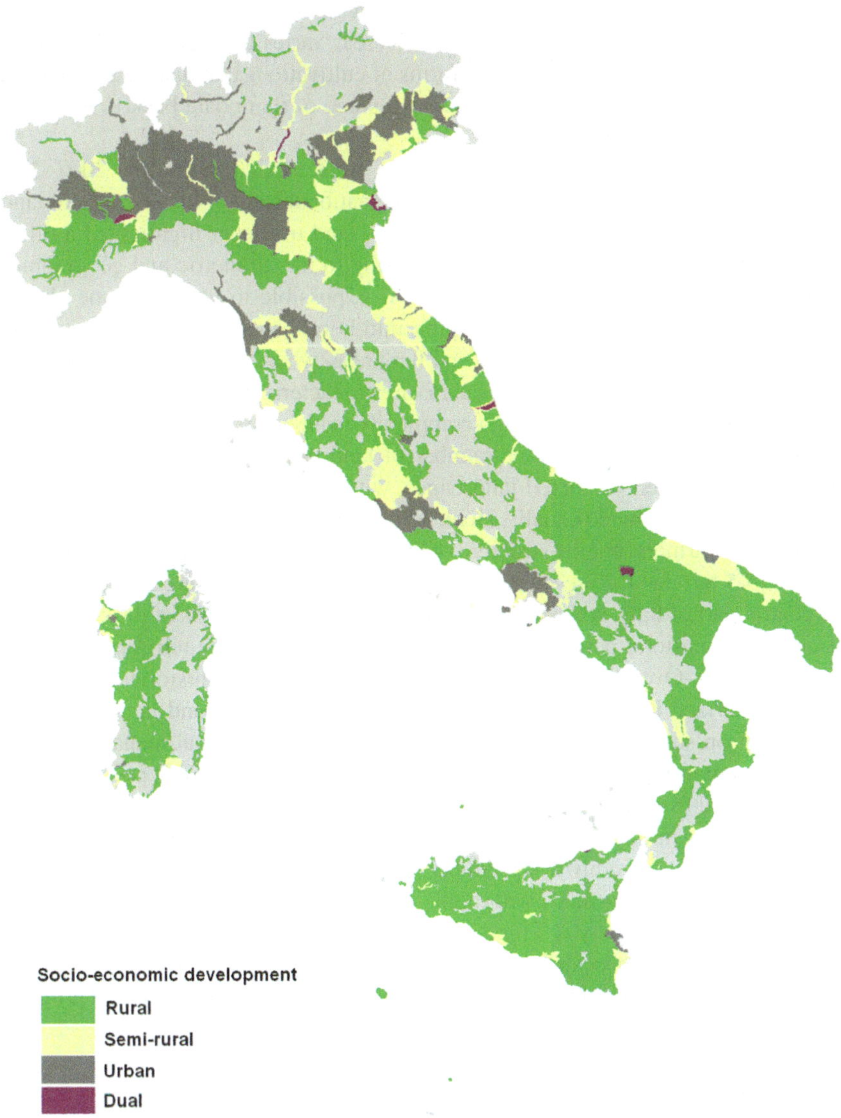


**Fig. 2.3** In prevalently agricultural areas (see map on Fig. 2.1), the development model is no longer entirely rural. There are now vast zones with diffuse urbanization and semi-rural settlement patterns. All this emphasizes the many functions of the rural landscape and the value assigned to it today, adding to the difficulty of adequate planning, but making it all the more necessary

## 2.5 The Analysis of the 123 Study Areas

The research has identified 123 areas, distributed in all the Italian regions, characterized by the presence of historical landscapes. The aim of the research was not to carry out a complete survey of the Italian historical landscapes but to give an idea of their wealth and variety. The size of the 123 areas varies from 218 to 5,750 ha. All these landscapes are characterized by forms of cultivations that date back to ancient times, most of them to the Middle Ages, but some of them date back to Roman or even pre-Roman times (Fig. 2.4).

The analysis of the landscapes of the 123 areas carried out on the 2007–2010 orthophotos represents the first database at the national level of historical rural landscapes. The data collected developing the first land use layer highlighted a differentiation in landscape characteristics, according to the altimetric and geographical localization and to the land use typology. It is possible to identify some characteristics, which allow a possible grouping into homogeneous classes. Agricultural activities are mainly located in hilly and flat areas, where agricultural activities are more economically profitable. These areas are characterized by a fine-grain structure of the landscape mosaic and by high complexity, with an average size of the patches equal to 1.12 ha and the average number of land use equal to 22. Mountains are mainly characterized by mixed landscapes, with grazing, forestry and agricultural activities often distributed equally. Here, agriculture is practiced on small surfaces by small-holder farmers, but the structure of the landscape consists on average of larger patches (5.5 ha) due to forests and pastures. Mixed landscapes are also common in the central part of Italy, where traditionally the landscape consists of the coexistence of agro-silvo-pastoral activities due to the organization of the territory into small farms (*poderi*). Overall, it is possible to state that the Italian historical landscapes are characterized by a high number of different cultivations and land uses, often carried out on small patches, as a consequence of the traditional management and high fragmentation of the properties. The high-quality products that come from this cultivation and from animal husbandry, which in some cases are found only on some tens of hectares in the whole national territory, justify and guarantee their maintenance, even if they are always products that risk disappearing in near future. The results are landscape mosaics with high complexity and diversity of the landscape structure. This complexity is a fundamental component of the bio-cultural diversity expressed by these landscapes, which include animal and vegetal species related to the traditional agricultural practices, as also described by the FAO GIAHS program.

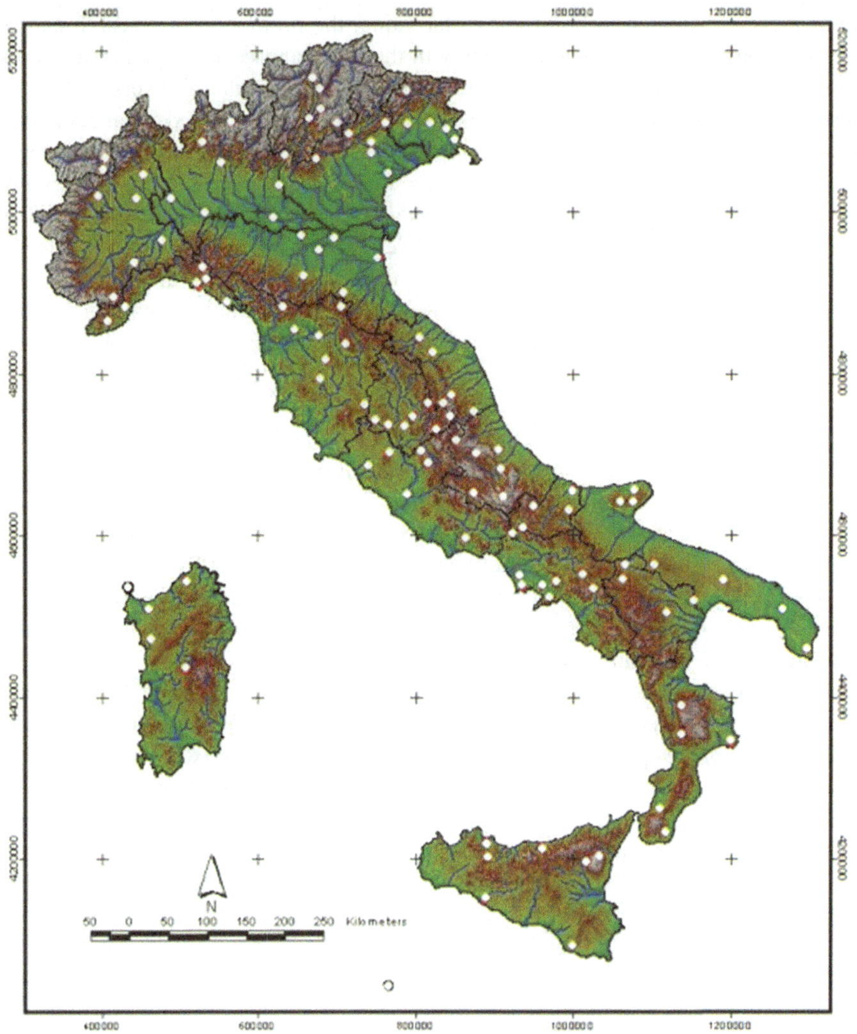


**Fig. 2.4** The localization of the 123 areas (white points) selected for the National Catalogue of Historical Rural Landscape

## 2.6 The National Observatory and the Register of Historical Rural Landscapes

One of the main results of the previous study is the fact that the Italian Ministry of Agriculture Food and Forestry Policies established in 2012 the "National Observatory of Rural Landscape, Agricultural Practices and Traditional Knowledge" (Decree n. 17070 of 2012). Among the tasks of the National Observatory of Rural Landscape can be found the surveying of landscape, of agricultural practices and of traditional knowledge considered to be of particular value, the promotion of research activities for studying the values associated with the rural landscape, its preservation, its management and planning, even in order to preserve bio-cultural diversity. It must also develop general principles and guidelines for the protection and enhancement of the rural landscape with particular reference to action taken under the Common Agricultural Policy. In addition to the landscape, the decree is aimed at the preservation and enhancement of "agricultural practices and traditional knowledge", defined as "complex systems based on ingenious and diversified techniques, on local knowledge expressed by rural civilization, which have made a major contribution to the construction and maintenance of traditional landscapes". This institution finally acknowledged the threats to the conservation of these cultural landscapes, as also occurring worldwide due to land abandonment, agricultural intensification, afforestation and urbanization which constitute threats to their diversity, coherence and identity (Antrop 2005). Rural areas losing their traditional landscapes, characterized by a small spatial scale, mixed cultures, limited technology, low use of fertilizers and pesticides and high biodiversity (Vos and Klijn 2000), require effective intervention, while also requiring dynamic conservation as suggested by many researchers (Farina 1998; Green and Vos 2001; Grove et al. 1994; Naveh 1993, 2005).

The same decree has also established the "National Register of Rural Landscape, Agricultural Practices and Traditional Knowledge". Through this Register, the Ministry identify and catalog "the traditional rural landscapes or landscapes of historical interest present within the national territory and connected traditional practices and knowledge, defining their significance, integrity and vulnerability, taking account both of the opinion of scholars and of the values ascribed to these landscapes, practices and knowledge by the concerned communities, subjects and populations". The Observatory, through the Register, has also the task of managing the "collection, analysis and classification of the data, ensuring its conservation for future generations and accessibility to potential users through a dedicated website as well as other means".

There are currently 27 landscapes and 3 traditional practices inscribed in the Register.

The Register is also the first step to access international programs, such as the Globally Important Agriculture Heritage Systems (GIAHS) program developed by FAO, the UNESCO World Heritage List and the UNESCO Network of Biosphere Reserves (MAB Program).

## References


Agnoletti M (2002) Il paesaggio agro-forestale toscano, strumenti per l'analisi la gestione e la conservazione. ARSIA, Firenze

Agnoletti M (2005) Osservazioni sulle dinamiche dei boschi e del paesaggio forestale italiano fra il 1862 e la fine del secolo XX, Società e Storia, n. 108. 377–396.

Agnoletti M (2007) The degradation of traditional landscape in a mountain area of Tuscany during the 19th and 20th centuries: implications for biodiversity and sustainable management. For Ecol Manag 249:(1/2)

Agnoletti M (2010) Paesaggio rurale. Strumenti per la pianificazione. Edagricole-Gruppo 24 Ore, Milano

Agnoletti M (ed) (2012) The Italian historical rural landscape. In: Cultural values for the environment and rural development. Springer Verlag, Dordrecht

Antrop M (2005) Why landscapes of the past are important for the future. Landsc Urban Plan 70:21–34

Brandolini P, Cevasco A, Capolongo D, Pepe G, Lovergine F, Del Monte M (2018) Response of terraced slopes to a very intense rainfall event and relationships with land abandonment: a case study from Cinque Terre (Italy). Land Degrad Dev 29:630–642

Farina A (1993) Bird fauna in the changing agricultural landscape. In: Bunce RGH, Ryszkowski L, Paoletti MG (eds) Landscape ecology and agroecosystem. Lewis Publishers, pp 159–167

Farina A (1998) Principles and methods in landscape ecology. Chapman & Hall, London

Gambino R (1994) Ambiguità e fecondità del paesaggio. In: Quaini M (ed) Il Paesaggio fra attualità e finzione. Cacucci, Bari

Green B, Vos W (eds) (2001) Threatened landscapes: conserving cultural environments. Sponpress, London, New York

Grove AT, Ispikoudies J, Kazaklis JA, Moody JA, Papanastasis V, Rackham O (1994) Threatened Mediterranean landscapes, west Crete. Research Report to EU. Department of Geography, Cambridge University, Cambridge, UK

Lear E (1964) Edward lear in Southern Italy: journals of a landscape painter in southern Calabria and the kingdom of Naples. William Kimber, London

McNeill, JR. (2001) Something new under the sun: An environmental history of the twentieth-century world (the global century series). WW Norton & Company

Moreira F, Pinto MJ, Henriques I, Marques T (2005) Importance of low-intensity farming systems for fauna, flora and habitats protected under the European '"Birds"' and '"Habitats"' directives: is agriculture essential for preserving biodiversity in the Mediterranean region? In: Burk AR (ed) Trends in biodiversity research. Nova Science Publishers, New York, pp 117–145

Moreira F, Queiroz I, Aronson J (2006) Restoration principles applied to cultural landscapes. J Nat Conserv 14:217–224

Moreno D (1988) Il paesaggio rurale fra storia e attualità. Monti e Boschi 1:3–4

Naveh Z (1993) Red Books for threatened Mediterranean landscapes as an innovative tool for holistic landscape conservation. Introduction to the western Crete Red Book case study. Landsc Urban Plan 24:241–249

Naveh Z (2005) Epilogue: toward a transdisciplinary science of ecological and cultural landscape restoration. Restor Ecol 13(1):228–234

Ploeg JDVD (2006) Oltre la modernizzazione. Processi di sviluppo rurale in Europa. Rubettino, Cosenza

Prussi PIETRO (1990) Continuità e trasformazione del paesaggio forestale: problemi e metodi della storia ecologica dei boschi. In: Cavaciocchi S, Istituto Internazionale di Storia Economica "F.Datini", Atti della XXVII Settimana di Studi: L'uomo e la Foresta, secc. XIII-XVIII, Prato 8–13 Maggio 1995. Firenze, Collana Atti delle settimane di studi ed altri convegni

Rackham O (1986) The history of the countryside. J.M.Dent & Sons Ltd., London

Romero Díaz A, Marín Sanleandro P, Sánchez Soriano A, Belmonte Serrato F, Faulkner H (2007) The causes of piping in a set of abandoned agricultural terraces in southeast Spain. CATENA 69:282–293

Sereni E (1961) Storia del Paesaggio agrario italiano. Laterza, Bari

Trechmann EJ (1929) The diary of Montaigne's journey to Italy in 1580 e 1581. Harcourt, Brace and Co

Vos W, Klijn J (2000) Trends in European landscape development: prospects for a sustainable future. In: Klijn J, Vos W (eds) From landscape ecology to landscape science. Kluwer Academic Publishers, Dordrecht, pp 13–29

# Chapter 3
# Cultural Heritage in the Region of Eastern Slovakia

**Slavomír Marcinčák, Boris Semjon, Peter Turek, and František Zigo**

**Abstract** This chapter examines the diverse relationships between cultural heritage and traditional food and agriculture products in the east part of Slovakia (Košice and Prešov region). In different regional contexts, heritage politics are encouraging the revitalization and the promotion of traditional food products with different aims, such as cultural recognition and market exploitation. We explore the multiple forms of cultural heritage that the region has occasioned, such as regional food, other agricultural products, and breeds of animals farm-raised in this region. In the nineteenth century here, there was extensive agriculture and animal production. Local food production significantly adds to food diversification by originating raw materials and specific processing conditions, which both contribute to food quality characteristics, including enhancement of a wide range of sensory properties. An important area of the eastern part of Slovakia is the Tokaj wine region. In Slovak Tokaj, traditional grape varieties are grown and produce specific wines typical of the area. The benefits of local food and agriculture products are highlighted in this research, assuming traceability of origins and relevant veterinary control.

**Keywords** Eastern Slovakia · Spiš sausage · Tokaj region · Traditional milk products

---

S. Marcinčák (✉) · B. Semjon · P. Turek
Department of Food Hygiene, Technology and Safety, University of Veterinary Medicine and Pharmacy in Košice, Komenského 73, Košice, Slovakia
e-mail: slavomir.marcincak@uvlf.sk

B. Semjon
e-mail: boris.semjon@uvlf.sk

P. Turek
e-mail: peter.turek@uvlf.sk

F. Zigo
Department of Animal Nutrition and Husbandry, University of Veterinary Medicine and Pharmacy in Košice, Komenského 73, Košice, Slovakia
e-mail: frantisek.zigo@uvlf.sk

© Springer Nature Switzerland AG 2022 
J. Hernik et al. (eds.), *Cultural Heritage—Possibilities for Land-Centered Societal Development*, Environmental History 13,
https://doi.org/10.1007/978-3-030-58092-6_3

## 3.1 Region of Eastern Slovakia

Two territorial units of the Košice and Prešov regions represent the Eastern region of the Slovak Republic. The Košice region accounts for 14% of the territory of the Slovak Republic. The region in the north and west is bordered by the Prešov and Banská Bystrica regions. In the south, it borders Hungary and Ukraine in the east. The Eastern part of the region is bordered by the "Vihorslatské vrchy", and the north by the "Slovenský raj" and the "Slovenské Rudohorie". The hilly area in the southeast extends into the "Východoslovenská nížina", which is separated from the "Košická kotlina" by the "Slanské vrchy". The highest point of the Košice Region is the 1,476 m high "Stolica" in the "Slovenské rudohorie" mountains, and the lowest point is the outflow of the "Bodrog" River from the Slovak Republic (94 m above sea level).

The region of Prešov is situated in the northeast of the country and represents approximately 18% of the area of the Slovak Republic. The northern border of the Prešov region is the state border with Poland. In the east, the region is bordered by Ukraine; in the south by the Košice region; in the southwest in a small section, it is bordered by the Banská Bystrica region; and in the west by the Žilina region. The northern point of the Prešov region is located in the village of "Becherov", the southern in the village of "Sečovská Polianka", the western in the village of Vysoké Tatry, and the eastern point, not only the Prešov region, but also the eastern point of Slovak Republic in the village of "Nová Sedlica", which represents the border point of the Kremenec crossroads (1208 m above sea level) between Slovakia, Poland, and Ukraine. Kremenec is located in the unspoiled nature of the "Stužica" forest (part of Bukovské vrchy), which is included in the UNESCO World Heritage List.

The Košice Region is represented of the former historic regions of "Abov", "Gemer", the southern part of "Spiš", and "Zemplín". The region of Prešov region covers the historical territory of central and "Horný Spiš", Šariš, and "Horný Zemplín". The region of the Eastern part of the Slovak Republic is extremely rich in cultural and historical sights, representing a part of the cultural heritage of the Slovak Republic.

The center of the Košice region is the second largest town in the Slovak republic, Košice. The first written mention of the southern suburbs dates back to 1230. Thanks to its commercial and strategically advantageous location, Košice had a rapid rise in the past. In 1369, Košice was the first city in Europe which had its own municipal coat-of-arm.

The historical center of the Košice city is the largest urban monument reserve in the Slovak Republic, in which stands the "Cathedral of St. Elizabeth ", the easternmost Gothic cathedral in Europe. Since 1924, the International Peace Marathon, which is Europe's oldest and the second oldest world marathon, is held in Košice every first Sunday in October. In 1993, the UNESCO World Heritage Site included a locality of: "Spiš Castle" with its surroundings and the town of "Spišské Podhradie", and the church town of "Spišská Kapitula". In 2000, the historical town of Bardejov which was considered to be the most Gothic city in Slovakia was also entered on the list. In 2009, the historical Spiš town Levoča was registered too.

The impact on the cultural heritage of the Eastern part Slovak region is also reflected in the national composition of the population, which, in addition to the prevailing Slovak nationality, also includes Hungarian, Roma, Ruthenian, Ukrainian, Czech, German, and Polish nationalities. The national influence was reflected in the cultural and folk traditions of the individual regions. The original breeds of animals and plant species, which were used to prepare food in this area, still retain their distinctive flavors.

## 3.2 Slovak Wine Region Tokaj

Tokaj is an exceptional wine-growing region in the heart of Europe located in two countries: Hungary and Slovakia. In Slovakia, Tokaj is located in its south-eastern part and in Hungary in the north-eastern part of the territory in the Bodrog basin. In the northern part, the border represents Zemplinske vrchy and in the south, Tokaj is bounded by the confluence of the rivers Tisa and Bodrog in Hungary. The Tokaj area is one of the oldest legislative Wine delineations in the world, as it was clearly defined and settled in 1737 by the royal decree of Charles III, and the Tokaj wine could only be grown and finished in the 21 sites listed in the decree. During Austria-Hungarian hegemony, respectively, Hungary was Definitive of Tokaj. After the First World War and the establishment of Czechoslovakia and “Trianon”, the state border divided this region into the Hungarian and Slovak parts. Since then, the influence of two states and hence two legislatures has been manifested in the Tokaj region. Slovakia received a smaller part, representing only 10% of the total area of Tokaj vineyards. According to the Act on viticulture and winery 313/2009 (2009), the Tokaj wine-growing area represents in Slovakia a closed wine-growing area bordered by cadastral territories of Tokaj villages Bara, Čerhov, Černochov, Malá Tŕňa, Slovenské Nové Mesto, and Veľká Tŕňa a Viničky. The information says that it has an area of 907 ha. The oldest wine-growing village is Malá Tŕňa, which belongs to the original Tokaj villages defined in the decree in 1737. Wine from this village got on the tables mainly of Polish nobility. Polish buyers worked in pairs. One tasted and the other handled money. If the wine suited “color, odor, and sapor”, they bought it. It is no coincidence that one of the wines is called self-made. The name itself is not of Hungarian origin, but of Polish origin.

The vine was probably brought to the territory of Tokaj by the Romans and the Slavs already knew it too. When the Hungarian tribes came to this territory in the ninth century, they already found the planted vine. Vineyard cultivation became more widespread after Christianity was adopted and even more after the Tatar invasion. As we know from history, few people remained after the Tatar ravages. King Bela IV invited settlers from Italy to the area. The village Bara, in the Slovak part of Tokaj, bears the name of an Italian town with the same name. New settlers brought new vine varieties. The variety Furmint, which became the basic Tokaj variety, was also among them. King Bela IV. called this land the Italian country. Wine production has grown in importance. Varieties such as Gohér, Balafant, Lipovina, Nutmeg, and Polyhoz

**Fig. 3.1** Tokaj region, Veľká Trňa, Slovakia

were also grown there. In addition to the Italian colonists, the region was occupied by French and German settlers. Wine from this area has long been considered a medium-quality wine. So how was the wine of such a quality born? (Fig. 3.1).

The Protestant pastor Máté Lackó Szepsi postponed grape harvesting because of the war. Autumn in 1630 was favorable for the development of mold and the wine prepared from such grapes was the first aszú. In 1655, the collection of raisins was legalized. The technology and legislation of wine have been unified and the glory of Tokaj wines spread to the world. During the looting of the Tatars, Turks, and royal troops, wine was sought. Tokaj´s winemakers were digging masked cellars to save their property. They hid people, property, and wine. And all this has proved to be advantageous for the ripening of wine. The corridors were narrow, excavated in tuff rocks. Corridors are often on the floors above each other. Various molds have settled on the walls of these cellars, in particular Cladosporium cellarae. A microclimate which contributed to the desired course of wine maturation was created here. Because the corridors were small, only small barrels with a volume of 136 L could fit in them. Their large surface, in terms of volume and the fact that they were not poured to the top, allowed the oxidative course of ripening and thereby created a unique taste called "chlebovinka" (bread-like). The main product of the Tokaj region was heavy wines. Table wines were produced in bad years.

The natural protection of the Carpathian Mountains is a special feature of the Tokaj wine region. It is open from the west by climatic air currents from the great Hungarian lowland. Autumn is mild and sunny. The grapes can mature well. The soil was formed on volcanic rocks. Andesites, ryolites, and their tuffs are present. These rocks contain many minerals, especially potassium necessary in increased amounts for the vine in vegetation (Kraus and Laštovcová 2011). Vineyards are planted at an altitude of 105–320 m above sea level. Only south, southeast,t and southwest oriented slopes are included in the Tokaj wine region. This type of slope orientation causes the vine shrubs to be exposed to the sun during warm autumn all day and sufficient amounts of natural sugar and aromatic substances in grape berries can be produced (Furdíková et al. 2015). Appropriate soil and climatic conditions are the basis for Tokaj wines quality. In good years, mold Botrytis cinerea grows into a noble form. Its hyphae outgrow on grapes, water evaporates, and the content of berry is concentrated. The whole berry shrinks, the sugar content increases significantly, the acid content is maintained, and "cibeba" is developed. In their juice, the fine aromatic substance that gives its special quality to the Tokaj wines is formed.

### *3.2.1 Specific Crops (Grape Varieties) in the Tokaj Region*

It is allowed to grow and produce wine from only three traditional Tokaj varieties: Furmint, Lipovina, and Yellow Muscat in the Slovak part of Tokaj. No more than 10% of the admixture of other varieties particularly of Riesling was tolerated. However, planting other than three Tokaj varieties is prohibited today. In addition to the three traditional varieties, three new varieties, Zeta, Kövérszölö, and Kobar, have only recently been approved in Hungary. These are varieties that ripen rather than furmint and are able to form cibeba more. The Hungarians were able to classify them among the Tokaj varieties even though they were not declared as such in earlier laws.

**"Furmint"**

The origin of the Furmint is not known exactly. According to some sources, it penetrated into Hungary in 1241 from Italy. Other authors consider that the Furmint comes from the Romanian region Tirnava (Pavloušek 2008). The most likely account seems to be that it originated from seeds and was later multiplied in the Tokaihegyalja area, from where it also penetrated into other areas of Hungary. It is widespread in several European countries, especially in Hungary and Slovakia. In the Tokaj wine region, forming the basis of Tokaj wines occupies up to 65% of the area Tokaj vineyards. In our country, it is exclusively used in Eastern Slovakia, especially in the Tokaj wine region. Thanks to its exceptional characteristics, the variety has become the basis for the production of Tokaj wines. Wines are characterized by higher acidity, which guarantees a long wine life. Modern reductive wines of fresh, fruity aroma, full of persistent taste with higher extract content and fresh acids are increasingly produced from the traditional Tokaj grape variety. Fresh apples as well as citrus fruits are found in its aroma.

**"Lipovina"**

It is probably a Hungarian variety, which originated randomly from seeds. It is widespread mainly in the countries of Eastern Europe, especially in Hungary, where it is grown in all wine-growing areas, especially in the Tokaj region. It is also known in Croatia, Romania, and Krym (Pavloušek 2008). Slovakia is the only area of the Tokaj wine-growing region, which represents approximately 29% of plantings. Lipovina comprises about 25% of Tokaj wine. The variety has been registered since 1952.

**"Muškat žltý"**

Varietas of Muscates (e.g., yellow muscate, violet, red, and black muscate) are a group whose origins date back to ancient antiquity and belong to the oldest vine varieties. Their old homeland is Syria, Egypt, or Arabia (Pavloušek 2008). Thanks to the Romans, Muscates were spread throughout their empire. After the reconstruction of the vineyards, economic reasons pushed Muscates into the background, but to this day they remained represented throughout Western Europe, the Balkans, Turkey, Greece, and the former Soviet republics, especially in the Krym. They are also grown in South Africa and North and South America. It grows mainly in the Tokaj region and represents 5–10% of the total planting of vineyards in the Tokaj wine region. "Muškát žltý"—Yellow Muscate—can be produced in a reductive way as a variety wine. The wine is strong, aromatic, and with a strong muscat aroma and fresh acid, which is missing in other muscat varieties. Especially naturally sweet wines made from overripe and tibed grapes are excellent and delicious. Most of the variety, however, is used together with Furmint and Lipovina to produce special Tokaj oxidative wines.

### *3.2.2 Specific Animals in the Region of Eastern Slovakia*

Livestock farming belonged to the basic components of agricultural production in eastern Slovakia. At the end of the eighteenth and in the nineteenth century mainly in the regions Zemplín, Šariš, Abov, Zamagurie, Spiš, and Gemer, the most important source of livelihood was tillage and the breeding of sheep, horses, cattle, pigs, and poultry. The peasant family formed a compact social and production unit, oriented to the preservation and development of the economy, which regularly worked on the allocated part of the land and paid the fees and part of the harvest to the lord or priest. The people who worked the soil did all the work manually with the use of horse power and often even cattle. In addition to tillage, sheep and cattle were grazed in most land overseen by the local shepherds. For this purpose, foothills and mountain pastures were used. Lower parts of inhabited areas such as valleys, fields, and meadows were used for cattle grazing after collecting the harvest and hay. The extent of grazing was limited not only by the corresponding area of pasture land but also by the amount of hay stock postponed for winter as a prerequisite for livestock breeding (Tančin et al. 2013).

Significant changes in agriculture and livestock production occurred in the first half of the twentieth century when gradual mechanization began to replace the pulling power of domestic animals. The connection of mechanical power engines with the electrification of agriculture had a huge impact on the organization, productivity, intensity, and profitability of agricultural production in all regions of eastern Slovakia. During the Second World War, there was a decline in agricultural development as a result of fighting on the Eastern Front and later as a result of the retreating German army, which confiscated especially horses and cattle for their needs. During the fighting events in the last months of the Second World War, there was a decrease in the number of livestock in the eastern regions of Slovakia by more than 60%. Livestock breeding and the agricultural economy stabilized until the establishment of United Agricultural Cooperatives on socialist foundations in the 50 and 60 s of the twentieth century (Sabol 2016).

#### 3.2.2.1 Traditional Breeds of Cattle Reared in Slovakia

The importance of cattle breeding is mainly about the production of milk and meat as well as other specific products (cheese, yogurt, and other dairy products). Many of them can be classified as functional foods containing significant levels of biologically active components that provide specific health benefits. Consumer increasing interest is mainly for maintaining or improving their health by eating these specific food products. In view of the increasing trend in markets for functional and traditional products, it is necessary to maintain and search for suitable animal breeds, which, by their production potential, can meet the requirements for food production of guaranteed quality (Zajác et al. 2012).

In addition to the production potential of animals, it is also important to take into account their resilience and adaptability to external conditions given the difficult geographical makeup of the country. Among the traditional breed of cattle with adequate utility and good adaptability belong Slovak Pinzgau cattle. This combined breed is characterized by its longevity and good constitution for cattle breeding in foothill and mountain areas. The first imports of Pinzgau purebred animals were organized a long time ago before 1894 when a system of cattle recording started in the territory of Slovakia. Slovak Pinzgau cattle originated by the crossing of Carpathian and Red cattle reared on the territory of north and central Slovakia with the Pinzgau breed imported from Austria. It is a small breed of cattle but with a strong constitution, suitable for mountain and extreme breeding conditions especially in Orava, Kysuce, Liptov, and Spiš regions. The Slovak Pinzgau cattle is a dairy-meat utility type, which is destined for the production of ecologically attentive milk and meat mainly in pasture farming conditions (Kasarda et al. 2009).

The live weight of the cows is 480–550 kg with a height at the withers 128–131 cm. The bulls can reach a live weight of up to 1050 kg. Typical color for this breed is the predominant chestnut-red to red-brown color with typical white badges as a white belt on the back and lower body, which also creates white cuffs on the limbs. The hoofs are dark brown and there are dark yellow horns that are dark in the upper third (Fig. 3.2).

In primary conditions, cows including older cows have a milk yield that ranges from 4300–6000 kg per lactation, respectively. Although the milk production of Slovak Pinzgau cattle does not reach the level of specialized dairy breeds, the milk

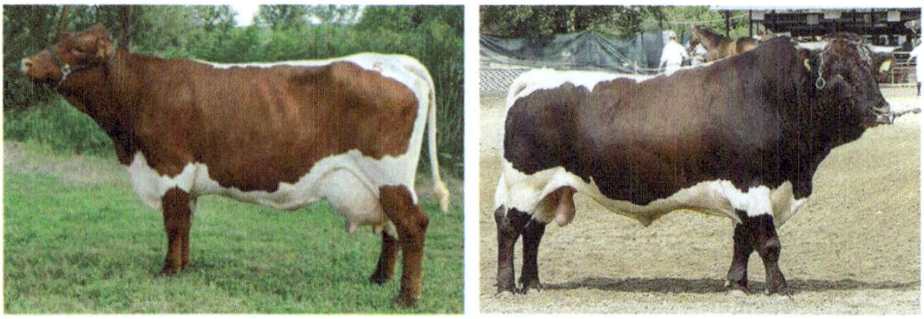

**Fig. 3.2** Slovak Pinzgau cattle

contains a high proportion of fat (4.0%) and protein (3.5%) and is ideal for Emmental type cheese production. The feeding used is mainly grazing with a combination of high-quality meadow hay. Most of the population of Slovak Pinzgau cattle are kept in herds of cows in 100 pcs. These farms are mainly specialized in the production of milk and local specialties such as cheeses, e.g., "Oravské korbáčiky". In addition to high milk quality, the meat from Pinzgau cattle is delicate and tasty. In countries with a long tradition of breeding, this breed in such nations as Austria, Slovakia, and Germany is for meat production mainly ensured by suckler cows breeding or breeding cows without market milk production based on the economical use of permanent grazing at relatively low investment and labor costs. In this system, a herd-breeding method is applied where cows are kept together with calves and calves suck milk throughout lactation. The market products of breeding are weaned calves aged 8–10 months under the weight of 250–300 kg for slaughter or for further fattening of bulls to a slaughter weight of 450–500 kg. The average daily gain in fattening bulls is 1000–1300 g and the carcass yield ranges from 55 to 58% (Kadlečik et al. 2013). Another system of beef production of Pinzgau cattle is based on pasture fattening of steers. The Pinzgau steers in extensive farming achieve high-quality meat with a live weight of 580 kg at the age of two years. The average daily gain of steers is 550–700 g with a carcass yield of 56% and 55 kg of meat in a half carcass part (Kasarda et al. 2009).

Due to areas of permanent grassland available in Slovakia especially in foothill and mountain areas, breeding and grazing of Pinzgau cattle are an appropriate economic sector of rural regions. In addition, the use of grasslands for cattle grazing contributes to landscaping and maintaining the countryside in mountain and foothill areas (Kadlečik et al. 2013).

#### 3.2.2.2 Traditional Breeds of Sheep Reared in Slovakia

Slovakia is one of the countries where sheep breeding has a long tradition. The beginning of sheep breeding dates back to the ninth century. Their breeding for centuries figured in several areas of economic, social, and cultural development of the country. Especially sheep grazing is one of the oldest, most natural, and popular breeding methods, which has not only biological and health benefits but also very important economic advantages compared to the intensive breeding method. The main terrain of continuous grazing areas for sheep is located in the mountain and foothill parts of northern, eastern, and central Slovakia. Breeding and grazing of sheep consist mainly of two functions, namely production and non-production. The production function consists in supplementing the food base, mainly by producing milk and meat, which enrich nutrition. Due to the chosen purposive focus and the difficult geographical makeup of the country on most farms in Slovakia, there are kept breeds of sheep with dual-purpose production or specialized breeds for milk production (Tančin et al. 2013).

Sheep milk is mainly used for the production of fine cheese varieties, yogurt, and whey cheeses. The high levels of protein, fat, and calcium by the casein unit make it an excellent matrix for cheese production. Also, it provides a number of health benefits, including the ability to lower cholesterol levels, strengthen the bones, boost the immune system, stimulate growth and development, prevent birth defects, reduce inflammation, and lower blood pressure (Balthazar et al. 2017).

Just for the purposes of milk production, there are notably Improved Valachian and Tsigai (Fig. 3.3), the most numerous sheep breeds on Slovak farms. These breeds are more suitable for extensive breeding systems because they are relatively appropriate given the breeding environment. However, the milk utility of both breeds is lower compared to specialized dairy breeds such as Lacaune or East Friesian sheep (Margetín et al. 2013).

The Improved Valachian variety is reared mainly in foothill and mountain areas of eastern Slovakia at an altitude of more than 600 m. An important stage of the breeding process concerning the Valachian sheep occurred after the Second World War with the relatively massive use of various forms of combination crossing in farms at all levels of the purposive policy control. In the first crossbreeding phase, with a combination cross was an improvement of primal rough-haired Native Valachian with Texel breed which has a good dense fleece of medium quality. In the next stage of the breeding process, there were used the English breeds Lincoln and Leicester with a semi-closed type of fleece. To a lesser extent, there was the use of Hampshire

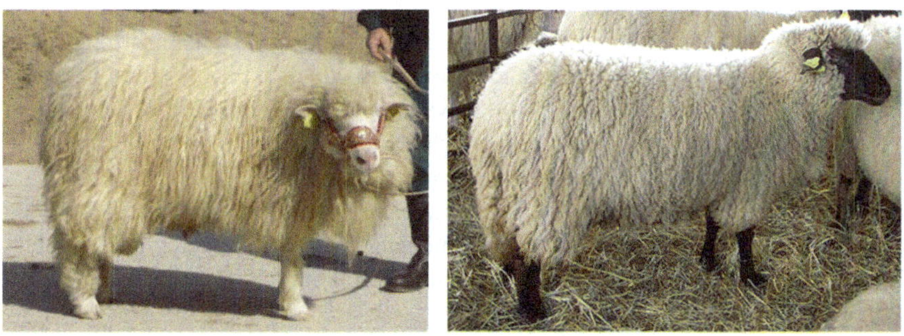

**Fig. 3.3** Improved Valachian and Tsigai

and Cheviot rams in the breeding process. The breed was recognized in 1982, as a new semi rough-haired breed of sheep with milk and wool productive potential. Ewes weigh from 45 to 50 kg, rams from 65 to 75 kg, with wool production from ewe 3.0–3.5 kg and for rams 5.5–6.0 kg. The wool is white in quality of grades C/D-D/E, with flowing character, the undercoat has at least a level of 60% of the amount of real hair. The prolificacy of ewes is again compared with previous breeds reaching 120–130% according to data over the past 10 years. Milk yield is 110–130 kg for a standard 150 day lactation period on farms included in the data. The average daily gain of lambs before weaning is 220–280 g. Depending on the breeding conditions, the production parameters are often significantly different and can exceed the breeding standard by 10–50% (Margetín et al. 2005, 2017; Mačuhová et al. 2019).

The second most numerous breed of sheep in eastern Slovakia is Tsigai. The breed originated from Asia Minor and is one of the oldest black-headed breeds which are reared predominantly in the foothills and mountain areas of eastern and central Slovakia especially in the regions Šariš and Spiš. Their breeding process and stabilization of the standard and useful properties were refined in the 50 and 60 s of the twentieth century. Later in the 80 s, there were tightened selection criteria for ewes with good exterior properties of the udder for the milking machine. This dual-purpose breed with a semi-closed type of fleece has a medium body frame, and its head and legs are covered with black hairs. Wool is white, sporadically there occurs in the fleece undesirable black hair. The live weight of ewes is 50–55 kg according to the breed standard. The wool production from ewes is 3.0–4.0 kg and from rams is 5.0–6.0 kg with wool quality is in the range of BC to CD. In purebred sheep, reproductive parameters and milk yields are very similar to those of the Improved Valachian. Fertility per ewe ranges from 115 to 140% and variability in this indicator is higher than for Improved Valachian. As regards milk production, the level of milk yield is very similar to the breed Improved Valachian while in some herds it exceeded 150 l per standardized lactation levels and the fertility per ewes gained to more than 150%. The average daily weight gain of lambs prior to weaning is 220–280 g (Margetín et al. 2005, 2013).

Breeding of sheep and other ruminants in addition to the defined production function plays an increasing role also in the non-production area. The non-production function of ruminant breeding contributes significantly to employment and thereby increases the standard of living and quality of life in rural areas. It has an irreplaceable role in the development of tourism and agritourism which, unlike Western European countries, did not have a tradition in Slovakia and is in the phase of developing. No less important is landscape creation. The extensive grazing method of farming allows farmers to maintain and cultivate hardly accessible areas of agricultural land with machines. Grazing creates favorable conditions for the growth of crops and herbs thus positively affecting the environment and cultural character of the country. Formation of the landscape environment and cultural heritage are closely associated with all this.

## 3.3 Traditional Cuisine

### *3.3.1 Specific Recipes of Tokaj Wines*

It is possible to produce varietal wines (Lipovina, Furmint, and Yellow Nutmeg) or typical Tokaj wines, mainly natural sweet wines produced from a mixture of three Tokaj grape varieties: Furmint, Lipovina, and Yellow Muscate in the Slovak part of Tokaj. Typical Tokaj wines include: "Tokajské samorodné" (dry or sweet), "Tokajské výbery", "Tokajská esencia", "Tokajská výberová esencia", "Tokajský mašláš", and "Tokajský forditáš". Variety wines are produced in an oxidative form, but at present, reductive varietal wines are becoming increasingly popular. Tokaj special wines are typically oxidative, maturing in 136 l barrels with oxygen access under a layer of skin-forming yeast called flor in Tokaj tuff cellars (Furdíková 2019).

The grapes for the Tokaj wine must be healthy, undamaged, exclusively from recognized Tokaj varieties, and grown only on qualified Tokaj terrain. The technological process of Tokaj wine production is based on differentiated harvesting. Above all, it is a late harvest. The grapes are harvested in late October and November. In particular, berries infected with Botrytis cinerea Persoon are harvested, which, through its metabolism, significantly interferes with the physiology of the berries, increases the sugar content, changes the chemical composition, and creates a sensory expression of the berries. These raisins, called cibebas, are concentrated in vats. Cibebas in the higher or smaller volume are an important part of Tokaj's special wines. Juice that flows under its own weight is essential.

The Tokaj self-made wine (tokajské samorodné) was named after the Slovak term "szamorodné", which at the end of the eighteenth century was taken over by the Polish who were buying this type of wine at that time. These wines are produced by a mixture of three Tokaj varieties with or without a low content of cibebas. "Tokajské samorodné" wine is produced in years when there are not favorable conditions for creating a sufficient quantity of cibebas. The grapes are eaten all together, cibebas are not selected separately.

"Tokajské samorodné suché" Tokaj, a self-made dry wine, contains a maximum of 10 g of sugar per liter and at least 12% of alcohol. The grapes used in the production are without cibebas with a sugar content of at least 21°NM.

"Tokajské samorodné sladké" Tokaj, a self-made sweet wine, is produced without a specific added cibebas selection, they are processed with the totality of grapes, while the content of sugar should be 24 °C at least. The grapes ferment 12 or 24 h, aromatic and color substances are extracted and must be oxidated partially. After fermentation, the wine matures oxidatively in 136 L Gönc barrels for 2–3 years. The color of the wine is pale yellow to golden yellow with a brownish shade. The aroma of the wine is the Tokaj character—overripe fruits, nuts. The taste is typical "Tokaj bread", honey to caramel tones and overripe fruit flavor.

"Tokajské výbery" Tokaj selections are made only in good years when a sufficient amount of cibebas is created. According to the old tradition, cibebas were put in special thin bags and stepped on with bare feet. Today, it is a special grinder that grinds onions without breaking the seeds. The crumbled matter is slippery and only the seeds remain in the palm when pressed. The rinds have almost disappeared during ripening. The old measure for the volume of cibebas is a "putna". One "putna" is about 25 kg of cibebas covered with 136 L of one-year Tokaj wine and represents a 1-putna aszú. "Tokajské výbery" are made only as 3–6 "putna" which means 3, 4, 5, or 6 "putna" of cibebas and 136 L of wine. After mixing, the cibebas are fermented for 18–36 h, decanted, and pressed. This liquor is poured into barrels and fermented for various periods, even half a year. It matures for at least 3 years in wooden barrels under a layer of skin-forming yeast. The usual maturation period of Tokaj selection wines is 5–10 years. The wine is yellow to deep amber in color, with a distinctive Tokaj aroma, honey tones, nuts, and overripe fruit flavor. The taste is typical Tokaj bread, honey to caramel tones, and an aspect of overripe fruits. These are natural sweet wines, where the sugar content in the wine ranges from at least 60 g.l-1 (3 putna selection) to at least 150 g.l-1 (6 putna Tokaj selection).

"Tokajská esencia", the Tokaj essence, is compared to the nectar that the gods drank at the Greek Olympus. It has a lot of sugar and extract. It is obtained from ground cibebas, from which it slowly (even over several weeks) flows out of its own weight without pressing a gold-brown liquid containing 40–60% sugar. This essence is produced only in years rich in cibebas. For years even decades, it is stored in tiny barrels stored in tufa cellars, and during this time, it may spontaneously ferment slowly. "Essence" is a diamond among wines and usually contains 2–3% of alcohol and 450–900 g of sugar per liter. It is characterized by a powerful aroma of honeycombs, dried fruit, and other divine fragrances.

"Tokajská výberová esencia", Tokaj selection essence, is obtained by the alcoholic fermentation of cibebas from qualified hunts. When harvested, grapes are selected which, after processing, are covered by a must from a defined vineyard or a Tokaj wine of the same year. The choice essence comprises at least 180 g.l-1 natural sugar and 45 g.l-1 sugar-free extract. The selection essence may be marketed after three years of aging, of which at least two years are in a wooden barrel.

"Tokajský fordítáš" is a wine for the preparation which will be used in the production of a Tokaj selection. Cibebas contain lots of sugars and extracts, so they are poured on must or young wine of the same vintage, mixed with the moldings, and allowed to ferment. Yeast sludge will be used in "Tokaj mašlaš" after the fermentation of Tokaj selections. After fermentation of the Tokaj selections, a considerable amount of sugar and extract remains in the yeast sludge. The fermented Tokaj wine is poured onto the sludge, mixed, and left to stand for 4–6 weeks. We get wine enriched with aromatic features, bouquets, and extract substances. The wine matures at least two years, of which not less than one year must require it being kept in tufa cellar. The color of "forditáš" and "mašláš" is yellow to amber yellow. The aroma is distinctive; Tokaj character–mead, noticeable fruits.

## 3.4 Specific Recipes of Meat Products

### 3.4.1 *History of Traditional Meat Production*

The Eastern Slovak region was historically part of the agrarian territory of the Austro-Hungarian Monarchy and the first Czechoslovak Republic after 2018. A more significant change occurred after the Second World War when the industrialization of the territory occurred. In the past, the important regions for the development of the meat sector were Spiš, Šariš, Horný Zemplín, and the whole area under the High Tatras. This was supported by extensive grazing of cattle and sheep, which corresponded to the mountain character of the area. Pig breeding has been widespread locally by small farmers for their own use and for the production of local meat products. Larger pig breeders and hence pig production were more widespread in the lowland area of Lower Zemplin.

An important region that historically enriched meat production was Spiš, by the production of Spiš sausages. There has been much description of the tradition of Spiš sausage production. The mentioned literature sources prove the fact that this product was produced in the region of Spiš, namely in Spišské Podhradie over 100 years ago and was made by butcher masters Štefan Varsányi, Karol Grieger, and Michal Blaško. The granddaughter of Štefan Varsányi in her book Spišské párky (Salamon 2013) states that the successful chronicle of Spiš sausages began to be written around the turn of the century at 1899–1900. The recipe for Spiš sausages, originally called "Podracké Viršle", was never written down on paper. Master butchers communicating in other ways passed on knowledge from generation to generation. From a historical perspective and from content expressed by their descendants, for example, we have some evidence to rely on. Igor Varsányi (Kurtanská 2017) and personal testimony by the grandson of Karol Grieger, Celestine Grieger, who experienced aspects of the production of Spiš sausages in the period 1938–1948 consistently state that Spiš sausages were made only from pork (lean and fat), salt, quality red ground paprika, and water, filled into ram casings. During the flourishing trade during 1938–1948, Spiš sausages required many butchers' masters in Levoča, Spišské Vlachy, and the like. The end of this era came after the nationalization and liquidation of customary trade after 1948. Based on these historical facts, the product was entered in the register of guaranteed traditional specialties of the European Union in 2011.

In the interwar period, butchery and smoked meat production developed as a separate part of the meat industry. New procedures, modern techniques, and technologies were introduced to mass-produce the meat. Under the conditions of Slovakia and most of all in Eastern Slovakia, the development of meat processing and the production of sausage and other meat products were limited by the low level of population consumption, both in the total volume and in the proportion of the overall meat industry. More than 60% of the working population worked in agriculture. Farmers provided mostly meat for their subsistence. Meat processing in the conditions of interwar Slovakia was the activity of several types of businesses and trades. In addition to slaughterhouses and operations for the primary processing of animals for slaughter, butchery production was concentrated in separate butcheries. They were the largest group of production units in the meat and meat processing industry (Hallon 1999).

Butchers and smokers were part of a craft business, whose legal framework was defined by the Trade Act passed in 1924. This trade was one of the traditional and firmly fixed branches in the structure of the production craft. Given his irreplaceable role in the production and sale of meat products, the basic food link for the population, the butcher was a respected personality of the town and municipality (Zelenák 1999).

After 1948, the change of ownership relations was decisive. After the expropriation of owners of a larger production, there followed the expropriation of approximately three thousand small traders–butchers-smokers. Already at the time of the first Czechoslovakia, then under the Slovak state, and until the 90 s of the twentieth century, the town of Prešov became the center of butchery (Fig. 3.4) and subsequently the center of the meat industry in eastern Slovakia.

The emerging national enterprises took the knowledge for the production of meat products from former private companies and tradesmen. Thus, in the new conditions of national enterprises, in the 1960s, new products emerged from the knowledge of former tradesmen or employees of nationalized companies, which were subsequently translated into technical economic standards in Czechoslovakia.

"Spišská klobása"—Spiš sausage

The Spiš Sausage was an important product for the Eastern Slovakia region, which was subsequently registered by the Czechoslovak State Standard (ČSN) 57 7270. It was one of the durable heat-treated meat products, of which there were very few on

**Fig. 3.4** The head of the bull, created by the master Bernat in 1905 for the butchers' guild. It was built into the front wall of the slaughterhouse in Prešov

the market and its popularity in the summer months at the time of agricultural work was very high for its shelf-life and palatability. The recipe was as follows:

Beef rear meat 32.5%
Beef front meat 20.0%
Pork lean meat 18.5%
Pork production meat (fat component) 62.0%

The meat dish was seasoned with a mixture of spices consisting of ground black pepper, sweet pepper, sugar, garlic, and coriander. The beef front was ground on a chopper with a board of 2.5 mm. Other meats were ground on a 3-hole plate. All meat ingredients were mixed in the mixer with the addition of spices and 1% water and subsequently ground on a chopper with 6 mm Ø holes. The meat dish was stuffed into pork thin intestines to form 30 cm long sausages. The product was smoked and then heat treated. After cooling, the products were transferred to an oven where they were smoked with cold smoke. Removal took place when the amount of water was reduced to the prescribed value.

The product was characterized by a high beef content, which was available in a sufficient quantity in this area. Nowadays, this traditional product is from a practical point of view no longer produced.

**"Prešovský kabanos"—Prešovský sausage**

Another meat product produced in this period was "Prešovský kabanos", classified under the Trade Standard (ON) 57 7224. It belonged to soft thermal treated meat products subject to cold storage. In terms of raw materials, beef meat was again an ingredient in a high percentage.

The recipe was the following:
Beef rear meat 55.8%
Pork production without skin 55.6%

The meat dish was seasoned with a mixture of spices: black ground pepper, caraway ground, sweet pepper, garlic, and hot pepper. Beef and pork were cut on a three-hole chopper. Individual meats were salted for 1–3 days. The salted beef was placed in a mixer and the mixed spices were added to water. Salted pork was then added. The mixed meat raw material was ground on a cutter with an 8 mm diameter plate and filled into continuous 22 mm diameter collagen casings in 50 cm strands. The heat treatment was carried out by placing it in a heated smokehouse where the product was baked for 90–100 min. After baking, the product was quickly cooled with cold water and prepared for distribution. At present, the product with this recipe is not on the market.

The region itself had many important meat products, which originated in meat production under the High Tatras (e.g., "Levočská klobása", "Sigordská klobása", "Prešovská saláma", "Čingovská saláma", "Tatranská saláma", and "Košická salâma"). All products showed a high content of beef, which was always available in sufficient amounts, due partly to the mountainous character of the region.

**Specific recipes of milk products**

The Slovak Republic has registered in the Quality policy system of European Union for agriculture and food products the following dairy products: "Slovenská parenica", "Slovenská bryndza", "Slovenský oštiepok", "Oravský korbáčik", "Tekovský salámový syr", "Zázrivský korbáčik", "Zázrivské vojky", and "Klenoveský syrec". Each of theese dairy product has a protected geographical indication. However, "Slovenská parenica", "Slovenský oštiepok", "Oravský korbáčik", and "Klenovecký syrec" do not have any registered producers, and it is impossible to find these dairy products in the market.

History of traditional milk production

Milking and milk processing in the region have been documented a long time ago. The fact that cattle, sheep, and goat breeding has since long ago been in Central Europe and has existed directly in our territory of the Slovak Republic has been proved by archeological excavations investigating evidence from ancient times. The beginnings of the dairy industry in the current territory of the Slovak Republic are also closely linked to the history of the dairy industry in the whole European region. In the early days, the processing of sheep milk predominated more than cow milk, which was not used until later. Breeding of cattle began to concentrate in our territory also around the middle of the sixteenth century (Keresteš et al. 2016), while sheep farming was one of the basic forms of agricultural production. Its significance in Slovakia increased relatively in the sixteenth and seventeenth centuries after colonization under Wallachian law when elements of mountain sheep fields penetrated here from eastern regions of the Carpathian mountains (Zuskinová, 1999). Until the second half of the nineteenth century, sheep were the most important source of meat and milk and these products were the basis of the daily human diet. Sheep provided products that were often the only market item of the peasant (Keresteš 2008). During World War II, the numbers of dairy cows, sheep, and goats radically decreased. Until the second half of the nineteenth century in Slovakia, there was not any progress in the dairy sector. Dairy cows started to concentrate in livestock centers after policies promoting the collectivization of peasants into unified peasant cooperatives, more particularly, dairy cooperatives. Mechanization of production as well as introduction of new methods of cleaning and disinfection increased the health safety of produced raw milk. In the nineteenth century, the industrial production of Edam cheese, mold, and various pasta filata cheese and smoked cheese was established in Slovakia (Keresteš et al. 2016).

The main sheep milk product in Slovakia was bryndza cheese, which is made by processing sheep lump cheese. In the family archives of the first industrial processors of bryndza cheese (Vagač and Molec family), it is stated that traders of cattle, butter, and other agricultural products, who came from the neighborhood of the Stará Turá trade in the middle of Slovakia, especially in the surroundings of Zvolen, found that significant oversupply of lump cheese could be sold in large cities (Mikuš et al.

1967). It is assumed that the first production of bryndza falls within the period of the last 20–250 years. The first bryndza cheese manufacture was founded by the butcher and trader Ján Vagač from Stará Turá in Detva in 1787 (Keresteš 2008). Ten years later, in 1797, bryndza cheese manufacture in Zvolen Slatina was established and subsequently in Zvolen. The first bryndza cheese manufacture was established in Liptov in 1850 in Ružomberok. The major bulk of the produced bryndza cheese was transported on carriages to Vienna and by rafts to Budapest (Mikuš et al. 1967). According to Teodor Valla, older traders taught sheep breeders how to preserve sheep's cheese by mixing it with a high amount of salt. A home-made and long-life bryndza cheese was thus made possible. This bryndza cheese was not spoiled during transport to distant markets. The preservation technique of sheep meat with salt was successfully adapted to the production of sheep cheese.

**"Slovenská parenica" cheese**

One of the traditional stretched pasta filata cheeses in the Slovak Republic is named "Slovenská parenica". On the other hand, the most famous traditional Polish steamed cheese is "Oštiepok" (Majcher et al. 2015). The authenticity of "Slovenská parenica" cheese is confirmed by the cheese strip, which can be unraveled into its fibers. Only long-term stored parenica cheese can have a very soft and loose consistency of cheese strips. Industrial dairy production of the Slovenská parenica cheese was introduced in Brezno in 1967. The Czech dairy professor Dr. Otakar Laxa in the book "Syrářství", Praha 1915, p. 340 writes: Sheep cheese of a special kind is produced around Zvolen and Brezno in Hungary by Slovak people. The strange shape of the parenica cheese is reminiscent of a decorative motif used frequently by the ancient Slavs. The same author in the periodical named "Mlékařský letáček" 10, no. 6, 193 in 1918 writes that on no other cheese is so obviously imprinted—a Slovak character as on parenica cheese, which has to be regarded as unique among cheese products!" (Keresteš and Selecký 2005). Commission Regulation (EC) No. 1049/2001 656/2008 from July 10th, 2008 registered the "Slovenská parenica" cheese in the Register of Protected Designations of Origin and Protected Geographical Indications. "Slovenská parenica" cheese is steamed lightly, smoked cheese wound into two connected rolls in the shape of "S", with the cheese strips with a diameter of 6–8 cm and 5–8 cm height. The rolls are bound with cheese string or a cheese chain. The cheese strip has before wrapping a thickness of 2–3 mm, a width of 5–8 cm, and a length of 4–6 m. "Slovenská parenica" cheese has a delicate taste and the odor of sheep's milk and a smoky smell. It is known for the characteristically pronounced fibrous structure of the curd.

Parenica cheese was originally produced in kitchens of sheep breeders, where temperature and space conditions were created for this purpose and there was a smoking and a storage place. When the shepherds learned how to build more enclosed and more spacious dwellings, the shepherd's huts, the production of the parenica cheese was transferred to the huts. The original mountain farming method has been preserved in households up to the present day. Raw sheep milk or a mixture of sheep and cow's milk with a minimum of 50% sheep milk is subject to renneting at temperatures of 29–32 °C. The rennet is processed in a wooden tray, which is a traditional way of making cheese, or in a double bottom stainless steel container, which is a modern way. The milk coagulates approximately half an hour after the addition of rennet. The curd is cut with a cheese knife and mixed, and a cheese harp is used to produce cubes with a size of 0.5 or 1 cm. Then, a stirred curd is left to sit and by hand, it is molded into a lump shape. The lump is picked out from the whey using a cheesecloth and left to hang up, while the curd continues to drip. The weight of the lump is from 3 to 5 kg (Official Journal of the European Union C 249/26).

After dripping and solidifying (from 4 to 10 h), the lump is placed on a wooden or stainless steel shelf, where the lump cheese is fermented. The process of fermentation takes about 24 h at a temperature of 20–23 °C, until the pH of the cheese reaches a value of about 5.3. Properly fermented lump cheese is cut into smaller pieces weighing about 0.5 kg and used in producing parenica cheese. The piece of lump cheese is placed in warm water with a temperature of 60–70 °C and spread on the inner side of the pot when the lump undergoes a plasticizing and kneading treatment until the cheese is formed. The cheese is taken out by hand and compressed to remove unnecessary amounts of water, stretched, and folded several times. The cheese strip is pulled out and formed by hand on a wooden board with a length of 4–6 m, a width of about 6 cm, and a thickness of 2–3 mm.

Subsequently, the cheese strip is placed in a prepared cold saline solution. Then the strip is removed from the solution, and after withdrawal, in the remainder of the saline solution and with the middle of the strip, it is folded on the board and rolled from the two ends into a full "S" shape. After that, the parenica cheese is tied with a cheese string or chain and placed on a wooden board to be left to drain. After drying, they are put into a smokehouse and they are smoked for about 2 h with hardwood smoke (Official Journal of the European Union C 249/26).

"Slovenská parenica" should contain at least 53% of dry matter, at least 50% fat in dry matter, and a maximum NaCl content of 3%. "Slovenská parenica" has a characteristic smell of sheep's milk and smoke aroma obtained by smoking with hardwood smoke. The taste should be soft, pleasantly salty from raw sheep milk. The consistency should be elastic, with a fibrous structure, and threads form when the cheese is broken open. The cheese rind color should be from yellow to brown after smoking and the color of the inside should be white or butter yellow (Official Journal of the European Union C 249/26).

At present, the consumer cannot find the product marked as "Slovenská parenica" cheese in the usual trade network. Industrial production offers vacuum packaged parenica cheese on the market, but those that are produced at shepherd huts are mostly packaged in bags. However, in the commercial network, it is possible to find only parenica cheese made from cow pasteurized milk in non-smoked and smoked varieties (Fig. 3.5).

**Traditional Easter lump cheese**

One of the traditional dairy products in Slovakia is the traditional "Syrec", "Syrek", or "Easter lump" or "Easter cheese". The period of Easter is the most important and most important feast of the year in the Christian religion. An important part of Easter was also the preparation of meals. The typical Easter meal includes Easter lump cheese.

**Fig. 3.5** Parenica cheese made from pasteurized cow's milk, unsmoked and smoked variety. *Source* Semjon et al. (2019)

The origin of the lump is said to be in Ukraine, but it is also prepared in Russia or Hungary. Fragrant and tasty cheese made from milk and eggs is being prepared in this period in most households of the Eastern region of the Slovak Republic.

"Syrec" could be made at home in different ways, sweet or salty. In the salty variety, there is added much more salt than in the classic variety, but chopped fresh green onion, chives, and parsley can be added too. In some households, there is added sliced Easter ham. The sweet variety is made with the addition of refined and vanilla sugar and in the process raisins could also be added, which enhances the sweet taste, and also enhances the cheese's appearance.

The process of making the Easter lump is very simple. We need 10 medium-sized eggs, one liter of milk, and two straight and aligned tablespoons of salt, for the production of approximately one kilogram of the lump. We have to break the eggs and whisk the egg yolks and whites together. Then we need to add the milk, preferably raw cow's milk, which we have previously pasteurized, or we can buy pasteurized whole milk.

If we are preparing a salty variety of the Easter lump, we need to add to the mixture two aligned tablespoons of salt to taste. If we want to add chopped chives, onions, or parsley, we can do this in this step. To prepare a sweet variety, add one tablespoon of salt to the egg mixture with milk and 100 g of crystal sugar again to taste. We can also add vanilla sugar and raisins.

Subsequently, we have to pour the mixture into a double bottom pot and start cooking on the stove. The mixture has to be warmed slowly, with a continuous stirring process. After a couple of minutes, we may notice that curds of egg mixture

**Fig. 3.6** Easter lump cheese made in the traditional way. From the left: classic variety; sweet variety; salty variety

with milk are being formed. It is important to stop cooking at the right time when the formed curd is still supple; otherwise, it will become hard and dry, and it would not create a consistent lump with a smooth surface.

After the coagulation, we pass the result through the cheesecloth, tie the corners of the cloth to each other, and twist it. We will create a lump with the shape of a ball and we hang it in a cool suitable place to drip. The drained and chilled lump can be consumed directly, or it could be stored in the refrigerator until the next day, which is done in many households during Easter time (Fig. 3.6).

## 3.5 Conclusion

This chapter discusses about cultural heritage in the region of Eastern Slovakia, which consists of two territorial units, Košice and Prešov regions. It explores the uniqueness of the Tokaj wine-growing region in the heart of Europe, with grape varieties like "Furmint", "Lipovina", and "Muškat žltý". Animals kept in this region, like Slovak Pinzgau cattle, Improved Valachian, and Tsigai sheep, have played an important role in livestock and food production for many years. The last part refers to a cultural tradition of sharing traditional meat, milk, and wine recipes with skills and techniques of their production including the history of their production.

The role of cultural heritage is examined with respect to promoting biocultural diversity and improving the capacity for livestock production among citizens. The food production from traditional raw materials, which are specific and characteristic for this region of Eastern Slovakia, is an important component of human nutrition here and can be an indicator for a culturally diverse region.

# References


Balthazar CF, Pimentel TC, Ferrao LL, Almada CN, Santillo A et al (2017) Sheep milk: physico-chemical characteristics and relevance for functional food development. Food Science Food Saf 16:247–262

Furdíková K (2019). "Special wines". Flor wines (In Slovak). Vinič a Víno 19(4):126–130

Furdíková K, Kakaš M, Malík F (2015) Technology of the Tokay wins (In Slovak). Vinič a Víno 15(1):16–18

Hallon E (1999). Butchery and smoked meat production in Slovakia in 1918–1938 (In Slovak). In: Mäsiarstvo a údenárstvo v dejinách Slovenska. Gradus, Martin, Slovakia, pp 173–182

Kadlečík and E., Pavlík I., Kasarda R. , 2013. Kadlečík OH, Hazuchová E, Pavlík I, Kasarda R (2013) Diversity of cattle breeds in Slovakia Slovak. J Anim Sci 46(4):145–150

Kasarda R, Kadlečík O, Ryba Š, Bučko O (2009) Estimation of genetic parameters of meat production traits of Pinzgau cattle in Slovakia. Acta Fytotechn Zootechn 12:37–40

Keresteš J, et al (2008) Sheep breeding in Slovakia: history and technology (In Slovak). Eminent Ltd., Považská Bystrica, Slovakia, p 592

Keresteš J, Selecký J (2005) Cheese processing in Slovakia—history and technology (In Slo-vak). Eminent Ltd., Považská Bystrica, Slovakia, p 368

Keresteš et al (2016) Milk in human nutrition (In Slovak). CAD PRESS, Bratislava, Slovakia, p 635

Kraus I, Laštovcová J (2011) The Slovak vinegrowing and its "terroir" from the point of view of the geologist (In Slovak). Vinič a Víno 11(2):43–45

Kurtanská K (2017) Potomok tvorcu spišských párkov Igor Varsányi má ťažké srdce na dnešných výrobcov: Jasný odkaz! (In Slovak). Nový Čas 11

Mačuhová L, Tančin V, Mačuhová J, Uhrinčať M, Margetin M (2019) Effect of weaning system and type of milk flow on milk production of crossbred ewes improved Valachian and Tsigai with Lacaune. Potravinarstvo Slovak J Food Sci 13(1):275–279

Majcher M et al (2015) SPME-MS-based electronic nose as a tool for determination of authen ticity of PDO Cheese Oscypek. Food Anal Methods 8:2211–2217

Margetín M, Milerski M, Apolen D, Čapistrák A, Oravcová M (2005) Morfology of udder and milk-ability of ewes of Tsigai, Improved Valachian, Lacaune breeds and their crosses. In: Physiological and technical aspect of machine milking. ICAR Technical Series, Rome, Italy, pp 259–263

Margetín M, Oravcová M, Makovický P, Apolen D, Debrecéni O (2013) Milk ability of Im-proved Valachian, Tsigai and Lacaune purebred and crossbred ewes. Slovak J Anim Sci 46(3):100–109

Margetín M, Oravcová M, Huba J, Janíček M (2017) Formation and characterization of Slovak dairy composite sheep breed: description of the process: a review. Slovak J Anim Sci 50(4):139–143

Mikuš M (1967) Sheep dairy and cheese processing (In Slovak). Slovenské vydavateľstvo pôdohospodárskej literatúry, Bratislava, Slovakia. p 166

Pavloušek P (2008) Encyclopedia of grape vines (In Czech) 2nd edn. Computer Press, Brno, Czech Republic, p 316

Sabol M (2016) Machine and tractor stations and mechanization of agriculture in Slovakia in 1945–1953 (In Slovak). Forum Hist 10:75–87

Salamon M (2013) Spiš sausage (In Slovak). GG, Kežmarok, Slovakia, p 159

Semjon B et al (2019) Sensory profile of Parenica cheese varieties made from pasteurized cow's milk. Potravinarstvo Slovak J Food Sci 13(1):76–82

Tančin V, Apolen D, Botto Ľ, Brestenský V, Brouček J et al (2013) Livestock farming in marginal areas (In Slovak). Press Group, Ltd., Banská Bystrica, Slo-vakia, p 174

Zajác P, Tomáška M, Murárová A, Čapla J, Čurlej J et al (2012) Quality and safety of raw cow's milk in Slovakia in 2011. Potravinárstvo Slovak J Food Sci 6:64–73

Zelenák P (1999) Butchers as a trade in the years 1945–1951 (In Slovak). In Mäsiarstvo a údenárstvo v dejinách Slovenska. Gradus, Martin, Slovakia, pp 201–207

Zuskinová, 1999. Zuskinová I (1999) Sheep breeding and farming in Liptov (In Slovak). Liptovský Mikuláš TeLeM, TLM Ltd., Slovakia, p 142

# Chapter 4
# Agricultural and Food Heritage of the Moravian Region

**Martin Král, Matej Pospiech, Lucia Hodulová, and Josef Kameník**

**Abstract** This chapter describes food and agriculture in the Moravian region (South Moravia and the Moravian–Silesian Region of the Czech Republic) with regards to cultural heritage. The main interest is focused on traditional crops, animals, and national cuisine of this region and it deals with their development through a decades-long period. The beginning of this period is marked by the work of Johann Gregor Mendel (1822–1884), the most prominent scientist related to the Moravian region. The first available records of crops in the area nowadays located within the Czech Republic date back to 1846. Results of the long-term development of culinary technologies and changes in agriculture are closely connected with the traditional national cuisine. The region's food base and cultural habits also include typical flavours of prepared dishes (seasonings and condiments). The decisive influence was clearly attributable to the cuisine of the pre-industrial rural self-supplying households or to the gradually expanding bourgeois households. Local food sources were used primarily in rural cuisine. Cereals (wheat, rye, and barley) were the base, as a source of protein supplemented by legumes (peas and lentils). Since the eighteenth century, the use of potatoes was expanding gradually and this trend has increased even more strongly during the nineteenth century, and potatoes have partly replaced the importance of cereals in people's diet. To summarize, our research highlights the advantages of cuisine and agricultural products typical for cultural heritage with a focus on bioactive compounds and their health benefits and other related aspects.

M. Král · M. Pospiech (✉) · L. Hodulová
Faculty of Veterinary Hygiene and Ecology Department of Plant Origin Food Science, University of Veterinary Sciences Brno, Palackého tř 1946/1, 612 42 Brno, Czech Republic
e-mail: mpospiech@vfu.cz

M. Král
e-mail: kralm@vfu.cz

J. Kameník
Faculty of Veterinary Hygiene and Ecology Department of Animal Origin Food and Gastronomic Sciences, University of Veterinary Sciences Brno, Palackého tř. 1946/1, 612 42 Brno, Czech Republic
e-mail: kamenikj@vfu.cz

© Springer Nature Switzerland AG 2022

J. Hernik et al. (eds.), *Cultural Heritage—Possibilities for Land-Centered Societal Development*, Environmental History 13,
https://doi.org/10.1007/978-3-030-58092-6_4

**Keywords** The Czech Republic · Fruit trees · Prunus domestica · Folk cuisine · Production

## 4.1 Czech Republic—Basic Characteristics and Agriculture

The Czech Republic is a landlocked country in the central part of Europe with an area of 78 866 km$^2$. It is located on the border of two mountain systems—the Bohemian Massif and the Western Carpathians—and it is surrounded by Austria, Slovakia, Germany, and Poland. The territory of the Czech Republic is characterized by varied soil and climatic conditions (Forchtsam and Prchal 1960). The oldest ethnic group was the tribes of Boii, and based on that, the territory was named *Bohemia*. Agriculture is an important sector of the country. The main role of agricultural production in the national economy of the state is the production of food for human consumption and raw materials needed for industry (Forchtsam and Prchal 1960). In 2016, the agricultural land fund made up 53.4% of the country's total area (URL 1). Development and changes in the agricultural and food industry in the Czech Republic are influenced by weather, market conditions, agricultural policy, and primarily by overall societal, especially governmental support of the Czech agrarian sector (URL 1). According to the EU typology, the most important type of production focus of Czech agriculture is the breeding of animals on a forage diet, field production, and mixed production. Enterprises specialized in these issues represent 98% of the agricultural land fund of the Czech Republic (URL 2) (Picture 4.1).

**Picture 4.1** Fields of the South Moravian region

**Picture 4.2** View of Moravian landscape

The Czech Republic is divided into 14 regions. The South Moravian Region with its area of 7 188 $km^2$ is the fourth largest one. Within the Region, there are located extensive cave complexes. The south is mostly flat and consists of fields, meadows, vineyards, and floodplain forests. An important part of South Moravia is the Pálava and Lednice–Valtice Complex. The dominant feature of the eastern part is the White Carpathians (URL 2). Agriculture is an important component of the Region with a long tradition. Agricultural land accounts for up to 60% of the Region's area. The most cultivated crops are cereals, rape, and sugar beet. The Region has excellent prerequisites for activities centred on viticulture, fruit, as well as vegetables. Vineyards in this Region make up 90% of the entire vineyard area of the Czech Republic. An important sector of animal production is pig and poultry farming (URL 2). Agricultural production, forestry, and fishing employ about 98 thousand workers. In agriculture alone, the estimate is approximately 84 thousand. In 2015, agricultural production accounted for 1.68% of total GDP, and food production for 2.19% of GDP (URL 1) (Picture 4.2).

## 4.2 Historical Changes in Animal Production and Animal Output

The oldest data on the agrarian sector in the Czech Republic and its regions date back to the 1950s. The development of agriculture since 1949 can be characterized as the transition from private small-scale farming—so-called “family farms”—to successful socialist large-scale production. This trend continued for 25 years, during which agricultural production steadily increased (in 1936, there were 1,530,000 private farms in Czechoslovakia; in 1976, there was a total of 2206 collective farms and 230 state farms) (Grolig 1979). Concurrent with to the intensification of agriculture, the spectrum of plants and animals central to conventional agriculture was significantly reduced. Profitable plant varieties and highly productive animal breeds came to the fore. The low level of diversity posed a risk to the future sustainability of agriculture and that was the reason to establish a national programme for the conservation and use of plant, animal, and micro-organism genetic resources important for food and agriculture (Ministry of Agriculture 2017). The difference between livestock farming in the past and present centuries has been in the approach to grazing. In the last century, grazing was an essential component of livestock breeding not only for health reasons, but also in terms of reduced production costs (Forchstam and Prcahala 1960). The year 1995 was an important period in agriculture. As shown by Chart 4.1, there was a sharp decline in cattle breeding after 1995, a decline of up to 38–40%. In 1996, the number of animals was expected to stabilize. However, this expectation did not come true and the significant decline in the number of animals continued until 2005 (Charts 4.1 and 4.2).

These sharp declines occurred in the Czech Republic over five years as a result of a reduction in previous state intervention in milk and meat prices, price liberalization, and a significant reduction in product consumption. At the same time, a complicated process of agricultural transformation began. Over the years, more than 90% of the almost exclusive forms of state and cooperative ownership were transferred to the private sector (Sereda et al. 1997).

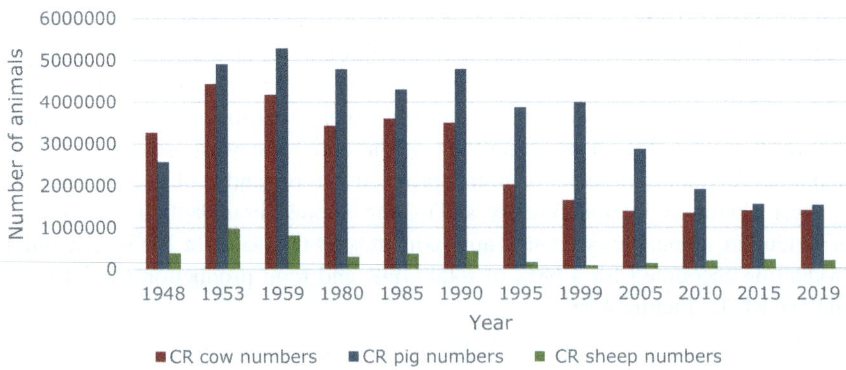


**Chart 4.1** Trends in the number of cows, pigs, and sheeps during 1948–2019

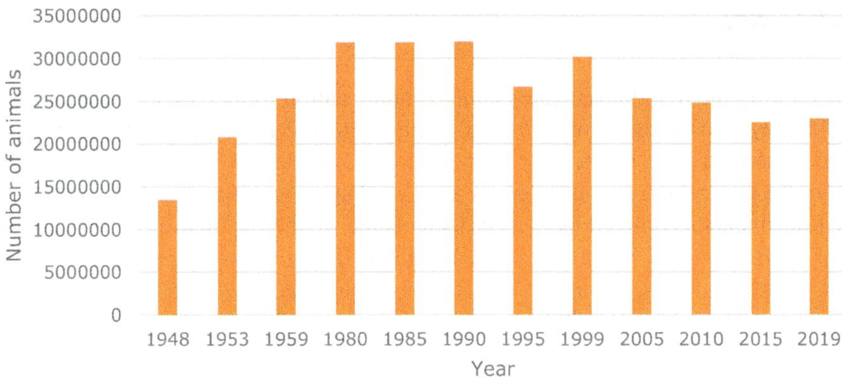


**Chart 4.2** Trends in the number of poultry during 1948–2019

## 4.3 Specific Animals in the Region

### *4.3.1 Cattle*

The original cattle breed in the territory of the Czech Republic was represented by brachycerous red cows. In the nineteenth century, various cattle breeds and varieties started to enter Bohemia and Moravia in particular from Austria-Hungary and they formed the basis of future Czech Fleckvieh Cattle (twentieth century). During the 1960s and 1970s, this breed was improved at first by Ayrshire and later by Red Holstein sires with the objective of increasing its milk production. (Jandurová 2005). Czech Pied Cattle originated from the original cattle by crossbreeding mainly with Simme and Bernese Cattle (Forchstam and Prcahala 1960). Czech Pied Cattle belong phylogenetically to the group of breeds of European Pied Cattle which is the largest and most powerful world bovine population of bilateral production orientation. It is one of the most widespread dairy breeds in the Czech Republic and it has been bred for a long time for both milk as well as meat yields (Hřeben 2019). The milk yield of Czech Pied Cattle ranged from 2200 to 4500 kg around 1960 with an average fat content of 3.8–4.0% (Forchstam and Prcahala 1960). Over the past 60 years, an average of 7344 kg of milk has been achieved with a fat content of 4.05% and a protein content of 3.55% (Andrýsek et al. 2018).

The *Bohemian Red Cattle* breed (the original breed in Bohemia and Moravia was called Bohemian Red Cattle, Silesian Red Cattle, Líštňany Red Cattle, etc.) should be considered as a cultural–historical part of breeding in the Czech lands. It is a so-called "rural" breed with a combined yield. The current population has a medium–strong skeleton, wedge-shaped head, rather short light horns sometimes with dark tips. The fur is red, sometimes with a yellowish tinge. It is characterized by constitutional strength and long life. The milk yield of the current population found on the first lactation is 1500–3000 kg, on subsequent lactations up to 4400 kg at a fat content of 3.6–3.8% and a protein content of 3.4–3.6% (Majzlík 2019).

### 4.3.2 Sheep

Sheep were the first domesticated animals in the nation formerly called Czechoslovakia. The oldest sheep in Bohemia belonged to the mouflon-like type of Ovis staderi, termed also as "sheep of the copper era". The first publication dealing with sheep breeding in our country was issued in Prague in 1561. The high standards of Bohemian sheep breeding can be documented by the fact that in 1769 breeding ewes were delivered from Bohemia to upper Austria to improve the flocks there. The reduction of wool prices after 1990 strongly influenced the decrease of the sheep population (Sereda et al. 1997).

***Wallachian sheep*** got into the Czech territory together with the Wallachian colonization of the Carpathians, which began in the fourteenth century and reached Silesia and Moravia in the Beskydy area in the fifteenth–sixteenth centuries. At the turn of the 40–50 s of the last century, the process of grading up Wallachian sheep began and it was completed in 1982 by recognizing the Wallachian Improved Sheep breed in Slovakia. The breed is well adapted to the conditions of breeding in mountain areas. The characteristic feature is a dual fleece with a short and fine undercoat, and long coarse guard hair. Limbs under the wrist and hock as well as head have no wool, except the tiny scalp of wool on the top of the head and forehead. Wallachian sheep can be used primarily in the context of the maintenance and use of mountain pastures by extensive grazing. They produce specifically traditional products, such as mountain cheese, lean meat with tender muscles, and skin with attractive long hair (Milerski 2019).

The new breed named ***Improved Wallachian sheep*** was publicly announced in 1982. In 1949, an intensive cross-breeding programme started to improve meat and wool production (Jandurova 2005).

### 4.3.3 Goat

White shorthaired goats were derived by crossbreeding of domestic farm goats between 1895–1950 with a white Saanen goat from Switzerland and Germany, first in Moravia and later in the entire territory of the Czech Republic. According to regionalization, it was intended for intensive production conditions. The breed is durable and suitable for both individual and intensive systems of breeding, and it is pure white coloured with significantly short, smooth hair. It achieves a high milk yield, an average of 710 kg in 280 days, and the best animals produced by individual studs can reach around 1400 kg of milk (Mátlová et al. 2019).

## 4.4 Traditional Crops in the Moravia Region

Since time immemorial, the main aim of agricultural production has been to achieve the highest yields of cultivated crops. Over the ages, the production process has gradually improved and, in addition to high yields, environmental friendliness is increasingly emphasized. Crop rotation and sowing procedures play a key role in both agricultural production goal-directed methods. With regard to the settlement of the Czech Republic, all basic methods of farming were gradually replaced within the framework of agricultural activities. In the early days of settlement, it was mainly a system of deforestation, slashing, and burning steppe vegetation. The gradual increase of population in the territory of the Czech Republic resulted in a situation such that this system was no longer viable and a fallow system was introduced in the eighth–ninth century. From the eighteenth–nineteenth century, this production system was replaced by a crop rotation system. New crops (potatoes, maize) as well as fodder crops and other improving crops, such as oilseeds and sugar beets, were incorporated into the sowing procedures. (Procházka 2009).

Even the crop rotation system underwent a number of changes and, from complicated 9- or 8-field sowing procedures (Forchstam and Prcahala 1960), crop rotation was gradually stabilized with the 4-field sowing procedures using also intensive exogenous inputs in the form of mineral fertilizers and plant protection products. Climate change also has an impact on crop rotation; it makes the growing of thermophilic crops, such as corn and soybeans, possible especially in the South Moravian Region. The changing subsidy policy and the fluctuating purchase price of agricultural crops also have a negative impact on crop rotation, where farmers are forced by economic pressure to fail to comply with appropriate crop rotation (Šmikmátor 2009), and thus, for example, the disappearance of legumes from the sowing procedure. Growing vegetables fertilized with organic fertilizers can also be considered as improving crops. Their inclusion in the sowing procedure is supported by the fact that the interest in self-harvesting crops is increasing, especially in suburban areas; vegetables are an ideal commodity for this form of agricultural activity.

The main crop in Czech agricultural production is cereals. They are mainly used for human nutrition (wheat, rye, and barley) and for feeding livestock (barley, oats, and corn). Traditionally, cereals are grown in winter varieties or less frequently in spring varieties (Forchstam and Prcahala 1960). The cultivation of varieties is constantly changing and the emphasis is on higher production. For example, the original winter wheat varieties of Kaštická osinatka, Židlochovická osinatka, Hadmerslebener Qualitas, Fanal, Diana, and Pavlovická 198 were grown until 1965 yielding 2.5–3 tonnes/ha and were gradually replaced by Nela, Brea, and Niagara yielding from 7–8.7 tonnes/ha in South Moravia (Haná area) (Kulovaná 2001). Lists of recommended varieties of agricultural crops for each year are issued by the Central Institute for Supervising and Testing in Agriculture (ÚKZÚZ) under the authority of the Administrative Council of the Community Plant Variety Office (CPVO) to carry out tests of variance, uniformity, and stability for the purposes of granting Community Plant Variety Rights in compliance with the Council Regulation (EC) 2100/94 in selected plant species. In the Czech Republic, proceedings on the registration of a variety are carried out in compliance with Act No. 219/2003 Coll. (Act on the Circulation of Seeds and Planting Material), as amended by Act No. 444/2005 Coll. and Act No. 178/2006.

In addition to cereals common at present, we can also come across growing of alternative cereals and pseudo-cereals to a lesser extent. The traditional alternative cereals include millet which is characterized by a short growing time and low demands for moisture. In the past, it was one of the main Slavic cereals (Kulovaná 2001). Traditionally grown varieties included Hanácka Mana (registered 1940) and Slovenské červené, while current newly registered varieties include Rubikon and Rupro. For warmer areas, sorghum is a suitable traditional crop, but it has higher fertilization requirements than millet. In the past, however, no varieties were registered in the Czech Republic for sorghum and mainly Hungarian and Soviet sorghum varieties were grown. Since 2013, new varieties of Great Millet and Súdánská tráva or their hybrids were gradually approved. Buckwheat has been cultivated in the Czech Republic since the sixteenth century. It was especially popular in Těšín, Moravian Wallachia, and Beskydy areas. The greatest boom of cultivation was until the seventeenth century. In the eighteenth century, interest in growing buckwheat was declining and its cultivation remained only in Moravian Wallachia. The original varieties include the variety called Doksanská (Moudrý 2005; Kulovaná 2001). At present, its cultivation is increasing mainly within organic farming and new varieties of Zita (registered in 2009), Zoe (registered in 2010), Zamira (registered in 2014), and Fages 4 rosa (registered in 2017) are also grown (Central Institute for Supervising and Testing in Agriculture 2019).

The situation in the cultivation of fruit trees and small fruits is different. Fruits are a product of several years old trees or herbs, thus by their very nature, it is not possible to grow them on a rotational basis. Therefore, requirements on the quality, soil type, and climate in the given area are often more important for planting them than for annual crops. The selected planting location should provide suitable conditions for growing dwarf trees (very small rootstock) for twenty-five years, for growing standard trees and espaliers (medium and large rootstock) for 50 years. (Forchstam and Prcahala 1960). It is the longevity and vegetative form of propagation that makes it possible to preserve the original varieties of fruit trees that have been preserved in the Czech Republic even considering the import of new fruit tree varieties. Fruit cultivation has a long tradition in the Czech Republic. The first evidence of fruit growing dates back to the Middle Ages. In the period between the thirteenth–fifteenth centuries, a number of apple and pear varieties began to appear, not only in the chateau and monastery gardens, but also as products of individual cultivation. In the seventeenth century, the first fruit nurseries were established. An important development occurred in the eighteenth century when the fruit growing industry developed organized fruit associations. Since the nineteenth century, we can talk about intensive fruit growing in the Czech Republic, which was supported by a number of pomological guides. After 1918, new forms of fruit cultivation were introduced into large-scale fruit production and the peak of intensive fruit growing was in 1970–1980 (Neč 2019). Plums have probably the most important history and foreign reputation in Moravia.

Plums represent 13% of all intensively grown fruit species and rank second, while the first rank is held by apple trees with 51% (Nesrsta 2016). A plum tree requires fertile soils; alluvia are suitable. It ripens to consumer maturity in areas up to 450 m above sea level. Plum trees tolerate heavier and very moist soils, but they can also be grown in poorer and drier soils on myrobalan rootstocks. The differences in soil requirements are also determined by the individual cultivars. Greengages are rather demanding in habitat selection compared to mirabelles and most other plums which have high yield even under deteriorated conditions (Forchstam and Prcahala 1960). Generally, it is recommended to grow varieties suitable for conditions in the selected geographical area. The most widespread original plums of Moravia are the varieties of Domácí velkoplodá and Wangenheimova registered since 1954. The old varieties can also include the Greengage of Althanova and Zelená renklóda as well as Mirabelle de Nancy, likewise registered since 1954. The cultivation of new varieties such as Čačanska lepotica (registered since 1991) is also widespread (Central Institute for Supervising and Testing in Agriculture 2019).

## 4.5 Use of Ruminant Meat in Folk Cuisine of Moravia and Silesia

Although the beef was the most frequently consumed type of meat among the population of the Lands of the Bohemian Crown in the Middle Ages, after the seventeenth century, its consumption gradually decreased. In the nineteenth century, butchers were still classifying the following as beef: (a) ox meat (the meat from oxen 3–4 years old was considered very good); (b) cow meat; (c) bull meat (it is reported that bull meat is not as tasty as ox meat or cow meat) (Rošický 2015).

In the folk kitchen, beef was mainly bought for soups. The beef was often consumed cooked. (The traditional recipe for beef rib preparation was as follows. Pour water into the pot, add salt, and bring to a boil. Rinse the rib, knock it a little, put it in the boiling water, cover it with a lid, and boil slowly until soft. Remove the clot that formed on the surface. About half an hour before the meat softens, add cleaned vegetables and whole onions, as well as spices, and the cooked rib should be brittle but not too soft).

In certain regions of Moravia, the influence of neighbouring Slovakia, which was part of Hungary as part of so-called Austria-Hungary for centuries, manifested itself. Goulashes from Hungary gradually became a popular meat dish in the Lands of the Bohemian Crown as well. To prepare beef, mainly ground sweet paprika and onion are used. The use of sweet paprika requires caution, it must not be rendered in hot fat, as the sugar in the paprika caramelizes, turn brown, and the paprika loses not only its nice colour but also gets a bitter taste.

Goulash preparation takes time (up to 3 h) since the collagen in the connective tissue needs to dissolve into gelatine. (Clean meat from stiff membranes and tendons, cut it into rather large pieces, and roast it in hot fat in a pot. Add salt, diced onion, crushed garlic, paprika, and occasionally water/broth to stew the meat until it is semi-soft. Let the juice then partially clear off, sprinkle with flour, fry it, add some more broth, and stew until soft).

In the foothill and mountain areas of Moravia, sheep were most commonly bred. Mutton with bone (e.g., breast) was used to prepare soups. In addition to mutton meat, root vegetables, onions, potatoes, dried mushrooms, garlic, and pepper were used to cook Wallachian mutton soup (Faktor and Žantovská 2017). Meat dishes from the Wallachian region include mutton stew. For four servings, one needed 1 kg of mutton with bone, root vegetables, 0.5 kg of potatoes, an onion and garlic, and for spices, ground sweet and hot paprika, allspice, bay leaf, ground pepper, and marjoram. (Cook the meat with the mix of spices in water until semi-soft. Prepare the paprika base. Add the cooked meat, vegetables, potatoes, paprika, and marjoram. Cook everything until soft). In Wallachia, mutton stews were often cooked in cabbage.

The popularity of fruit growing was also reflected in traditional Czech cuisine. The processing of ripe plums gave rise to a special plum jam called powidl. The consistency of powidl is only achieved by evaporating water from the overcooked fruit mash. The traditional method used is to overcook and thicken the fruit mash in copper cauldrons or stone pots (Pisch 1902). This method of processing has been preserved to the present day. However, contemporary sources also mention other methods of plum processing, such as the preservation of sloes in vinegar or the steaming of fruit, but those traditions have gradually disappeared.

Fruit with a high content of pectins (apples) has traditionally been processed into so-called jelly. Typically, this method of processing utilized red fruit, inferior quality fruit, and fruit trimmings processed for other purposes. The processing mainly involved boiling the fruit most often in copper cauldrons, followed by filtration (through a canvas or sieve made of horse hair) and adding sugar. Further cooking resulted in a reduction of the water content and reaching the desired consistency. Egg white was traditionally added for clarification and, if desired, the jelly was stained red with alkermes juice, green with chlorophyll, or blue with indigo (Pisch 1902).

## 4.6 Conclusion

The Czech Republic is a landlocked country in the central part of Europe. It is characterized by varied soil and climatic conditions. For much of history, primarily local food sources were used in rural cuisine. Cereals were the base, as a source of protein supplemented by legumes. From the eighteenth century onward, the use of potatoes expanded gradually. Fruit cultivation has a long tradition in the Czech Republic. The first evidence of fruit growing dates back to the Middle Ages. In the period between the thirteenth-fifteenth centuries, a number of apple and pear varieties began to appear. The most important foreign reputation about fruit arose from plum trees. At the present time, apple trees and plum trees are the most important fruit species. The most notable difference between livestock farming in the past and present has been in the approach to grazing. In the last century, grazing was an essential component of livestock breeding for health and economic reasons. In 1995, there was a sharp decline in cattle breeding but the number of these animals has stabilized at present. The original cattle breed on the territory of the Czech Republic as the brachycerous red cow. The Bohemian Red Cattle breed should be considered as a cultural–historical component of combined yield breeding in the Czech lands. The traditional goat derived by crossbreeding of domestic farm goats is the white shorthaired goat. The oldest traditional breed is the Wallachian sheep which is well adapted to the conditions of breeding in mountain areas. The breed had an impact on food, e.g., in traditional dishes like Wallachian mutton soup as well as cooked meat with paprika and a cabbage base. In addition to the role of local food sources in cultural heritage, attention to local food sources may suggest possible assets for purposes of tourism and export, beyond solely local significance.

## References

Act No. 219/2003 Sb (2005a) o uvádění do oběhu osiva a sadby pěstovaných rostlin a o změně některých zákonů (zákon o oběhu osiva a sadby). In Sbírká zákonů ČR, vol. 79/2003, 4053–4085

Act No. 444/2005 Sb (2005b) o územních finančních orgánech, ve znění pozdějších předpisů, a některé další zákony. In Sbírká zákonů ČR, vol. 155/2005, 8130–8168

Act No. 178/2006 Sb (2006) o uvádění do oběhu osiva a sadby pěstovaných rostlin a o změně některých zákonů (zákon o oběhu osiva a sadby), ve znění zákona č. 444/2005 Sb a některé další zákony. In Sbírka zákonů ČR, vol. 61/2006, 2069–2096

Central institute for supervising and testing in agriculture. plant variety rights and national list database. Ministry of Agriculture 2019. http://eagri.cz/public/app/sok/odrudyNouQF.do

Council regulation (EC) No 2100/94 of 27 July 1994 on community plant variety rights. Official Journal L 227, 01/09/1994, p. 0001–0030.

Faktor, V. Žantovská, K. (2017): Tradiční česká kuchyně. Práh, Prague, third edition. 344 pp. (in Czech).

Forchtsam V, Prchal Zemědělská výroba v kostce. 1960

Grolig A (1979) The development of agriculture in Czechoslovakia following world war II. Agric Adm 6(1):31–42

Hřeben F (2019) Metodika uchování genetického zdroje zvířat, Plemeno: Český strakatý skot 2017, qtd. on 26 July 2019. available online at http://genetickezdroje.cz/wp-content/uploads/2019/01/2.-Metodika-uchov%C3%A1n%C3%AD-GZ-%C4%8CESTR.docx

Andrýsek J, Král P (2018) Vznik a vývoj českého strakatého skotu. Náš chov 9/2018 volume 79 ISSN 0027–8068

Jandurova OM et al. (2005) Genetic relationships among Šumava, Valachian and improved Valachian sheep. Small Rumin Res 57 157–165

Kulovaná E Významné etapy ve vývoji sortimentu a výnosů ozimé pšenice – 2. polovina 20. století. www.uroda.cz

Majzlík I (2019) Metodika uchování genetického zdroje zvířat, Plemeno: Česká červinka 2017, qtd. on 26 July 2019. available online at http://genetickezdroje.cz/wp-content/uploads/2019/01/2.-Metodika-uchov%C3%A1n%C3%AD-GZ-%C4%8C%C4%8C.doc

Mátlová V, Konrád R (2015) Metodika uchování genetického zdroje zvířat, Plemeno: Koza bílá krátkosrstá, 2015, qtd. on 26 July 2019. available online at http://genetickezdroje.cz/wp-content/uploads/2019/01/2.-Metodika-uchov%C3%A1n%C3%AD-GZ-KB.doc

Milerski M (2019) Metodika uchování genetického zdroje zvířat, Plemeno: Valašská ovce 2016, qtd. on 26 July 2019. available online at http://genetickezdroje.cz/wp-content/uploads/2019/01/2.-Metodika-uchov%C3%A1n%C3%AD-GZ-OV.doc

Ministry of Agriculture (2017) Národní program konzervace a využívání genetických zdrojů rostlin, zvířat a mikroorganismů významných pro výživu a zemědělství–strategic and programme document for 2018–2022, Prague 1, Czech Republic 2017 ISBN: 978–80–7434–385–8.

Moudrý (2015) Jan. Pohanka a proso. Ústav zemědělských a potravinářských informací.

Neč V (2019) Ovocinářská unie České republiky, qtd. on 19 August 2019. www.ovocnarska-unie.cz

Nesrsta D, Tomáš J (2016) Přehled odrůd ovoce. Central institute for supervising and testing in agriculture, Brno. 154

Pisch O (1902) Zahradní hospodářství. Brno, Moravská akciová knihtiskárna v Brně 1902. 200

Procházka V (2009) Strukturální skladba plodin, Zemědělec, 3

Rošický V (2015) Řeznictví a uzenářství. Ing. Václav Šedivý–nakladatelství OSSIS, Tábor. Second edition, 300 (in Czech).

Sereda L, Urban F, Vachal J, Prazak C (1997) Breeding strategies for cattle, pigs, and sheep in the Czech Republic in breeding strategies for cattle, sheep and pigs in eastern Europe: task force on livestock production in Eastern Europe. In: Proceedings of a meeting held on 21–22 Jan 1996, ICC, Berlin, FAO, 1997, 151, ISSN 1024–2368, online Series 47 http://www.fao.org/3/AD250E/ad250e0a.htm#bm10

Šmikmátor J (2009) Osevní postupy v součastném zemědělství. bakalářská práce, mendlova zemědělská a lesnícká universita v Brně. Brno 57.

(URL 1) http://eagri.cz/public/web/mze/zemedelstvi/

(URL 2) https://www.czso.cz/csu/xb/charakteristika_jihomoravskeho_kraje

# Chapter 5
# Wine Quality as a Part of Cultural Heritage Affected by Its Different Geographical Origins

**Martina Fikselová, Andrea Mendelová, and Ján Gažo**

**Abstract** Regional products more and more often attract attention which as a result contributes to growing producer income and to the development of the whole region as well. Local products can be offered through innovation in restaurant services, enhanced production, farm visits and tastings related to tours of the vineyards. Regional products which can be awarded by Protected Designation of Origin (PDO) or Protected Geographic Indication (PGI) are those whose quality is connected with the places of their origin and local manufacturing methods. There is registered 15 415 ha of total vineyard area in Slovakia and production of quality varietal wines with the "Designation of Origin" or "Geographical Indication" has increased recently. In our work, the quality of Slovak wines within two selected varieties of different geographical origins was analyzed. Because in Slovakia there are six wine-growing regions, we tried to find some characteristics that differentiate the wines that originated from different wine-growing regions. Firstly, we used mainly chemical (FT-IR) analysis of 13 Sauvignon wines from three regions. Secondarily, Chardonnay variety (15 samples of wine) from five different growing regions was used and another chemical analysis based on GC–MS analysis was performed. By processing of results with statistical evaluation (PCA analysis), we detected differences among growing regions.

M. Fikselová (✉)
Faculty of Biotechnology and Food Sciences Department of Food Hygiene and Safety, Slovak University of Agriculture, Tr. A. Hlinku 2, Nitra 949 76, Slovakia

A. Mendelová
Faculty of Biotechnology and Food Sciences Department of Technology and Quality of Plant Products, Slovak University of Agriculture, Tr. A. Hlinku 2, Nitra 949 76, Slovakia
e-mail: andrea.mendelova@uniag.sk

J. Gažo
Faculty of Agrobiology and Food Resources, Department of Genetics and Plant Breeding, Slovak University of Agriculture, Tr. A. Hlinku 2, Nitra 949 76, Slovakia
e-mail: jan.gazo@uniag.sk

© Springer Nature Switzerland AG 2022

J. Hernik et al. (eds.), *Cultural Heritage—Possibilities for Land-Centered Societal Development*, Environmental History 13,
https://doi.org/10.1007/978-3-030-58092-6_5

## 5.1 Introduction

*Vitis vinifera* (the grapevine) is one of the oldest cultivated plants grown by man. Wine, as the most famous product of grapes produced by alcoholic fermentation, has been known to our ancestors. In many countries, wine has become part of the culture. The character and potential of wine are determined by the quality of the grapes. The quality of the finished wine is affected by a number of factors such as climatic conditions, rainfall and soil. The technology of grape processing then completes the parameters obtained from the basic raw material (Vnuková 2018). If the same grape varieties are grown in different areas, it is possible to obtain wines that will differ in their quality. This is due to an effect called terroir, by which a defined area whose geographical location, climate and physico-chemical natural conditions make it possible to produce specific wine products (Fic et al. 2015). Local products can be offered through innovation in restaurant services, enhanced production, farm visits and tastings related to tours of the vineyards. Regional products more and more often attract attention which as a result contributes to growing producer income and to the development of the whole region as well.

In Slovakia, legislation describes the rules of grape cultivation and wine production up to detailed specifications of wines from individual wine-growing regions. According to the wine labeling rules under the legislation, wine must be designated as non-geographical, Protected Geographical Indication (PGI) or Protected Designation of Origin (PDO) wine. There are six wine regions recognized in Slovakia: Nitrianska, Malokarpatská, Južnoslovenská, Stredoslovenská, Východoslovenská and Tokaj.

"Designation of origin" means the name of a region, a specific place or, in exceptional cases, a country used to describe a product that complies with the following requirements: its quality and characteristics are essentially or exclusively due to a particular geographical environment with its inherent natural and human factors; the grapes from which it is produced come exclusively from this geographical area; its production takes place in this geographical area and it is obtained from vine varieties belonging to *Vitis vinifera* (Council Regulation (EC) No 491/2009). "Geographical indication" means an indication referring to a region, a specific place or, in exceptional cases, a country, used to describe a product that complies with the following requirements: it possesses a specific quality, reputation or other characteristics attributable to that geographical origin; at least 85% of the grapes used for its production come exclusively from this geographical area; its production takes place in this geographical area and it is obtained from vine varieties belonging to *Vitis vinifera* or a cross between the *Vitis vinifera* species and other species of the genus Vitis.

The wine industry has a long history of labeling wines based on varietal, regional or age (vintage)-related properties. Misleading-labeled wines, by including them in a higher quality level class of wines, have been noticed. Even if the wine contains more than 88–90% of a single grape variety but the synthetic colorants and flavors are detected, it cannot be named according to the controlled denomination of origin.

In this case, the wine is considered misleadingly labeled and adulterated (Fikselová and Kunová 2012).

Protecting consumers from poor-quality products with fake ingredients or descriptions, while protecting honest and traditional traders from unfair competition, are critical issues in food quality control processes. It should be ensured that wine and all other foodstuffs are subject to the authenticity of the ingredients and particulars in the product description (Arvanitoyannis 2010). Research on wine traceability has yielded a number of new chemical markers that could not be identified without adequate analytical techniques combined with liquid chromatography and mass spectrometry, which have a high resolution and detection power. The identification of variety, geographical origin and type of wine can be successfully performed by profiling or analyzing chemical markers such as volatile compounds, polyphenols, carbohydrates and mineral elements. Chemical markers are affected by environmental conditions and wine production technology (Montet and Ray 2017). The combination of instrumental analyses with multivariate statistics has allowed the successful classification of different products by variety, geographical origin and certain aspects of the wine-making process. Multivariate statistical methods are a powerful tool for the study and classification of grapes and wines, based on complex sets of chemical and sensory data (Versari et al. 2014).

The aim of this work was to test selected basic chemical parameters of two wine varieties which originated from different wine-growing areas of Slovakia in order to differentiate these growing regions and wines from them as local products.

## 5.2 Material and Methods

### *5.2.1 Material*

Firstly, 13 samples of dry quality wines of Sauvignon variety (2015 year) were tested originating from three different wine-growing regions of Slovakia (Fig. 5.1):

- Južnoslovenska (JS): 5 samples.
- Nitrianska (NR): 5 samples,
- Malokarpatska (MK): 3 samples,

Basic chemical parameters by FT-IR method, such as total acids, tartaric acid, malic acid, glucose and fructose content, were observed.

Secondary, 15 samples of dry quality wines of Chardonnay variety (2016 year) from five different wine-growing regions of Slovakia were analyzed for volatile compound content by GC–MS analysis:

- Južnoslovenska (JS): 6 samples.
- Nitrianska (NR): 2 samples.
- Malokarpatska (MK): 5 samples.
- Vychodoslovenska (VS): 1 sample.

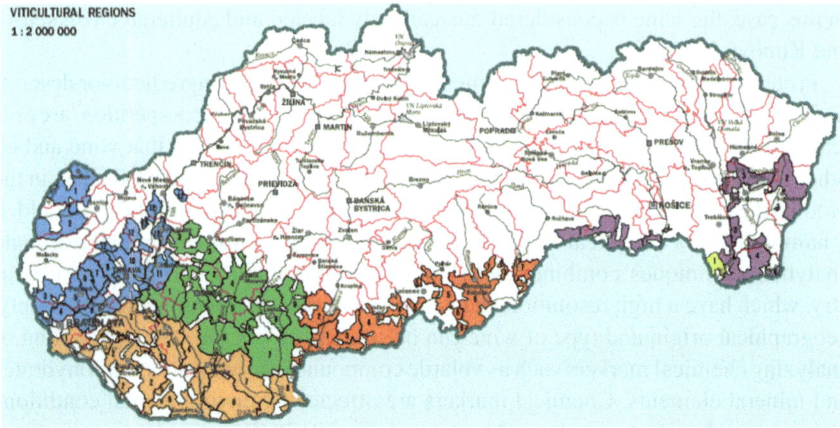


**Fig. 5.1** Viticultural regions in Slovakia (www.svapezinok.sk) *Notice* blue (MK), brown (JS), green (NR), red (SS), purple (VS), yellow (Tokaj)

- Stredoslovenska (SS): 1 sample.

## 5.2.2 *Methods*

### *FT-IR (Fourier Transform Infrared Spectroscopy) Analysis*

The determination of the selected chemical parameters in Sauvignon wines such as total acids, tartaric acid, malic acid, glucose and fructose content was performed using the ALPHA Bruker Optik GMBH analyzer. It is an integrated system for rapid analysis of musts and wines by FT/ IR infrared spectroscopy. Infrared spectroscopy provides useful information on the physical and chemical characteristics for quality control monitoring (Teixeira dos Santos et al. 2018). The sample of wine was applied to the analyzer measuring head using an applicator, where the sample was allowed to warm to 40°C. The sample was evaluated within 70–100 s and read.

### *GC/MS* **Analysis**

A set of wines (15 samples) of Chardonnay variety was analyzed by GC/MS. The concentration of the individual volatile substances was determined by extraction with methyl t-butyl ether (MTBE). Into the 25 ml volumetric flask was pipetted 20 ml of wine, 50 μl of 2-nonanol (500 mg/dm$^{-3}$) was added, together with cyclopentanone (25 g /dm$^{-3}$) in ethanol serving as an internal standard with 5 ml of saturated solution $(NH_4)_2SO_4$. The flask content was mixed and 0.75 cm$^3$ of extraction solvent was added (MTBE with 1% neohexane). After shaking and separating the phases, the top organic layer was taken into a microtube also with the resulting emulsion. The centrifuged and clear organic phase was dried with anhydrous magnesium sulfate. The extract was subsequently used for GC/MS analysis (Marko 2019). Shimadzu

GC-17A equipment was used (autosampler: AOC – 5000, detector: QP – 5050A, software: GC solution).

***Separation Conditions***

Column: DB-WAX 30 mm × 0.25 mm, 0.25 μm stationary phase (polyethylene glycol).

Sample injection volume: 1 μl split ratio 1: 5.

Gas flow rate He: 1 ml / min (linear gas velocity 36 cm/ s).

Injection chamber temperature: 200°C.

The final temperature was maintained for 5 min. The total analysis time was 30 min. The detector operated in SCAN mode with an interval of 0.25 s in the range of 14–264 at a detector voltage of 1.5 kV. Individual substances were identified on the basis of MS spectrum and retention time. Quantification was performed by comparing the sample peak area and the external standard with a correction for the internal standard.

***Statistical Analysis***

Results were processed by the Principal Component Analysis method (PCA). Statistical processing of the results was performed using STATISTICA 10 CZ software.

## 5.3 Results and Discussion

The European Union remains the world's biggest wine producer, producing around 60% of world production of wine. Wine is not an undifferentiated commodity: each type of wine even produced within the same area has specific particularities. The quality of wine produced in another year can differ from the one produced this year. Appreciation and consumption of a certain type of wine also depend on cultural aspects (European Commision 2015).

Firstly, in evaluating the quality of wines from one year, within the basic chemical parameters of Sauvignon wines, we selected those originating mainly from raw material (grapes) and potentially less affected by the grape processing.

The total acid content in our wines of the Sauvignon variety ranged from 5.33 $g.dm^{-3}$ from one region Južnoslovenska to 7.03 $g.dm^{-3}$ in the sample originating from the Nitra region.

By monitoring selected individual acids, malic acid content ranged from 1 to 2.9 $g.dm^{-3}$. It is one of the active intermediates of grape metabolism. In young wine after the must has been fermented, its content decreases, transforming it into sensory-acceptable, finer and harmonic lactic acid by the action of lactic bacteria. Reduction of malic acid is desirable because its higher concentration affects the taste and aroma of wine (Poláček et al. 2018). Excess amounts of malic acid may be present in the grapes harvested during exceptionally cold months (Volschenk et al. 2006). After the must has been pressed, malic acid undergoes further catabolic reactions, its content decreasing rapidly in fermentation, which according to Ilc et al. (2016) involved *Oenococcus oeni*, whose role is to reduce the acidity of wine by converting malic acid into lactic acid.

Tartaric acid is the main acid in wine. It occurs in grape juice as tartaric acid in the range of 1 to 7 $g.dm^{-3}$ (Margalit 2012). Tartaric acid can be synthesized from glucose by various metabolic pathways in grape berries, on which its content in the final product depends (Volschenk et al. 2006). All investigated samples of wines showed tartaric acid in content from 1.77 to 3.2 $g.dm^{-3}$. The tartaric acid content remains almost constant during the ripening of grapes (Silva et al. 2015). Within wine production, tartaric acid is not metabolized, but its amount can be reduced through physico-chemical mechanisms such as precipitation (Waterhouse et al. 2016).

The sugar content of wine depends on the cultivar, vintage, soil, weather and degree of maturity (Margalit 2012). It is affected also by variety, ripeness and climate conditions (Michlovský 2017). The sugar utilization by yeast *Saccharomyces cerevisiae* during fermentation is largely affected by the transport of sugars; glucose is usually consumed more rapidly than fructose (Dumont et al. 2008). The fructose content in our samples of wines varied from 0.77 $g.dm^{-3}$ to 10.13 $g.dm^{-3}$ and glucose content from 0.13 to 3.53 $g.dm^{-3}$.

Experimental techniques such as Gas Chromatography with Mass spectrometry (GC/MS), Nuclear Magnetic Resonance (NMR) and Fourier Infrared Spectroscopy (FT/IR) are used for wine authentication (Arvanitoyannis 2010). Another approach is based on the analysis of several components of a larger sample set, with the measured values processed by statistical methods, most often by the Principal Component Analysis (PCA) method or by cluster analysis. Samples that differ in composition are considered to be non-authentic or group samples are formed according to the origin, technological process, age, etc. (Čížková et al. 2012).

By evaluating our results by PCA analysis of basic chemical parameters of Sauvignon wines, the degree of similarity for individual wine-growing areas was shown between factors 1 and 3 (Fig. 5.2). Factors describe 69.81% of the total variability. The first homogeneous group with the degree of similarity was represented by five samples from the Južnoslovenska region. The second homogeneous group consisted of three samples from the Malokarpatska region. The last group consisted of five samples from the Nitra region and two samples from the Južnoslovenska region. Overlapping of homogenous groups can be affected by relatively close geographical distances among observed wine regions.

Plot variable factor coordinates (Fig. 5.3) show four formed groups at selected basic parameters. Selected chemical parameters of wines by different levels supported different variability of tested factors 1 and 3. The discriminating power of factors 1 and 3 revealed that total acid content is separated from tartaric acid and malic acid. It seems that there is a possibility that other organic acids can be used for the future separation of wines as well.

Secondly, to assess markers from GC/MS results of Chardonnay wines, PCA analysis of the main components for groups of compounds such as higher alcohols (isoamylalcohol, isobutylalcohol, 2-phenylethanol, 1-propanol, 1-hexanol,

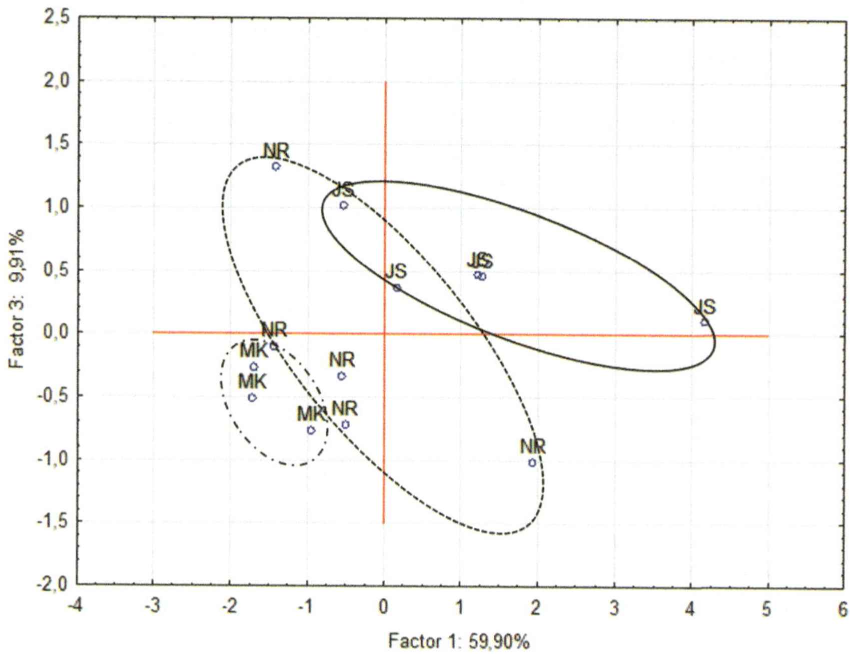


**Fig. 5.2** Plot of case factor coordinates in basic chemical parameters of Sauvignon wines of different geographical origins *Notice* Južnoslovenska (JS), Nitrianska (NR), Malokarpatska (MK)

**Fig. 5.3** Plot of variable factor coordinates in basic chemical parameters of Sauvignon wines

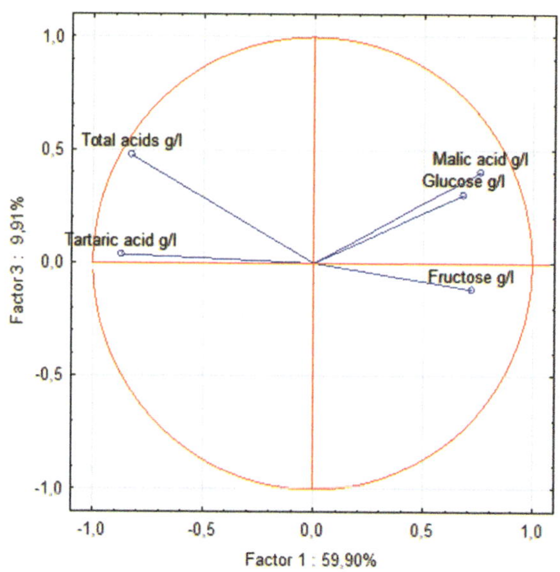

1-butanol, benzyl alcohol, methionol); terpenes such as linalool, ho-trienol, α-terpineol, nerol, geraniol) and volatile phenols (4-vinylguaiacol, 4-vinylphenol, 4-ethylguaiacol, 4-ethylphenol) were selected (Marko 2019). By evaluating volatile compound content of wines by PCA analysis, the degree of similarity for individual wine-growing areas was shown between factors 1 and 3 (Fig. 5.4). Factors describe 44.9% of the total variability. The largest group with some degree of similarity of volatile compounds consisted of all five samples from the Južnoslovenska region and two samples from the Malokarpatska region. The second group consisted of five samples from the region of Malokarpatska and the sample from the Južnoslovenska region. The third group consisted just of the Nitra wine region. The samples from the Nitra region are significantly different from the other two groups. It can be seen that there is high variability within Nitra region samples that can be explained by their opposite geographical location within this region.

Vinylphenols are the main phenols of white wines. Technology in wine production does not affect the volatile phenol content of the final product (Ayestarán et al. 2019). There were formed four groups of discriminating variables in volatile compounds of wines (Fig. 5.5). The most interesting seems to be the group consisting of 1-hexanol and benzyl alcohol. This is confirmed also in the study of Oliveira et al. (2006), who investigated the possibilities of distinguishing monovarietal wines. Hexanols

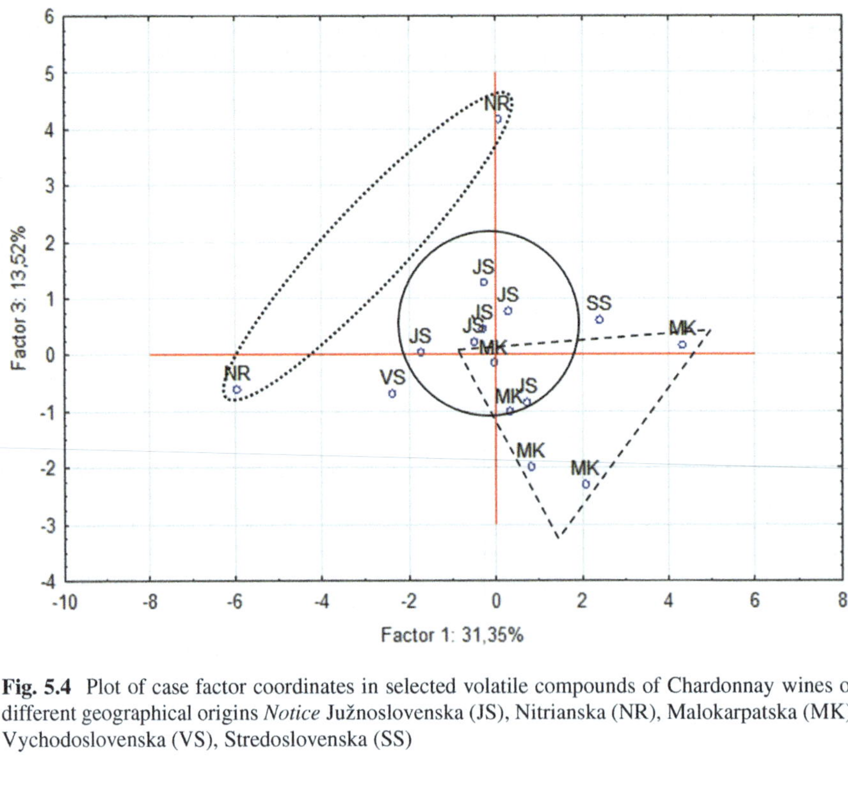


**Fig. 5.4** Plot of case factor coordinates in selected volatile compounds of Chardonnay wines of different geographical origins *Notice* Južnoslovenska (JS), Nitrianska (NR), Malokarpatska (MK), Vychodoslovenska (VS), Stredoslovenska (SS)

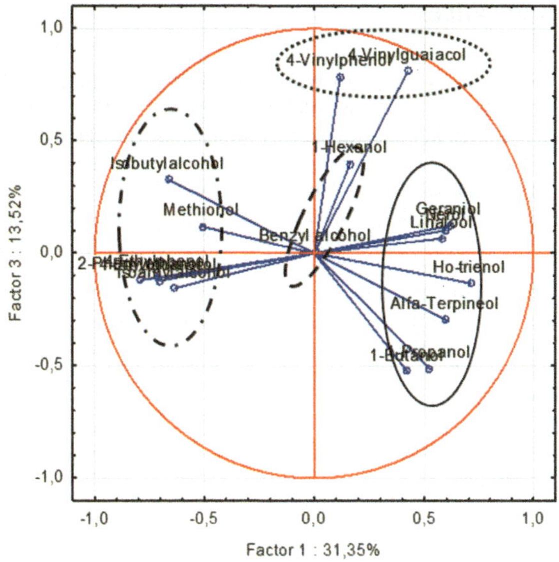


**Fig. 5.5** Plot of variable factor coordinates in selected volatile compounds of Chardonnay wines

are saturated higher alcohols containing six carbons in the chain. They are volatile substances affecting primary wine aroma. One of the most important is 1-hexanol, whose aroma is irritating, ethereal, green to fruity (Malík et al. 2017). Oliveira et al. (2006) investigated whether there is a possibility to characterize and differentiate varietal wines from different regions according to the 1-hexanol representation, taking into account the different technological processes of production, different years and different origins. They concluded that despite the dual origin of 1-hexanol, both pre-fermentative and fermentative, it is possible to distinguish wines among wine-growing areas.

We can conclude that PCA analysis of volatile compounds makes it possible to use them as one of the authentication markers to distinguish wines among individual wine-growing areas.

## 5.4 Conclusion

Since the area of production raises issues about standards concerning originality and quality characteristics of products and is subsequently reflected in the final price, the determination of geographical origin is one of the primary requirements for wine authenticity that is important as a part of cultural heritage protection as well.

Our pilot study showed that selected chemical parameters that we used were helpful to separate samples of the wines by their region, but should be combined with the PCA method or other appropriate statistical method, as their combination has been shown as an effective tool in region separation.

# References

Arvanitoyannis IS (2010) Wine authenticity, traceability and safety monitoring. Managing Wine Quality, 218–270. https://doi.org/10.1533/9781845699284.2.218

Ayestarán B, Martínez-Lapuente L, Guadalupe Z et al (2019) Effect of the winemaking process on the volatile composition and aromatic profile of Tempranillo Blanco wines. Food Chem 276:187–194. https://doi.org/10.1016/j.foodchem.2018.10.013

Čížková H, Ševčík R, Rajchl A et al. (2012) Trendy v autenticitě potravin a v přístupech k detekci falšování. chemické listy. 106:903–910. http://chemicke-listy.cz/docs/full/2012_10_903-910.pdf

European Commission (2015) Wine–market situation. https://ec.europa.eu/agriculture/sites/agriculture/files/wine/statistics/market-situation-2014-07_en.pdf

Montet D, Ray RC (eds) (2017). Food traceability and authenticity: analytical techniques. Bosa Raton, CRC Pres

Dumont A, Raynal C, Raginel F et al. (2008) The ability of wine yeast to consume fructose. Wineland Mag 110–113

Fic V et al. (2015) Víno–analýza, technologie, gastronomie. 2 theta, Český Těšín

Fikselová M, Kunová S (2012) Food adulteration and authentication. Slovak University of agriculture.

Ilc T, Werck-Reichhart D, Navrot N (2016) Meta-analysis of the core aroma components of grape and wine aroma front plant. Sci 1–15. https://doi.org/10.3389/fpls.2016.01472

Malík F, Furdíková K, Ruman T et al. (2017) Abecedarium vini. Fedor Malík a syn

Margalit Y (2012) Concepts in wine technology: small winery operations. Board and Bench Publishing, San Francisco

Marko M (2019) Využitie vybraných markerov pre geografické rozlíšenie vína odrody Chardonnay. diploma thesis. Slovak University of Agriculture

Michlovský M (2017) Vinohradnictví. rakvice vinselekt michlovský

Oliveira JM, Faria M, SÁ, F. et al (2006) $C_6$-alcohols as varietal markers for assessment of wine origin. Anal Chim Acta 563:300–309. https://doi.org/10.1016/j.aca.2005.12.029

Poláček Š, Tomáš J, Vietoris V, Lumnitzerová M (2018) Vinárstvo, someliérstvo a enogastronómia. Slovak University of Agriculture, Nitra

Council regulation (EC) No 491/2009 of 25 May 2009 amending regulation (EC) No 1234/2007 establishing a common organisation of agricultural markets and on specific provisions for certain agricultural products

Silva FLN, Schmidt E, Silva F, Messias C. et al. (2015) Quantitation of organic acids in wine and grapes by direct infusion electrospray ionization mass spectrometry. Anal Methods 7: 53–62. https://doi.org/10.1039/c4ay00114a10.1039/c4ay00114a

Teixeira dos Santos C, Páscoa R, Porto P et al (2018) Raman spectroscopy for wine analyses: a comparison with near and mid infrared spectroscopy. Talanta. https://doi.org/10.1016/j.talanta.2018.04.075

Versari A, Laurie F, Ricci A et al. (2014) Progress in authentication, typification and traceability of grapes and wines by chemometric approaches. Food Res Int https://doi.org/10.1016/j.foodres.2014.02.007

Vnuková I (2018) Charakteristika akostných vín odrody sauvignon rôzneho geografického pôvodu. diploma thesis. Slovak University of Agriculture. http://svapezinok.sk/files/gallery/20091015134800_Odborne%20vzdelavanie%20v%20SR_SK.pdf

Volschenk H, Van Vuuren H, Viljoen-Bloom M (2006) Malic acid in wine: origin, function and metabolism during vinification. S Afr https://doi.org/10.21548/27-2-1613

WATERHOUSE, Andrew – SACKS, Gavin – JEFFERY, David. (2016) Acids. Understanding Wine Chemistry, pp. 19-33, ISBN: 978-1-118-62780-8

# Chapter 6
# Food and Meals in Czech Lands from a Cultural-Historical Perspective

**Josef Kameník**

**Abstract** The term *meal* means a structured event of eating (an eating occasion) organised by rules concerning time, place and sequence of action. A meal can also be considered a product (a dish), i.e. ingredients transformed by cooking and combined into a meal. In contrast with a meal, a snack is typically a relatively unstructured food event usually without any culturally understood rules of combination or sequence. Meals contribute to ordering our days into segments: morning—midday—afternoon—evening. Food preparation and meals involve interactions with work and other activities. This model was created during the Neolithic Revolution. In the Middle Ages, doctors believed that one should eat only when an earlier meal had left the stomach. According to this opinion, people were able to eat two full meals a day. The two-day meal pattern (lunch—dinner) became the standard for mediaeval Europe. The eighteenth and nineteenth centuries in Europe were a time when national cuisine was formed. There are three basic pillars of national cuisine—basic foods, culinary techniques and flavour principles. Three basic types of cuisines co-existed for centuries: the rural cuisine, the burgher cuisine and the aristocratic cuisine. The most widespread type was the rural cuisine. The rural cuisine in the Czech Lands was based mostly on local products and plants. Cereals consumed in the form of mashes or bread were the basis of the diet in the Middle Ages. Potatoes spread in the nineteenth century and replaced cereals to some extent in some regions. Since the seventeenth century, the proportion of meat in the diet has declined. In the nineteenth century, its proportion fell to twenty percent in comparison with the sixteenth century. From the Middle Ages until the eighteenth century, meals were prepared in open fireplaces. However, these were gradually replaced by a bread oven with a closed fireplace with a cooking plate. The flavour of dishes was influenced by the herbs and spices used, and by garlic, onion, horseradish, chervil and root vegetables. A long-term cultural-historical perspective is obviously important in understanding the cultural heritage regarding food at any given point in time.

J. Kameník (✉)
Faculty of Veterinary Hygiene and Ecology, Department of Gastronomy, University of Veterinary Sciences, Brno, Palackého tr. 1946/1, 612 42 Brno, Czech Republic
e-mail: kamenikj@vfu.cz

© Springer Nature Switzerland AG 2022 
J. Hernik et al. (eds.), *Cultural Heritage—Possibilities for Land-Centered Societal Development*, Environmental History 13,
https://doi.org/10.1007/978-3-030-58092-6_6

**Keywords** National cuisine · Culinary technology · Cereals · Potatoes · Cabbage

## 6.1 Introduction

As soon as the human race moved from the culture of hunter-gatherers to a settled agricultural way of life, the hours of the day began to be divided into work and rest (Elias 2014). In the intervals between work, i.e. at times of rest, space was also given for the consumption of foods prepared in the form of meals. The term "meal" is used to mean foods consumed as part of a certain structured activity (breakfast, lunch, dinner), and in this sense a meal is understood as both the act of consuming food and the content of the meal, i.e. the product itself (Kjærnes et al. 2009). In Anglo-Saxon literature, the word *meal* is differentiated from the term *snack* (an abbreviation of *snack foods*). A snack means a relatively unstructured activity involving the consumption of food with no rules (or only looser rules) of combination or sequence. To put it concisely, the term snack (or *snacking*) means the consumption of foods between the times of the main meals of the day. In this sense, the consumption of calorific drinks is also included among snacks (Hess et al. 2016).

Long periods of development over the centuries created certain specific properties of foods prepared in individual regions that led to characteristic national cuisines. National cuisines differ from one another in three areas: (1) the culinary techniques used, (2) the choice of basic foodstuffs and (3) the tastes and aromas of dishes resulting from the application of typical spices, herbs and other ingredients (Meiselman 2009). For example, the use of garlic, basil and oregano is characteristic of southern Italian (Neapolitan) cuisine, while spicy dishes seasoned with chilli and Szechuan pepper are typical of Chinese Szechuan cuisine from the Sichuan province.

Czech national cuisine is also the result of the centuries-old development of culinary technology, basic foodstuffs, supplementary ingredients and cultural customs influenced by the sharing of cultural information between the social and ethnic groups that lived in the area that today makes up the Czech Republic. Historically, Czech cuisine is related to Bavarian and Austrian cuisine, while it has also been influenced by Hungarian cuisine. Germans, Jews, Italians, Spaniards (under Rudolf II) and Russians (as a result of emigration after 1917) have all contributed towards the enrichment of Czech cuisine.

Three categories, or three types, of national cuisine can be observed historically in the majority of the countries of Europe. The most widespread type was comprised of the rural cuisines of the village population living a country way of life. A burgher type of cuisine predominated in the towns and cities, while the aristocracy enjoyed the cuisine of the elite. Each of the given cuisines had its characteristic features. Rural cuisine was the most widespread of the three categories and had a regional character. It was the most rigid (i.e. it was the least subject to developmental change), and was based entirely on local products. It had a subsistence character and was subject to pronounced seasonal effects. Foodstuffs for town cuisine were acquired largely by purchases made in town markets or stores (butchers, bakers, gingerbread bakers).

This was characterised by a more regular system (it was not influenced by changes in the length of the day during the course of the year or the associated seasonal work in the fields) and assimilated many influences from the surrounding area. Aristocratic cuisine (the cuisine of the elite) was based on the work of professional chefs and developed under the influence of other European cuisines (it was cosmopolitan). It was characterised by frequent sharing of recipes, and a wider range of foods and ingredients were available for this type of cuisine. While aristocratic cuisine took advantage of the skill and experience of male chefs, the preparation of food dishes was performed by female cooks in the kitchens of the towns and villages.

The aim of this chapter is to describe the development of the national cuisine in Moravia in the Czech Republic. The text focuses primarily on the rural type of cuisine, which was depended on the basic regional foods of its time. This chapter considers the development of kitchen technology, ingredients used and the predominant food dishes traditionally prepared in previous centuries.

## 6.2 The Development of Culinary Technology

The first use of fire for heat and cooking is credited to *Homo ergaster/erectus* (Outram 2007). Specific archaeological evidence of the use of fire is not, however, always conclusive (Svoboda 2014). The ability to make, maintain and use a fire was one of the accompanying and, evidently, contingent factors in the migration of Man from Africa to Eurasia. Archaeological finds point to the use of fire in a locality in Israel (0.79 million years ago) and in China (0.4–0.5 million years ago), though they are, however, rather circumstantial. More consistent evidence of the existence of localised fireplaces comes from the period 0.40–0.23 million years ago in a number of European settlements.

Energy from cooked food can be used earlier by the body and cooking significantly reduces the amount of energy consumed by digestion itself. This is particularly true of animal proteins—meat. Man prepared food directly on an open fire for thousands of years. Fundamental change in food preparation from the technological point of view came in the Middle Ages when homes were gradually equipped with bread ovens with a firebox and smoke vent (Petráňová 2014). The body of the oven protruded into the living room where it acted as a source of heat (Fig. 6.1). This combined body with an oven and a fireplace for cooking proved so successful that it served in various modifications until the eighteenth and nineteenth centuries. This system can be considered specific to central Europe, as food was cooked in a fireplace beneath a chimney in Southern and Western Europe, while a furnace with no designated space for cooking was used in Eastern Europe (Fig. 6.1).

In the summer, a fire was made in the chamber of the furnace and the smoke went up the chimney. Pots stood on a tripod, and pans on their own earthenware legs (Fig. 6.2). In the winter, a fire was made in the depth of the furnace and the flame taken so that it warmed the body of the oven. The cook had to put cooking pots with food in through the opening of the oven with the use of a pole stuck into the hollow

**Fig. 6.1** In the Middle Ages homes were gradually equipped with bread ovens with a firebox and smoke vent

handles of the pots. An adequate amount of dry wood (as much as 2 $m^3$!) was needed for the regular baking of bread and occasional baking of white baked goods. The size of the oven depended on the size of the household. It could hold as many as 16 loaves and baking was generally performed every other week. The shape of the pots also corresponded to the method of cooking over an open fire. Ceramic pots were narrowed towards the bottom to enable the flames to heat them freely on an open fire.

The transition to cooking on an enclosed fire (the "stove revolution") was the second fundamental milestone in the development of communal catering (Petráňová 2014). This took place from the last third of the eighteenth century onwards in the towns, and from around the middle of the nineteenth century in the countryside. Pots with a wide flat bottom began to be used on flat stove hobs, initially also made of ceramics, and later made of cast iron and enamelled sheet metal. Enamelled pots became the most popular cooking vessels as they combined the advantages of iron and glass (Teuteberg 2007). They were resistant to acids, easily cleaned and practically unbreakable.

Cooks could place more than one pot on the stove plates at one time and this shortened the dish preparation time (Fig. 6.3). Cooking was also combined with the heating of water. The stove not only saved energy, but also meant safer kitchen operations. Most of the fires that destroyed large parts of towns and villages from the

**Fig. 6.2** The pots on tripods were used to prepare meals on open fireplaces

Middle Ages to the eighteenth century originated from kitchen fireplaces. Collected fat was sometimes added to the fire with a ladle to speed up cooking. If the flame rose too high, the soot in the smoke outlet caught fire and fire spread to the roof. The danger of fires was limited by the decree issued during the reign of Maria Theresa (1740–1780) on brick chimneys. The introduction of stoves also added new technical possibilities to the existing methods of cooking. Stewing meat in its own juices, improved frying and everyday baking either on hotplates or in the oven were all new developments.

Stoves made of cast iron and brass, in which coal was burned instead of wood, later began to appear in the towns. Coal stoves were gradually replaced by gas stoves

**Fig. 6.3** Stoves replaced open fireboxes for meal preparation. They were combined with an oven (on the right) to bake smaller dishes. The bread oven was used for baking bread (in the picture behind the stove on the top left). The method of heating and baking bread remained the same as in the Middle Ages

in the period 1880–1930 (Teuteberg 2007). Gas stoves were faster, didn't create any ash, and allowed more precise temperature regulation. Electric cookers that limited the source of heat to a single ring spread from the 1930s onwards. The hob was later separated from the main baking oven, and hot-air ovens were introduced.

### 6.2.1 The Range of Basic Foodstuffs

Czech cuisine is based on raw ingredients that could be cultivated near home, primarily cereals, legumes and certain kinds of vegetables such as cabbages, onions, garlic, root vegetables and, from the eighteenth century onwards, potatoes (Faktor and Žantovská 2017). Findings of carbonised plant remnants at Slavonic settlements in the Czech Lands of the sixth to seventh centuries testify to the fact that their inhabitants focused primarily on the cultivation of cereals—wheat, barley, rye, oats and millet. They were acquainted with the legumes peas and lentils and also hemp. Of fruit trees, only the wood of the plum tree has been found, and no traces of the cultivation of vegetables have been found. Archaeological proof of cucumbers has been found only from the ninth century onwards (Váňa 1983).

People in Europe were increasingly dependent on cereal crops from the ninth century until the period of the peak population boom in the early fourteenth century (Woolgar 2007). Mash from crushed cereal grains, cooked over a fireplace and bread (or rather leavened and unleavened flatbread baked on hot stones, in hot ash and in simple ovens) formed the basis of the diet in the prehistoric period. If there was a universal food, it was undoubtedly cereal mash (Anderson 2013). It was cheap and easy to prepare, filled you up and was nutritious. A thick cereal mash (*puls*) made from prehistoric wheat was popular with the Ancient Romans. The Ancient poet Ovid mentions *puls fabricia*—a cereal mash with fatty bacon. The mash *kykeon* from Ancient Greece was prepared from barley flour, ricotta and wine instead of water. Cereal mashes in the western world are generally a bowl of cereals cooked to varying degrees, sweetened with sugar, honey or syrup, spiced with cinnamon or nutmeg, and served with milk or cream and dried fruit such as raisins.

Farinaceous dishes in the Czech Lands were based on the cereals rye, wheat, barley, oats, millet and, from the fifteenth century, buckwheat. Watermills became widespread in the Middle Ages and replaced hand milling, the most time-consuming and laborious work, in the preparation of cereal food. Rotating hand mills, which were introduced to central Europe by the Celts in the last centuries B.C., were used before the development of watermills and windmills (Malinová and Malina 1982). Simple ovens for baking unleavened bread dating back to the Early Iron Age have been discovered in Austria. The bread ovens with a fore fireplace introduced in the Middle Ages enabled improved cooking of the prepared dough.

In Old Czech, the expression *vaření* referred to all mashy foods, not just farinaceous dishes, but also leguminous dishes, thickened soups (such as *kysela*), fruit mashes (*varmuže*) and meat mashes (Petráňová 2014). Bread baked in ovens became the main satiating food in the Middle Ages, and pushed this ancient mashy food off the menu to a certain extent. All breadstuffs made from leavened bread were known as bread for a long time. Bread has had an unambiguous meaning and taken the form of a round loaf since the sixteenth century.

There was a disproportionately large number of animals in relation to the number of people in Europe in the Late Middle Ages, which led to an increase in the consumption of meat (Woolgar 2007). Beef, mutton and pork meat was the main source of animal proteins for the town population in the Middle Ages. Beef accounted for more than 40 percent of the red meat consumed, while pork accounted for around 20 percent (Ervynck and Van Neer 2017; Páral and Pyszko 2015). Consumption of meat in Europe was governed by religious rules.

Christianity advocated abstinence beginning in the fourth century, as this was thought to provide spiritual benefits. Foregoing meat and dairy fats, and thereby carnality and the associated sins of gluttony and lasciviousness, eased the path to redemption of the soul (Woolgar 2007). In the Catholic Church, fasts are an expression of repentance and count as good deeds, similarly to alms and prayer. The lengths of fasts and their intensity are (were) prescribed. Abstinence meant foregoing meat, while fasting allows just a single meal a day.

The system of fasts was applied to the full in the High Middle Ages in the fourteenth century. The Roman Catholic Church prohibited its parishioners from eating meat for more than half the days of the year. Dairy products could also not be eaten for 51 days. The system of fasting was gradually relaxed by the Reformation, though most of the population continued to observe moderation in its eating habits in connection with the church calendar for centuries to come. The system of fasting was still observed in the nineteenth century both by households and on the menus in pubs, hospitals and barracks. It became customary not to eat meat on Wednesday and Saturday, and it was directly prohibited on Friday. Sunday, Monday (in order for the leftovers from Sunday to be eaten), Tuesday and Thursday remained meat days. During the fast neither meat nor eggs could be eaten, and lard and butter were not used for cooking, with only oil to be used (Petráňová 2014). Hempseed and poppy seeds, for example, were eaten on days of fasting. There were a number of recipes for making soup or mash out of these ingredients (Faktor and Žantovská 2017).

Long periods of the Middle Ages, during which liquid or semi-liquid mashy foods predominated, were followed in the Late Middle Ages by times in which bread with meat prevailed. The proportion of meat in the diet declined from the seventeenth century onwards and fell over the course of two centuries to a mere 20 percent of the amount consumed in the sixteenth century. The accompanying fall in nutritional and calorific value was compensated by an increasing proportion of the cereal component and legumes. Animal housing improved in the eighteenth century and greater milk production led to an increase in the proportion of dairy foods in the diet—what was known as a "black-and-white diet" (bread and milk). The basic foods used in folk cuisine in the Czech Lands were greatly transformed at the beginning of the nineteenth century by the spread of potato cultivation, coffee drinking and loaf sugar.

Potatoes were discovered by Spanish conquistadors in Peru in 1539, and isolated records of them can be found at the end of the sixteenth century in Spain, Italy, the British Isles and Central Europe (Cowan 2007). Potato cultivation in the Czech Lands was promoted by Maria Theresa (Faktor and Žantovská 2017). They did not become widespread here until the nineteenth century, when they became a basic foodstuff. Many folk names and regional names for potatoes can be found (e.g. bandúrek, bandor, bobal, brambola, erteple, grumbír, grumbolec, kobzol, kobzole, krumpel, krumpír, krumpolec, santák, šupák, zemčátko, zemák, zemlják, zemče, etc.). We also know the popular pancakes made of raw potatoes under many names in the Czech Lands, such as bramborák, cmunda, hanuška, sejkora, jabkanec, liťák, bandorka, strouhanec, drbanec, babčák, prskanec, křápanec, křapáč, kramfík, drh and pokroutka. In South Moravia (the Kyjov and Břeclav areas), potato pancakes were known as "přesňáky" or "patenty" (Hluk and the surrounding area). They were made from an extremely simple potato dough shaped from potatoes cooked and peeled in advance along with flour and salt. Flat cakes rolled from dough were baked dry without fat on a medium-hot plate. Today, they tend to be made in a sweet version covered with jams and sprinkled with poppy seeds (Faktor and Žantovská 2017).

Cabbage played a vital role in the popular diet in the Czech Lands. It was either eaten fresh or prepared by fermentation, which was the oldest and also the cheapest method of conserving. Grated cabbage was trodden in a pre-scalded wooden barrel lined with vine leaves, dill and sometimes slices of beetroot (Fig. 6.4). It was sprinkled with onion, salt and caraway seeds, and sometimes each layer was interspersed with apples or pears. The lid of the barrel was weighed down for around three weeks while fermentation took place and the cabbage matured. Cabbage was eaten practically the whole year round in the Silesian countryside. It was cooked in a large pot and thickened with a thickener and grated potato, and bacon was then added. It was eaten with a sausage or a piece of smoked meat after the pig-slaughter. The more time that passed after the pig-slaughter, the more often cabbage was eaten just with potatoes (Faktor and Žantovská 2017).

**Fig. 6.4** For the processing of food for centuries have been used wood, or ceramic containers. In the middle of a picture a grater for grinding cabbage for subsequent fermentation in barrels

Beans became established in Moravia in the eighteenth century. Buckwheat and millet gradually disappeared almost entirely from folk cuisine until the twentieth century. The spice saffron disappeared and the drink mead was replaced by beer and spirits.

Peas and peeled barley were often cooked together in the Middle Ages. More peeled barley was eaten than peas (Faktor and Žantovská 2017). Peas and peeled barley were basic foods for the poor in the Middle Ages.

The typical diet of country folk changed according to the season of the year (Petráňová 2014). Only two, occasionally three, meals a day were eaten in the short days of the winter months. As many as five meals a day were eaten at the time of seasonal work in the long summer days. Soup (beer soup, sour soup, garlic soup, mushroom soup, etc.) with bread was a hot breakfast for long generations. The main meal of the day was lunch. It was usually comprised of soup and sweet or savoury floury or potato dishes. If the main dish was a mash, the soup was generally eaten last.

Soups are, and were, a dish that comprised part of practically all three main meals of the day—breakfast, lunch and dinner—in the Czech Lands. In the Middle Ages and Early Modern Times, soups were cooked quickly from, for example, whey or sourdough and egg. Garlic soup was also quick to make. Soup used to be served from a communal bowl from which everyone ate with a spoon. The householder began and finished first, with the others taking his lead. Soup was prepared depending on the possibilities available to the housekeeper—the available ingredients and the preparation time. We can find soups in old recipes that are no longer made and that no one would probably want today. Těšín dog-rose soup was made from dried dog roses. Fifty grams of dog roses, 2 dl of milk, 40 g of medium ground flour, a little sugar and butter were needed for two portions. This soup was made in the mountainous regions of Silesia and in Moravian Wallachia (Faktor and Žantovská 2017). Some soups made use of the available dairy products. Hanakian cheese soup was made from milk and curd cheese and was served with boiled potatoes. Demikát soup comes from Moravian Wallachia. Soft sheep's cheese was added to thin slices of bread, and sieved water boiled with onion, caraway seeds, peppers and salt was poured over it. Hanakian cabbage soup with mushrooms was a more substantial soup. Pickled cabbage, dried mushrooms, dried plums, cream and Moravian sausage were needed to make it. Moravian cabbage soup was similar, though smoked pork rib was added instead of sausage. Potatoes were also added to cabbage soup. Fruit soups were widely made in Silesia from, for example, dried apples or pears. The fruit was peeled before being dried in the sun (Faktor and Žantovská 2017).

The sequence of meals and the composition of dishes were a matter of regional and family tradition. Snacks were taken only in summer, usually bread with milk or pickled cabbage water, though bread with bacon, cheese, jam or scrambled eggs would be eaten on the demanding days of harvest. The evening meal involved the gathering of the family around the table, though it was usually taken without any great preparations. The leftovers from lunch, some bread or potatoes washed down with sour milk usually sufficed (Petráňová 2014).

Economical management of supplies was an essential condition for existence for a rural family. Stored grain and potatoes, pickled cabbage, peas, some lentils and, in the warmer Moravian regions, some beans were a guarantee of survival. These stores were supplemented with dried herbs and mushrooms, fruit and nuts, melted butter in stone jars, bacon, cheeses, root vegetables, horseradish, garlic and onions, poppy seeds, linseed and hempseed, jams, honey and other foods according to local possibilities. The general rule was that a family should still have half of its supplies at the turn of January and February (Petráňová 2014).

Folk tradition devised a number of ways of surviving in times of need and shortage. As the main foodstuff was bread, these means of survival were aimed primarily at achieving ways of substituting for flour, by using ingredients of lower quality such as oats, millet and mannagrass (*Glyceria maxima*) as well as ground chestnuts, acorns, lichen, couch grass and later potato flour.

Regional differences in the composition, preparation and naming of dishes were shaped over the centuries in which a subsistence way of life prevailed. Communication and cultural factors, first and foremost catchment markets and the sharing of culture among neighbouring ethnic groups, contributed to the formation of regional differences in addition to the specific character of production areas (dependent on the climate and soil conditions). A role was also played by family traditions (Petráňová 2014). Sausages were made in Moravia that were never matched by Bohemian sausages. The very beginnings of cheese production in the Czech Lands are associated with primitive folk cheese making at farms and sheepfolds (Kopáček 2018) where housekeepers made curd cheese and simple acid cheeses in the shape of cakes and discs out of cow's milk. Preparation of flat round cheeses, which began to be exported to a large part of Europe before the middle of the seventeenth century, developed in the flat lands of the Haná region (Petráňová 2014). The first records of these dairy products date back to 1452 (Kopáček 2018). In Bohemia, however, domestic cheese making produced larger pointed cheeses from the same ingredient—acid milk—that were preserved by drying and that were used grated for cooking in the winter.

### *6.2.2 The Tastes and Aromas of Dishes Prepared in the Czech Lands*

Mediaeval cookery was characterised by a passion for seasonings and distinctive tastes (Woolgar 2007). Spices became an important trading article and some of them originated from India, Indonesia and China. Spices were not used merely for their aromatic and digestive properties, but also for the colour they added to food. People from lower classes of society used less in the way of spices and used the kinds of flavourings that were grown in the area in which they lived (e.g. mustard seeds, crab apples and saffron). Certain imported spices, such as pepper, became more readily available as the volume of trade increased over the course of time.

Rural housewives generally knew how to cook just five to ten simple dishes, which they prepared repeatedly on weekdays (Petráňová 2014). They seasoned food with fresh and dried herbs, mostly marjoram and caraway seeds, which grew wild between the fields. The greatest treat at home was honey and confections from fairs made from the sixteenth century onwards by confectioners and apothecaries from oriental recipes.

Innovations (such as the use of pepper, ginger and cinnamon) were brought by parish cooks and maids returning from service in aristocratic or town households. Housekeepers on large farms and at mills had the advantage of devoting themselves entirely to cooking, and thereby got more training, in view of the large number of people to be fed. They generally also worked at important events in village life at which large numbers of people had to be catered for, such as weddings and funerals. Ingredients for these irregular feasts were usually provided by neighbours or by the invited guests themselves who provided flour, butter, curd cheese, eggs, etc. in advance. The consumption of larger amounts of food of good quality accompanied festivities associated with the completion of harvesting work.

The main days of plenty for country people were Shrovetide and the time of pig-killing, when a system of mutual neighbourly food gifts made it possible to extend the consumption of meat dishes over a period of several weeks. There were also regular annual festivals on the main church holidays at which dishes dictated by tradition were served and which were taken into consideration in advance by the subsistence system of farming. The climax was the two-day Kermesse festival held in the autumn at the end of the agricultural work. The country folk only ate their fill of fresh meat during the Kermesse and at the time of pig-killing. The meat they did not eat at once was salted and smoked. Every butcher and farmer guarded the secret of their pork seasoning carefully through the generations. It was generally ground and mixed the day before to ensure it was as fresh as possible. White pudding always contained marjoram, pepper and new spice. Savoury or oregano was added instead of marjoram in Moravia and Silesia. A ground mixture of cloves, caraway seeds and ginger was used in Silesia (Faktor and Žantovská 2017). In Silesia, they also preferred to smoke bacon rather than render lard. Pork fat was always rendered over a gentle fire. It was basted with a little water or milk at the beginning to make sure it didn't burn. The fat had to be sliced evenly. Care was taken to ensure that the cracklings did not stick to the bottom of the cauldron and did not turn too brown. The housekeeper then poured the rendered lard into milk pots. The lard had to last the whole year until the next pig-killing.

Josef Tašl includes a recipe for Moravian sausages in his Recipe Book. The author comes from a family of butchers from Veselí nad Moravou where members of the Tašl family are recorded from the second half of the eighteenth century. One kg of sliced marbled pork is added to 4 kg of lean pork meat, and the meat is ground to the size of nutmeg. The mix is seasoned with ground white pepper, caraway seeds and nutmeg. Hanakian sausages were prepared in a similar manner, though new spice was added in place of nutmeg (Tašl 2019).

Smoked meats were added to soups (see Moravian cabbage soup) and were also part of main dishes such as "trnčenice" from the Haná region. Two handfuls of dried plums, 2–3 spoonsful of plum jam, sweet cream and butter are needed for its preparation (4 portions). The seasonings used are cloves, new spice and lemon peel. Plain flour served for thickening. "Šimlovice" with smoked knuckle was made in a similar way in Silesia. Beef stock was poured over a light thickener, and plum jam, dried plums and cinnamon added. Whisking was followed by boiling for half an hour. The cinnamon was taken out, milk added and the sauce sweetened. Boiled and boned pork knuckle was served with sauce and potato flour dumplings (Faktor and Žantovská 2017).

A programme of education of the agricultural communities in rural areas began in the second half of the nineteenth century (Petráňová 2014). It targeted women, and focused first on improvements in domestic dairy farming and later on housekeeping and cookery. The graduates of housekeeping schools (first established in Kroměříž in 1884) then made a significant contribution towards modernising rural cuisine before World War I. The standard of the peasant diet on the national level was, however, low round 1900 according to a survey conducted by the central authorities. Changes in the menu of the population of Central and Western Europe from 1850 to the end of the twentieth century included (Teuteberg 2007):

- Increased consumption of beef and pork meat
- Stagnation/decline in the consumption of veal
- Decline in the consumption of goat meat and mutton
- An increase in consumption of milk, cheese, butter, eggs, wheat flour, fruit and vegetables
- A fall in the consumption of potatoes and rye flour from 1900 onwards
- The greatest of declines was seen in consumption of legumes
- Increased consumption of poultry.

These changes came to a climax after centuries of progressive development of the national cuisines of central Europe, including the Czech Lands. National cuisines received significant new stimulus as a result of mass tourism, extensive information sharing and the easy accessibility of information, and the availability of foodstuffs from practically all over the world. While the rural cuisine of the Czech Lands retained some of its characteristic dishes at least in part until the nineteen seventies, the trends of modern business including catering, including the preparation of *"convenience"* foods, predominated on the menu of the later generations in the last third of the twentieth century and, first and foremost, in the new millennium. The preservation of cultural traditions, including the dishes made by our ancestors using original foodstuffs and ingredients, represents, however, an important element of our upbringing and can enrich the diet of practically all age groups of the population with products that have a positive influence on our health.

## 6.3 Summary

Until the nineteenth century, the rural population represented the largest part of the population of the Czech lands. Their cuisine was based on traditional food produced within their farms. The diet of the rural population was dominated by food of plant origin, namely cereals, cabbage and, from the nineteenth century, potatoes. The meal preparation of these foods was dependent on the available cooking technologies and techniques. Since the Middle Ages, households have been equipped with a bread oven, first with a firebox and smoke vent. Later, in towns since the last third of the eighteenth century, in the countryside around the mid-nineteenth century, the open fireplace was replaced by stoves. Cooking on an enclosed fire has brought a number of advantages and made it possible to extend some cooking technologies, such as stewing.

The development of agriculture since the nineteenth century allowed the expansion of cattle breeding, which increased the consumption of milk and cheese. Cheese making has traditionally been associated with the desire to extend the shelf-life of milk in the form of dairy products and has been carried out for centuries in individual households—small farms. It was only in the eighteenth century that the first manufactories were established, but their development did not occur until the nineteenth century.

If we compare the diet of the rural population until the end of the nineteenth century with the current diet, the most significant differences are the current high proportion of industrially processed food and the lower proportion of food of plant origin. Among plant-based foods, there is a noticeable decrease in the consumption of legumes, a lower proportion of potatoes and a higher proportion of wheat flour.

## References

Anderson HA (2013) Breakfast a history. Rowman & Littlefield, Maryland, USA, 227p.

Cowan B (2007) New worlds, new tastes. in: Freedman P (ed) Food: The History of Taste, University of California Press, 368 p.

Elias M (2014) Lunch a history. Rowman & Littlefield, Maryland, USA, 193p.

Ervynck A, Van Neer W (2017) Beef, pork and mutton: an archeological survey of meat consumption in medieval and postmedieval towns in the southern Low Countries (Flanders & Brussels, Belgium). Quarternary Int. https://doi.org/10.1016/j.quaint.2017.02.004

Faktor V, Žantovská K (2017) Tradiční česká kuchyně. Práh, Praha, 3rd edition. 344 p. (in Czech).

Kjærnes U, Holm L, Gronow J, Mäkelä J, Ekström MP (2009) The study of Nordic meals lessons learnt. In: Meiselman (ed) meals in science and practice. CRC Press. Woodhead Publishing Limited, Cambridge, UK, 681p.

Kopáček J (2018) Putování za sýry. Česká republika–země sýrů. Potravinářská Revue, 15, (6) 74–80. (in Czech)

Malinová R, Malina J (1982) Vzpomínky na minulost aneb experimenty odhalují tajemství pravěku. Profil, Ostrava, 1st edition, 277p. (in Czech)

Meiselman HL (2009) Dimensions of the meal: a summary. In: Meiselman (ed) Meals in science and practice. CRC Press. Woodhead Publishing Limited, Cambridge, UK, 681p.

Outram AK (2007) Hunter-gatherers and the first farmers. In: Freedman P. (ed.): Food: The History of Taste. University of California Press, 368 p.
Páral V, Pyszko M (2015) Kosti ze středověké kuchyně. Maso 26(1):18–24 (in Czech)
Petráňová L (2014) Strava In: Tyllner L (ed): Velké dějiny zemí Koruny české. tematická řada, sv. IV. Nakladatelství Paseka s.r.o. v Praze a Litomyšli, 1st edition, 803 p. (in Czech)
Svoboda JA (2014) Předkové. Evoluce člověka. Academia Praha, 1st edition. 479 p. (in Czech)
Tašl J (2019) Odborná nauka pro řezníky a uzenáře. Receptář. OSSIS, Tábor, 3rd edition. 180 p. (in Czech)
Teuteberg HJ (2007) The birth of the modern consumer age. In: Freedman, P. (ed.): Food: The History of Taste. University of California Press, 368 p.
Váňa Z (1983) Svět dávných Slovanů. Artia, Praha, 1st edition, 239 p. (in Czech).
Woolgar CM (2007) Feasting and fasting. In: Freedman P (ed) Food: The History of Taste. University of California Press, 368 p.

# Chapter 7
# Regional Gastronomy and the Preservation of Culinary Heritage

**Agnieszka Filipiak-Florkiewicz, Kinga Topolska, and Adam Florkiewicz**

**Abstract** Gastronomy is one of the oldest forms of human service activity, defined as "the art of preparing and serving dishes based on professional culinary knowledge, as well as knowledge about products, their nutritional value, appropriate preparation of dishes, composing of meals, culinary traditions as well as habits and customs related to food" (Makała 2015). Regional gastronomy deserves special attention. The specificity of local cuisine, traditional products from a given region, as well as culinary events are important components of the tourist "package", increasing the uniqueness of the experience and the extraordinary character of the place (Stasiak 2007). Appreciating the role and importance of regional cuisine as an extremely important element of cultural heritage, traditional foods and meals products have been specially protected in the European Union. They can fall into one of many categories, e.g., protected designation of origin, protected geographical indication, or a traditional specialty that is guaranteed as such. Moreover, in Poland (besides Registers of the European Union) products with quality or unique properties resulting from the use of traditional production methods can be found on the list of traditional products (Stasiak 2007); Internet source no. 1). Region-specific dishes, prepared and served in a special way, offered in facilities with unique local architecture and interior climate are undoubtedly a tourist attraction, and the more so, the more the cultural values of the area are expressed (Wieczorkowski et al. 2011). This is to emphasize that Polish cuisine has a real abundance of flavors which have been developed over centuries. Polish dishes can be both prestigious and full of surprising taste experiences (Purzycka 2018). Małopolska is one of the regions of Poland that meets the requirements described above. Undoubtedly, it is one of the areas in our country with a unique wealth of local gastronomic "treasures"; many places therein are associated with such gastronomic "calling cards". Often, the headliner of various tourist

A. Filipiak-Florkiewicz (✉) · K. Topolska
Department of Plant Products and Nutrition Hygiene, Faculty of Food Technology, University of Agriculture, Krakow, Poland
e-mail: agnieszka.filipiak-florkiewicz@urk.edu.pl

A. Florkiewicz
Department of Food Analysis and Quality Assessment, Faculty of Food Technology, University of Agriculture, Krakow, Poland

© Springer Nature Switzerland AG 2022

J. Hernik et al. (eds.), *Cultural Heritage—Possibilities for Land-Centered Societal Development*, Environmental History 13,
https://doi.org/10.1007/978-3-030-58092-6_7

offers is eating a meal in one of the regional restaurants, trying local specialties, as well as learning about local culinary customs. In addition, there are also fairs, demonstrations of the process of traditional products, and other types of culinary events. Many habits and traditional regional products and dishes are associated with regional cuisine, which was affected by such factors as the availability of natural resources, the possibility of farming and animal husbandry, historical events, the pace of economic development of the region, local traditions, customs and religious ceremonies or various types of rituals related to the preparation and consumption of meals. Regional products owe their special character primarily to the methods of production for which the region is famous for, the ingredients used for production and environmental factors that often have a major impact on the quality of local products (Berndt-Kostrzewska 2001). The chapter includes some culinary history as well as information concerned with traditional Polish products and dishes, especially these from the Małopolska region that are the most popular among visitors, as well as those that are only remembered by the oldest residents. This article is without a doubt worded as part of a journey meant to evoke tasty possibilities while following culinary heritage.

**Keywords** Regional gastronomy · Local cuisine · Culinary heritage · Traditional products · Dishes · Małopolska

## 7.1 Introduction

Interest in topics related to broadly understood culinary topics such as cooking, national and regional cuisines, food festivals and culinary routes is growing every year (Internet source no. 2).

The desire to taste new dishes or the possibility to enjoy favorite flavors at their place of origin is an additional motive of tourist trips. Tourists want to get to know local food products and traditions. This is why regional cuisine has become an important feature of choosing the destination of the journey (Internet source no. 2).The culinary aspect of traveling is thus essential (Charzyński et al. 2015). Food is a significant component of overall tourist spending, and it is a main element of the travel experience (World Tourism Organization 2012; Gheribi 2018). Nowadays, the increasing interest in culinary art and corresponding new trends in traveling are notable. As a term, "culinary tourism" was created in 1998 and introduced by Lucy M. Long (Kowalczyk 2008). Among English researchers, the term "culinary tourism" occurs along with a broader term—"food tourism" (Hall et al. 2003).

Most of culinary tourism definitions are associated with cultural tourism (Czarniecka-Skubina 2008; Mikos von Rohrscheidt 2008; Jędrysiak 2010). It means, *inter alia,* taking part in culinary festivities and festivals, participating in gastronomic competitions, traveling on so-called culinary trails. The most obvious reason for people to choose culinary traveling is the need to get to know the taste of traditional regional cuisine and dishes prepared by different ethnic groups (Kowalczyk 2008).

Food has therefore developed from being a basic necessity for tourist consumption to being regarded as an essential element of regional culture (Jones and Jenkins 2002). One of the most important factors stimulating the relationship between tourism and food experience is the role of these elements in local development (Gheribi 2018).

Poland is becoming increasingly attractive as a tourist destination. Its natural environment, modern infrastructure, and its improvements in access to culture and national heritage of exceptional value make the country popular to international tourists (Gheribi 2018). Polish cuisine is appreciated by people from many countries. The promotion of Polish cuisine is therefore an excellent possibility to introduce to tourists the attractiveness of our country, both based on the values of the natural environment (slow food, healthy food) and cultural heritage (traditional cuisine) (Internet source no. 2).

Bearing in mind the fact that this chapter describes issues related to culinary heritage, it is worth to mention that historical events have had a great impact on development of specifically Polish cuisine (Charzyński et al. 2015).

The fact that the Polish people for generations have cherished culinary traditions is evidenced by the List of Traditional Products kept by the Ministry of Agriculture and Rural Development (Internet source no. 1). The wealth of culinary history of Poland is an invitation to learn it. It is worth knowing that each region of the country surprises with its diversity of tastes and traditions. The List of Traditional Products contains products that are part of the cultural heritage of the region in which they are produced. Their quality or exceptional characteristics and properties result from the use of traditional production methods (used for at least 25 years). They are a part of the local community's identity. The List contains 1977 products (as of August 12, 2019) (see Table 7.1).

Among the raw materials and products being used in the kitchen, the perfectly pickled cucumbers and cabbage are among those deserving mention. If you want to make dill pickled cucumbers called "ogórki kiszone" in Polish, you should prepare cucumbers, water, salt, dill, garlic, and horseradish, put them in a jar and then leave it for several weeks. When the season for this vegetables begins, you can see people buying them in huge amounts and trying to do their best to prepare perfect pickles (Polish Food 2018).

In turn, cabbage (with the addition of water, salt, and optionally carrots, etc.) is used to make sauerkraut (Wójcik 2017). As a curiosity, it should be noticed that in the old days Krakow housewives used sauerkraut also ....to clean carpets (Krygowski 1980).

Małopolska is an excellent example of a region that has attempted to include regional foods in its tourism strategy. A vast variety of tourist products in this part of Poland are linked directly or indirectly with regional foods. For instance, monasteries and abbeys facilitate an intangible connection with the rich culinary tradition of the region through a range of food-related products and activities (Dominik et al. 2017). As regards to monasteries, they were specialized in baking bread. On the day of their patron, they distributed them to the people (Wójcik 2017). For instance, tourists are lured to Szczyrzyc in Beskid Wyspowy (charming and steeped in legends) by both the architectural heritage of a former Cistercian abbey, and by the role

**Table 7.1** Products on the List of Traditional Products kept by the Ministry of Agriculture and Rural Development (as at August 12, 2019)

| Voivodeship | Type of Product | | | | | | | | | | Total |
|---|---|---|---|---|---|---|---|---|---|---|---|
| | Cheese and other dairy products | Fresh meat and meat products | Fishery products, including fish | Nuts, seeds, cereals, vegetables, and fruits | Bakery and confectionery products | Oils and fats (butter, margarine, etc.) | Honeys | Meals and Dishes | Beverages (alcoholic and non-alcoholic) | Other products | |
| Dolnośląskie *(eng Lower Silesia)* | 4 | 11 | 3 | 4 | 9 | 1 | 6 | 8 | 6 | 0 | 52 |
| Kujawsko-pomorskie *(Eng. Kuyavia-Pomerania)* | 6 | 8 | 1 | 8 | 17 | 2 | 1 | 25 | 14 | 6 | 88 |
| Lubelskie (*Eng. Lublin)* | 11 | 29 | 5 | 18 | 48 | 8 | 8 | 48 | 36 | 3 | 214 |
| Lubuskie *(Eng. Lubusz)* | 6 | 17 | 4 | 8 | 13 | 2 | 7 | 8 | 14 | 1 | 80 |
| Łódzkie *(Eng. Lodzkie)* | 7 | 33 | 2 | 20 | 31 | 3 | 4 | 24 | 18 | 0 | 142 |
| Małopolskie *(Eng. Małopolskier or Lesser Poland)* | 13 | 78 | 0 | 15 | 42 | 1 | 8 | 40 | 17 | 6 | 220 |
| Mazowieckie*(Eng. Masovia)* | 5 | 39 | 5 | 13 | 14 | 5 | 8 | 30 | 23 | 5 | 147 |
| Opolskie (Eng. Opole) | 3 | 10 | 2 | 8 | 12 | 2 | 3 | 15 | 3 | 4 | 62 |
| Podkarpackie *(Eng. Subcarpathia)* | 23 | 79 | 3 | 10 | 47 | 4 | 3 | 49 | 16 | 1 | 235 |
| Podlaskie *(Eng. Podlasie)* | 11 | 12 | 2 | 6 | 13 | 3 | 3 | 11 | 7 | 3 | 71 |

(continued)

**Table 7.1** (continued)

| Voivodeship | Type of Product | | | | | | | | | | Total |
|---|---|---|---|---|---|---|---|---|---|---|---|
| | Cheese and other dairy products | Fresh meat and meat products | Fishery products, including fish | Nuts, seeds, cereals, vegetables, and fruits | Bakery and confectionery products | Oils and fats (butter, margarine, etc.) | Honeys | Meals and Dishes | Beverages (alcoholic and non-alcoholic) | Other products | |
| Pomorskie *(Eng. Pomerania)* | 4 | 27 | 17 | 16 | 30 | 5 | 3 | 52 | 20 | 5 | 179 |
| Śląskie *(Eng. Silesia)* | 12 | 13 | 5 | 12 | 33 | 2 | 5 | 55 | 7 | 1 | 145 |
| Świętokrzyskie *(Eng. Świętokrzyskie)* | 4 | 25 | 6 | 17 | 17 | 4 | 7 | 8 | 4 | 2 | 94 |
| Warmińsko-mazurskie *(Eng. Warmia-Masuria)* | 3 | 15 | 1 | 1 | 7 | 0 | 4 | 8 | 3 | 0 | 42 |
| Wielkopolskie *(Eng. Greater Poland)* | 66 | 33 | 1 | 7 | 6 | 6 | 2 | 20 | 10 | 2 | 153 |
| Zachodniopomorskie *(Eng. West Pomerania)* | 1 | 4 | 6 | 7 | 10 | 1 | 11 | 4 | 8 | 1 | 53 |
| Total | 179 | 433 | 63 | 170 | 349 | 49 | 83 | 405 | 206 | 40 | 1977 |

that it played—according to centuries-old chronicles—in the production of a thirst-quenching beverage called *cerewizja*, a mixture of malted grain, hops, and chicory. Other examples of successful agritourist products based on a culinary tradition are homemade strong alcoholic drinks (based on plums) produced in Łącko. The harvest of fruit (in September) remains a tourist attraction throughout the region, giving the opportunity to witness the production of fruit juices, such as those manufactured in the traditional Fruit Press Factory in Zarzecze (Dominik et al. 2017).

Nowadays, local chefs and restaurateurs go back to childhood flavors and tirelessly unearth old recipes used by their predecessors: chefs cooking for wealthy aristocratic families. We are witnesses of the rebirth of a tasty culinary history (Internet source no. 3; Książyk 2016).

## 7.2 Famous Dishes of Polish and Regional Cuisine

The culinary heritage of Poland is rich with many traditional ingredients and tastes. According to the List of Traditional Products kept by the Ministry of Agriculture and Rural Development (as of August 12, 2019) more than four hundred meals and dishes belong to the tradition ones. Indeed, Polish food can often be characterized beside delicious taste by extra additions (for instance butter, sour cream, roux), a lot of meat dishes with vegetables, mushrooms, and groats, and surprising connections of seemingly mismatched ingredients (like herring and cream) (Polish Food 2018).

One of the most recognizable Polish dishes are "pierogi" (dumplings). They were already known in Poland since the Thirteenth century. One of the legends says that they came to our country thanks to Hyacinth of Poland, the Dominican friar, who lived in poverty, and having only flour, cabbage and mushrooms, combined these ingredients into one, from which the Polish "pierogi" originated (Książyk 2016). The dough for "pierogi" is prepared from flour, water, and salt (sometimes eggs as well). Then, it is rolled and cut into wheels with a glass. The most common stuffing is ground-meat, a simple combination of potatoes and cottage cheese, sour mix of sauerkraut and mushrooms. The traditional dumplings stuffed with potatoes and cottage cheese are called "pierogi ruskie" (Ruthenian), and those stuffed with sauerkraut and dried mushrooms are a typical Christmas Eve dish (Książyk 2016). Most restaurants serve a wider selection of dumplings, for example, with spinach, lentils or tomatoes and cheese. **Sweet versions of this dish** are—in turn—dumplings usually stuffed with **wild blueberries, strawberries, or sour cherries.** These delicious little dumplings are served with **sugar and cream** (Polish Food 2018). Classic "pierogi" have their smaller cousins, we call them "uszka". Molded in a way that resembles ear-like shapes, "uszka" are typically stuffed with sauerkraut and mushrooms.

The king of **Polish dishes is pork chopped cutlet—"kotlet schabowy"** (schnitzel). It's a simple breaded pork served with a side of boiled, mashed, or fried potatoes (Polish Food 2018).

Potato dishes are also very popular. Potato flans—"Placki ziemniaczane" deserve special attention. These are potato pancakes, usually served with sugar, sour cream, or goulash—"gulasz" (Polish Food 2018).

"Kopytka" are a kind of **Polish gnocchi**. The only difference is another shape (they look like a horse's hoof ("*kopyto*" in Polish). "Kopytka" are often eaten plain or with butter and sugar or with a meaty sauce (Polish Food 2018).

Another special Polish dish is cabbage stew—"bigos", where the sauerkraut plays the main role. However, this meal is rich with many ingredients, being a combination of sauerkraut, mushrooms, sausage, bay leaf, sometimes even dried plums. **Cooking "bigos" is not easy**; it is prepared during a few hours of hard mixing with a wooden spoon (Polish Food 2018). The more times reheated, the better (more delicious) "bigos" is (Wójcik 2017).

One of the traditional cakes, known for centuries, is the "Kołacz" wedding cake from Jodłownik. Initially, it was baked for church and nuptial festivities. Today, the "Kołacz" much like those produced centuries ago, is sweet, with golden-brown crust and a thick layer of cheese and is produced also for everyday consumption (Internet source no. 1).

Another favorite dessert among the Poles are doughnuts ("pączki")—round treats made with yeast dough, deep fried in lard grease and sprinkled with a glaze of sugar or covered with icing. Their origins can be traced back to ancient Rome. To begin with in Poland they were made with bread dough and filled with different types of jams. The most common ones are made of plums and rose petals. In Poland no one can imagine Fat Thursday (the Polish equivalent of Mardi Gras) without "pączki" (Polish Food 2018; Tomaszewska-Bolałek 2016).

Engel wings or brushwood, called "faworki" or "chrust" in Polish, are another delicacy that is also eaten on "Fat Thursday". They are bits of fried dough coated with powdered sugar (Polish Food 2018). They were first created, probably by accident, when an inexperienced pastry chef threw a slice of doughnut dough into piping hot oil (Tomaszewska-Bolałek 2016).

## 7.3 The Heritage of Krakow's Cuisine

It is not easy to describe the rich history of Krakow, the city located at the intersection of main trade routes, populated by a variety of nations, nearly each of which has had an impact on the city's culture (architecture, cuisine, etc.) (Internet source no. 3).

In the **Middle Ages**, Polish cuisine was based on locally grown wheat, meat, fruits, honey, herbs, and spices. In those times Polish dishes were characterized by high energy value and were quite spicy. A very strong tradition of hunting was also not without significance. Polish cuisine has a great respect for forest fruit and mushrooms, and that is why Poles are still organizing family trips to the woods in their free time. Meat, fish, and vegetables were preserved with salt or by the sun-drying process (Polish Food 2018). The manner of meal consumption was also different from what is contemporary, because flatware was not in use; everyone relished the food only with their hands. At that time, Krakow belonged to the richest cities in Europe. The goods transported through the city reached the tables of Krakow's burghers. Until the middle of Fifteenth century, the cuisine of Polish burghers as well as meals distribution (they ate twice a day) did not differ significantly from practices of the nobles. The diet of the burghers was largely based on cereals and cereal products. Millet and barley were especially popular (Wójcik 2017). According to Jelicz (1966), at that time, Krakow's bakeries offered nine main types of bread and cakes. Fish (dried, smoked, salted and fresh, fried, etc.) were consumed very often. Among them, salmon and sturgeon were the most expensive. The former Krakow's inhabitants used to eat a lot of dairy products or used to include a lot of dairy products in their diet. At that time soups with eggs and cheese were popular. For years, the basic drink of the burghers was beer and honey, but wine liquors became widespread in the Fifteenth century. With the prosperity of the city, the medieval Krakow burghers began to reach for various "powidlings"—confections prepared with honey or even sugar (Jelicz 1966). The dishes were seasoned with dill, garlic, mustard (Wójcik 2017). It is worth mention that from the mid-Fifteenth century, dishes were richly decorated. And the ingredients used to prepare the dishes were often surprising (Wójcik 2017).

The cheapest sources of street food in Krakow as early as the Middle Ages were so-called kitchens under the sun run in market squares by cooking women called "faryniarki" in Polish. The pharines functioned almost unchanged for centuries and served such meals as sausages, borscht on the bone, pea soup, potato soup, as well as the cheapest Rumford soup. The last dish was served by the Charity Society to poor people (Estreicherówna 1968; Krygowski 1980).

In the Sixteenth century, imported luxury goods (such as the most expensive spices, sugar, rice, and wines) became more widespread (Januszek-Sieradzka 2009).

In the notes of Marcin Gilewski one can read that in the Seventeenth century, among bakery goods, new products called in Polish "kukiełka" (or "gżele", "gżegżółki") were served in Collegium Maius (Jagiellonian University—the oldest Polish university). They had a delicious taste, especially when they were eaten with butter (Wójcik, 2017). One of these special bakery "Kukiełka Lisiecka" is produced even now (Kowalik 2018).

The Eighteenth century was an important time for Polish culinary culture. One of the reasons was that in this time the cultivation of potatoes, which significantly modified the menu, started to be widespread. Bringing potatoes to Poland is attributed to King Jan III Sobieski. During the Viennese expedition he was to send Queen Marysieńka potatoes as a peculiarity of imperial gardens in Vienna. The royal gardener planted them in Wilanów, and served them at royal parties and they became fashionable as an exotic vegetable (Wójcik 2017). At this time potatoes were not popular in Poland. People were afraid of their potential harmful properties including the possibility of having a fever after potatoes consumption. Before potatoes appeared in Polish cuisine, another tuber (Jerusalem artichoke) was popular (Wójcik 2017). Once, at Christmas, Hawełka imported five kilos of potatoes from South Africa. A kilo cost the same as the monthly salary of an aspiring magistrate official (Broniewski 1970; Krygowski 1980; Homola-Skąpska 2007).

Among holidays or traditions celebrated in Krakow, "Babski Comber", "Ścinanie Mięsopusta", "Emaus", and "Rękawka" should be mentioned. The first one, taking place on the Krakow Market Square, is the first typical Krakow holiday. Starting on Fat Thursday and lasting until Tuesday preceding Ash Wednesday and thus Lent, Krakow traders (female) gave some participants the excuse to eat and drink street food. They caught men to dance, eat, and drink with them (Wójcik 2017).

During Easter, within a three day period people had to share eggs in all related and friendly homes. Guests were greeted with a holy egg and invited to a richly laden table. They have "to taste" the whole assortment of sausages in tartar sauce, a piece of roasted piglet with horseradish, a large selection of cakes, and a lot of sweet wine and liqueurs (Broniewski 1970). According to Grabowski (1909), the tables of rich men, at that time, bent under piles of hams, delicious cakes and other luxury goods.

During "Emaus", candies, gingerbread, rings of bagels, or bagels, piles of salty barbells and puffed cakes were sold (Homola and Łopuszański 1980; Krygowski 1980). Another holiday celebrated only in Krakow is "Rękawka". On Tuesday after Easter in Krzemionki, after a service in the church of Saint Benedict, the fun began, throwing rolls, apples and oranges, cookies and gingerbread from a group on high down to where an even larger crowd was gathered (Wójcik 2017).

On the Christmas Eve menu, there were included beetroot soup with "uszka", almond soup, fish patties, kale with chestnuts, baked pike, zander with eggs, noodles with plums. For dessert on Christmas Eve, poppy seed legume was served. The dinner ended with kutia, but also apples, nuts, and other delicacies (Wójcik 2017).

## 7.4 The Traditional Products and Dishes of Małopolska

Both residents of the individual regions and tourists increasingly appreciate the unique properties and high quality of local food. The former discover in it the tastes of their childhood, while the latter wandering around our country want to explore the tastes specific to the given region (Internet source no. 5).

The List of Traditional Products (Internet source no. 1) contains 220 products from the Małopolskie Voivodeship (see Table 7.1).This impressive number is largely due to local producers who care about the highest quality. Traditional foods also embody the concept of localization of production and consumption, which favors short, local food supply chains, thus limiting so-called "food miles"—i.e., the distance that food products need to be transported from the producer to the consumer.

One of the most famous culinary symbols of Krakow is the "Kraków Obwarzanek"—bagel. Already in 1496, King John Albert ordered what hopefully the European Union will soon "confirm": true "obwarzanek" cannot be bought outside Krakow. Today, these fresh and crunchy bagels are available in the Main Market Square and all through the city: sprinkled with salt, poppy seed, or sesame. Similar, but not the same, product is "Kraków Precelek" (pretzel); small, oval-shaped, with the ends twisted inside into a knot, hard, glazed, crumbly, and light. It keeps freshness for a long time, and is therefore rumored to have been ordered in the quantity of a few carts by King Jagiełło before he set off to the battle of Grunwald. Fortunately, the battle ended in Polish victory and the recipe did not fall into the hands of the enemy (Internet source no. 4).

Not only residents, but also tourists are delighted also with salted and ripened rennet cheese produced of ewe's milk ("Bryndza Podhalańska") or smoked ewe's rennet cheese salted and smoked ("Oscypek"). Although the names of traditional products sound mysterious to many, one thing is certain—their taste is unique and unrepeatable. These are real Małopolska unique products.

Traditional foods are a valuable element of tourist offer, a factor in protection of traditional cultural landscapes (Montanari and Staniscia 2009), and reportedly one of the key reasons why tourists choose to stay at agrotourist farms (Dominik et al. 2017).

As regards the wealth of our country's culinary history, Małopolska commands special attention. The Małopolska voivodeship land is not only the cultural heritage in itself but also includes an amazing culinary heritage, which is undoubtedly another attractions of our region. The fact that this region of Poland nurtures culinary traditions with forty dishes is evidenced on the List of Traditional Products kept by the Ministry of Agriculture and Rural Development (as of August 12, 2019)—see Table 7.2.

One of the most popular soups is **"Żurek po krakowsku"** (Polish Sour Soup Kraków Style); it is made on a base of sourdough and meat with plenty of margery. It includes pieces of white sausage and boiled egg and is sometimes served in a bowl made of bread (Polish Food, 2018). In Małopolska, the tradition of souring (fermenting) rye dates back to the Thirteen century. The Archives of the City hold the following recipe for Krakow Sour Soup: three handfuls of rye flour (good for "żurek"), a quart of boiled water, a small chunk of bread from Prądnik or Krakow, a clove of garlic for the taste, and a small jar of white (i.e., wheat) flour. Having softened all these in a stone jar, one should set it for two nights by the stove (Internet source no. 4).

The other traditional soup of Małopolska is Kraków Red Borscht ("Barszcz"). For centuries, to prepare this delicious dish, only beetroots which have been grown in the soft soils of the Prądnik and Szreniawa River valleys (even in small, home gardens) were used. Every good housewife would make "sour beetroot juice" to add to tasty soups or to drink straight as a healthy drink. The tradition continues to this day in the region. Many would find Christmas Eve supper without Red Borscht absolutely impossible. This soup can be served in a cup without anything floating about ("clean") or most popularly with little dumplings "uszka" (Polish Food 2018, Internet source no. 4).

One of the regional soups is cumin soup (Wójcik 2017). The Teslar' recipe for cumin soup is one of the traditional recipes but forgotten by younger residents. According to this recipe, you should take two tablespoons of butter, finely chopped carrots, celery, and parsley. Next fry two tablespoons of cumin for a while, add 4 tablespoons of flour and fry until yellow. Then add water, stir, and boil for half an hour. Salt and browned sugar should be used to improve the taste. Next, the soup should be strained through a napkin and served warm with croutons fried in butter.

**Table 7.2** The meals and dishes of Małopolska evidenced by the list of traditional products kept by the ministry of agriculture and rural development (as of 12 August 2019)

| Original name in polish | Date of the entry | Short characteristics |
|---|---|---|
| Żurek po krakowsku | 2007–02–28 | Thick soup on the basis on natural sourdough, cooked on smoked meat or sausage and served with potatoes or egg |
| Barszcz czerwony krakowski | 2007–02–28 | A thin soup prepared from natural beet sourdough, served in various ways: with potatoes and cream, with dumplings or croquettes |
| Sułkowicka krzonówka | 2010–12–01 | A bright, thick soup with horseradish, prepared with pieces of chopped cured meat, meat and sausage, as well as a boiled egg |
| Ziemniaki po cabańsku | 2011–01–17 | Multilayer dish prepared from potatoes, onions, carrots, and beets, baked on a fire in a cast-iron pot |
| Pierogi łomniczańskie | 2011–08–25 | Oval dumplings (cooked in water); two layers: ring of boiled potatoes and circle of white cheese which are visible in cross section |
| Moskol | 2011–08–25 | A cake (1 cm thick, up to 8 cm in diameter), made from boiled and mashed potatoes, flour, egg, fried directly on a baking sheet |
| Zupa zoproska | 2012–03–06 | Thick soup with visible ingredients, i.e., dried mushrooms, mint, carrots, parsley, onion, dill, cumin, butter and thicken with wheat flour |
| Krzonówka po orawsku | 2012–06–27 | Thick soup with horseradish made on the basis of juice from sauerkraut. Visible pieces of sausage, ham, and eggs |
| Sałata po orawsku | 2012–06–27 | Soup with visible, floating lettuce leaves, cooked in buttermilk or sour milk |
| Siuśpaj | 2012–11–26 | Thick dish made of cooked barley, with dried fruit, sweetened with sugar and honey |
| Strząska | 2012–12–14 | Dish consisted of chopped cured meat (preferably sausage), hard-boiled eggs, grated horseradish with spices, with a vinegar solution in water |
| Kapusta po góralsku | 2013–03–22 | Thick soup with visible stripes of cooked sauerkraut, with pork scratching and potato pieces |

(continued)

**Table 7.2** (continued)

| Original name in polish | Date of the entry | Short characteristics |
|---|---|---|
| Tarcioki—kluski scykane—bukty | 2013–03–22 | Small, oblong noodles from raw grated potatoes with flour, egg, spices, cooked in water and served with pork scratching |
| Kluski hałuski | 2013–03–22 | Irregular, jagged noodles made of wheat flour, eggs, and salt, cooked in salted water. They are served on milk or with pork scratching |
| Kaszana—kiszka z kapustą zasmażaną | 2013–04–24 | Kiszka (a type of black pudding) in barley groats, served with sauerkraut fried with pieces of carrot, parsley and celery, and green parsley |
| Prażone—prażonki polańskie | 2013–04–24 | The dish prepared in a cast-iron pot, where the chopped raw materials (potatoes, carrots, onions, bacon, sausages, parsley) put in layers and baked on the fire |
| Kapuśniorka z grzybami | 2013–06–14 | Soup with visible pieces of mushrooms and potatoes. Prepared from sauerkraut juice, dried mushrooms, cream and flour, served with boiled potatoes |
| Zolipskie pirogi | 2013–06–27 | Dumplings with a slightly undulating surface with visible stuffing. The taste is "soaked" with fried onion, sweet cabbage and sausage and noticeable black pepper |
| Strojcowskie zawijoki z ziemniaków | 2013–07–18 | Dish made from potatoes rolled in white cabbage |
| Rosół polski | 2013–08–26 | A clear soup with visible, shiny mesh of fat on the surface, with noodles, addition of carrots and chopped green parsley |
| Kasza pęcak z kapustą | 2014–02–25 | Loose-grain cooked barley groats combined with boiled, chopped cabbage, fried in fat with pork scratching, fried onion and spices |
| Fitka z Lelowic | 2014–10–07 | Soup based on vegetable stock from potatoes, onions, carrots and parsley, optionally celery or leek |
| Siuśbak—kasza z grochem | 2014–10-07 | A traditional dish derived from old, rural cuisine, with visible grains of groats and beans as well as pieces of bacon and poultry meat |

(continued)

**Table 7.2** (continued)

| Original name in polish | Date of the entry | Short characteristics |
|---|---|---|
| Wolańska kura nadziewana—peerelka | 2014–10-07 | Whole chicken, golden brown color, with brown, toasted skin, stuffed with meat |
| Gałki sadczane | 2015–04–16 | Small, round noodles with a porous surface, consist of two basic ingredients: grated, raw potatoes and wheat flour. Boiled in salted water, served with browned bacon |
| Imbramowska kapusta na żurze | 2015–09–25 | White, shredded, and short-pickled cabbage with a small amount of sour rye soup |
| Hulajdy | 2015–12–03 | Small, round noodles, with a slightly porous surface, made of grated raw potatoes, covered with milk or pork scratching |
| Pasztet z fasoli z marchewką lub grzybami | 2015–12-03 | Pate prepared from beans "Piękny Jaś" variety (cultivated in the Dunajec Valley), with carrot or forest mushrooms, onions, garlic, oil, breadcrumbs, potato flour, sour cream, egg, and spices |
| Zupa karpielanka | 2015–12–16 | Swedish turnip soup with visible vegetables: beans, carrots, and green parsley |
| Dziadki kokuszczańskie | 2015–12–16 | Round noodles, with a smooth or slightly rough surface, made from scalded mixture of potato and flour (wheat flour, often wholemeal), without additional cooking. Served alone or with pork scratching, cream, and sheep's cheese ("bryndza") |
| Mirowska parzybroda | 2015–12-16 | A concentrated soup with visible pieces of vegetables (cabbage, sauerkraut, potatoes, carrots, onions, tomatoes, and lovage) with pork scratching and herbs |
| Sakiewka ziemniaczana | 2016–09–14 | Dish made of white cabbage leaves, tied with green chives or a leek leaf. Stuffing prepared from potato and onion. Served with sauce with pieces of mushrooms |

(continued)

**Table 7.2** (continued)

| Original name in polish | Date of the entry | Short characteristics |
|---|---|---|
| Kładzionki ziemniaczane witanowickie | 2016–09-22 | Cake prepared from boiled potatoes with onion, flour, eggs, and spices (such as salt and pepper), dipped in flour and fried in fat |
| Pierogi z fasolą Piękny Jaś | 2017–03–03 | Dumplings with a semicircular shape with a characteristic frill in the place of sticking, stuffed with cooked beans "Piękny Jaś" variety (cultivated in the Dunajec Valley) with the addition of garlic, fried onion and pork lard or goose fat |
| Klęczańska zupa wigilijna—jasiek ze śliwką | 2017–04–20 | A clear soup with a sweet and sour taste, with dried smoked plums (with stones), beans "Piękny Jaś" variety, carrots and parsley slices |
| Pierogi kościeleckie | 2017–04–20 | Dumpling with a triangular or semicircular shape, with a characteristic frill in the place of sticking. Wheat flour cake, stuffing: vegetarian, cheese with curd, cheese with plum jam (sweet dumplings), cooked in water |
| Placek po rytersku | 2017–04–20 | Fried cake (10–15 cm in diameter), prepared from boiled potatoes, served with salty or sweet additives |
| Święconka (święcelina) spiska na ciepło | 2018–06–19 | Thick soup made on a stock of dried plums, with pieces of cooked eggs and cured meat |
| Lubecka zupa z lubczykiem | 2019–07–12 | Soup with visible pieces of potatoes, vegetables, meat, and lovage leaves. Vegetables are cooked on a stock of pre-fried smoked bacon, beef, pork, or poultry. The ready soup is thicken with cream cheese or cream |
| Rzodkie ziemnioki | 2019–07–12 | Soup with a unique taste (not very thick), prepared from potatoes |

Moreover, the most known traditional dumplings of Małopolska are "Pierogi łomniczańskie", "Zolipskie pirogi", "Pierogi z fasolą Piękny Jaś" as well as "Pierogi kościeleckie" (see Table 7.2).

Speaking of surprising old dishes, an interesting thing is that, in the past, Krakow, which, at that time, was a capital of Poland, had its own version of hamburgers. It was called "maczanka krakowska" (sippet soup), i.e., chuck steak or pork loin roasted in a baking pan served with bread and cumin sauce. No one knows when the dish was first served but it definitely was one of the favorite snacks among horsedrawn carriage drivers and students (Tomaszewska-Bolałek 2016).

In brief format, the short characteristics of other delicious dishes originated in the Małopolska region are presented in Table 7.2.

## 7.5 Krakow—European Capital of Gastronomic Culture 2019

Krakow (the capital of Małopolska) was selected as the European Capital of Gastronomic Culture 2019. The title was awarded for the first time by the European Academy of Gastronomy. It is an opportunity for the city of Krakow to organize a variety of initiatives where local culinary heritage is showcased and the famous Krakow hospitality can be experienced (Internet source no. 6).

Krakow has the resources to make this success happen—good recipes, highest quality products from local suppliers, as well as hospitality and friendly atmosphere (Internet source no. 7).

Krakow's culinary heritage includes street food and recently "re-discovered" dishes (Internet source no. 8).

One of such initiatives is the Krakow Capital Menu. As part of this event, in 21 restaurants you can sample dishes inspired by the city's history, historic recipes as well as local products. Restaurants that take part in the initiative will be marked with the ECGC 2019 logo (Internet source no. 9).

For years Krakow has promoted good gastronomy. Culinary festivals are the another way where you can find regional, traditional products and dishes. Quite a few have been organized since the 90 s, some—like the soup festival—have already disappeared from Krakow's calendar, while others: the Pierogi Festival or the Bread Festival, have a regular audience. For several years, smaller culinary festivals (once Foodstock, today Najedzeni Fest!, Art & Food Bazaar, or the Festival of Young Wine) as well as restaurant festivals (Restaurant Week, Fine Dining Week) have been regularly held. There are also events in Krakow that are part of the Malopolska Festival of Taste (Wójcik 2017, Internet sources 10–12). Street food fans can look forward to some intriguing menus during the Street Food Festival Poland, which is also hosted by Krakow (Internet source no. 3).

In Malopolska, culinary heritage features are promoted *inter alia* via an Internet portal "Wrota Małopolski" (Koskowski and Dominik 2017). It offers detailed tourist information, including separate sections dedicated to a regional food festival "Małopolski Smak", a culinary trail "Małopolska Trasa Smakoszy" and many others (Kosowski and Dominik 2017). Moreover, it is also an example of how marketing success can lead to protection and revitalization of culinary heritage (Kosowski and Dominik 2017).

Culinary events treat tourists to excellent local dishes in such agritourism-famous communities or municipalities (the smallest administrative unit in Poland) as Mszana

Dolna, Niedźwiedź, Dobra and Tymbark in the Limanów district, and Żegocin, Łapanów and LipnicaMurowana in Bochnia district of the Małopolska (Tomczyk-Miczka 2008). Other examples of culinary heritage promotion are the regional taverns which serve Old-Polish dishes and specialties of regional cuisine. These restaurants cherish an exceptional architecture, decoration, atmosphere, serving styles, etc. (Wieczorkowski et al. 2011).

Most regional cultures have the potential to become tourist attractions. It is especially true with regard to regions characterized by distinctive cultural features, tangible and intangible (Dominik et al. 2017).

## 7.6 Conclusion

Living culinary traditions enhance the tourist's experience by enabling visitors to get close to regional history and tradition. Above we describe a few examples to show that gastronomy plays a crucial role in tourism development in the region as well as in promotion of culinary heritage.

To protect valuable inherited food recipes, and habits of their production and consumption and to sustain the uniqueness of a small homeland as a whole the base of regulation in EU and each member country has been developed but the real support, on the day after day basis, is given by the autochthonic members of local communities.

## References

Berndt-Kostrzewska J (2001) Kuchnie różnych narodów. Format-AB, Warszawa

Broniewski S (1970) Igraszki z czasem czyli minione lata na cenzurowanym.Wydawnictwo Literackie, Kraków

Charzyński P, Łyszkiewicz A, Musiał M, Podgórski Z (2015) Culinary tourism in the Cuiavian-Pomeranian Province, Poland. In: Sahin S, Charzyński P (eds) The Cultural Heritage and its Sustainability in Europe. Pegem Akademi, Ankara, p 69–89

Czarniecka-Skubina E (2008) Culinary tourism, Gastronomic Review. Turystyka kulinarna, Przegląd Gastronomiczny, p 12

Dominik P, Grochowicz J, Koskowski M (2017) The influence of regional culinary traditions on the attractiveness of agrotourism offer. Tourist value of traditional foods in selected regions EJSM. 24(4): 5–10. doi: https://doi.org/10.18276/ejsm.2017.24-01

Estreicherówna M (1968) Życie towarzyskie i obyczajowe Krakowa w latach 1848–1863.Wydawnictwo Literackie, Kraków

Gheribi E (2018) The foodservice business in big Polish cities. Urban Dev Issues 53:57–63

Grabowski A (1909) Wspomnienia. Towarzystwo Miłośników Historyi i Zabytków Krakowa, Kraków

Hall CM, Sharples E, Mitchell R, Cambourne B, Macionis N (2003) Food tourism around the world: development, management and markets. Butterworth-Heinemann, Oxford

Homola I, Łopuszański B (1980) Kapitan i dwie panny. Wydawnictwo Literackie, Kraków

Homola-Skąpska I (2007) Co to jest Kraków? Ot, Sukiennice i Hawełka. In: Z dziejów Krakowa, Galicji i Śląska Cieszyńskiego, Księgarnia Akademicka, Kraków, p 264
Januszek-Sieradzka A (2009) Artykuły luksusowe na stole królewskim w późnośredniowiecznej Polsce. Studia Mediaevalia Bohemica 1:268
Jelicz A (1966) W średniowiecznym Krakowie. PIW, Kraków
Jędrysiak T (2010) Village cultural tourism. Wiejska turystyka kulturowa. PWE, Warszawa
Jones A, Jenkins I (2002) A taste of Wales—Blas ArGymru: Institutional malaise in promoting Welsh food tourism products. In: Hjalager A, Richards A (eds) Tourism and gastronomy. Routledge, London
Koskowski M, Dominik P (2017) The role of culinary heritage in regional tourism development strategies in Poland in the face of contemporary trends in tourism. Przedsiębiorczość i Zarządzanie XVIII 8:93–104
Kowalczyk A (2008) Present-day cultural tourism between tradition and modernity. Współczesna turystyka kulturowa—między tradycją a nowoczesnością. In: Krawczyk A (ed) Turystyka kulturowa. Spojrzenie geograficzne. Uniwersytet Warszawski, Wydział Geografii i Studiów Regionalnych, Warszawa
Kowalik A (2018) Ziemia Lisiecka. Ludzie historia tradycje. Wydawnictwo Księgarnia Akademicka Sp. Z o. o, Kraków
Krygowski W (1980) W moim Krakowie nad wczorajszą Wisłą. Wydawnictwo Literackie, Kraków
Książyk M (2016) The unforgettable flavour of the childhood. Pol Food, ISSN 1232-9541:19-21
Makała H (2015) Innowacyjne formy działalności gastronomii. Specjalizacja zakładów gastronomicznych. Zeszyty Naukowe. Turystyka i Rekreacja 2(16):193–202
Mikos von Rohrscheidt A (2008) Cultural tourism. Fenomen, potential, perspective. Turystyka kulturowa. Fenomen, potencjał, perspektywy. GWSH Milenium, Gniezno
Montanari A, Staniscia B (2009) Culinary tourism as a tool for regional re-equilibrium. Eur Plan Stud. https://doi.org/10.1080/09654310903141656
Polish Food (2018) Complete guide that will rock your tastebuds. https://discovercracow.com/polish-food/
Purzycka J (ed.) (2018) One Hundred Recipes for One Hundred Years of Independence (1918–2018). Polskie Skarby Kulinarne, Kulinarne Zjednoczenie Polski, ISBN 978-83-952306-0-8, https://www.polskieskarby.pl/polskie-skarby-kulinarne/conowego/ksiazka-sto-receptur-na-sto-lat-niepodleglosci
Stasiak A (2007) Gastronomia jako produkt turystyczny. Turystyka i Hotelarstwo 11:103–132
Tomaszewska-Bolałek M (2016) Polish culinary paths. Wydawnictwo Hanami, Warszawa
Tomczyk-Miczka E (2008) Małopolska—palce lizać. MOT, Kraków
Wieczorkowski J, Cieślik E, Topolska K, Horwath L, Łodziński J, Łodzińska P, Gądek K, Tomczyk-Miczka E (2011) Księga jakości gastronomii regionalnej czyli „O czym karczmarz wiedzieć powinien. Praca zbiorowa, Małopolska Organizacja Turystyczna, Kraków
World Tourism Organization (2012) Global report on food tourism. UNWTO, Madrid, http://cf.cdn.unwto.org/sites/all/files/pdf/food_tourism_ok.pdf
Wójcik M (2017) Kulinarne dziedzictwo Krakowa. Slow food—CE Culture, Heritage, Identity and Food, współfinansowanego ze środków programu INTERREG Europa Środkowa. https://www.bip.krakow.pl/plik.php?zid=206081&wer=0&new=t&mode

## *Internet Sources*

Internet source no 1
https://www.gov.pl/web/rolnictwo/lista-produktow-tradycyjnych12
Internet source no 2
https://www.pot.gov.pl/attachments/article/7780/Culinary%20campaign%20English%20version.pdf
Internet source no 3
http://culinary.krakow.pl/culinary_heritage/232067,2056,komunikat,krakow_s_culinary_gems_-_sample_the_finest_cuisine_html
Internet source no 4
http://krakow.pl/get_html.php?dok_id=200126
Internet source no 5
https://www.gov.pl/documents/912055/913531/PF_spring_2018.pdf/40c6520f-0158-a0fe-8247-e180b76b5d2b
Internet source no 6
https://polska.pl/tourism/urban-tourism/krakow-european-capital-gastronomic-culture-2019/
Internet source no 7
http://krakow.pl/krakow_open_city/see_also___/226642,339,komunikat,krakow_selected_the_2019_european_capital_of_gastronomy_culture_.html
Internet source no 8
http://www.krakow.pl/ccb_en/hot/221740,251,komunikat,krakow_selected_the_2019_european_capital_of_gastronomy_culture.html
Internet source no 9
http://kulinarny.krakow.pl/aktualnosci/229588,2030,komunikat,krakowskie_menu_stoleczne__sprobuj_specjalnych_dan.html
Internet source no 10
https://www.malopolska.pl/malopolskismak/malopolski-festiwal-smaku/malopolski-festiwal-smaku-2019
Internet source no 11
http://karnet.krakow.pl/33307-krakow-xvii-festiwal-pierogow
Internet source no 12
https://www.bip.krakow.pl/?dok_id=59397

# Chapter 8
# Ruthenian Culinary Traditions of Lemkivshchyna

**Marcin Łukasiewicz, Gabriela Zięć, Kinga Topolska, Wiktor Berski, and Adam Florkiewicz**

**Abstract** A part of the Carpathian Mountains, the Lower Beskid range that is located in eastern Małopolska, is a special region as to cultural heritage. Here, in the mountain valleys, for centuries a specific multicultural society was created by mixing together traditions, beliefs, and customs of various ethnic groups. The native Ruthenian culture (Lemko) was in continuous contact with the influences of the Poles (Lachs). Significant ethnic groups were also Jews and Roma people. In the culture of the region, Hungarian (and Austrian) and Slovak influence can also be found. An extremely important element affecting the fate of individual people and entire communities were also great military conflicts, which at the beginning of the twentieth century swept through the region (Gorlice–Tarnów Offensive—WWI and Battle of the Dukla Pass—WWII) as well as later expulsion of the Lemko population by Poles (operation Vistula). These events resulted in the bustling region being turned into an almost uninhabited desert, which had to wait a few decades for the return of its native inhabitants. This is the main reason the region is still one of the few

M. Łukasiewicz (✉)
Department of Engineering and Machinery for Food Industry, Faculty of Food Technology, University of Agriculture in Krakow, ul. Balicka 122, Krakow 30-149, Poland
e-mail: rrlukasi@cyf-kr.edu.pl

G. Zięć · W. Berski
Department of Carbohydrate Technology and Cereal Processing, Faculty of Food Technology, University of Agriculture in Krakow, ul. Balicka 122, Krakow 30-149, Poland
e-mail: gabriela.ziec@urk.edu.pl

W. Berski
e-mail: rrberski@cyf-kr.edu.pl

K. Topolska
Department of Plant Products and Nutrition Hygiene, Faculty of Food Technology, University of Agriculture in Krakow, ul. Balicka 122, Krakow 30-149, Poland
e-mail: rrtopols@cyf-kr.edu.pl

A. Florkiewicz
Department of Food Analysis and Quality Assessment, Faculty of Food Technology, University of Agriculture in Krakow, ul. Balicka 122, Krakow 30-149, Poland
e-mail: adam.florkiewicz@urk.edu.pl

© Springer Nature Switzerland AG 2022 
J. Hernik et al. (eds.), *Cultural Heritage—Possibilities for Land-Centered Societal Development*, Environmental History 13,
https://doi.org/10.1007/978-3-030-58092-6_8

in southern Poland, where traditional culture is still alive and is not just an artificial facade for tourist purposes, as is the case, for example, in the Podhale region. The specificity of this culture is related not only to the influences of various nations, but also to the characteristic terrain and the low level of affluence of the majority of settled and migratory populations. In this context, the culinary tradition, or the preparation of food, has focused on readily available products, including wild ones. A special place in the culinary culture of the region was occupied by vegetable products. Food which was eaten in the countryside was divided into ordinary and festive food. The daily meals included those prepared on the basis of various cereals (mainly oat and wheat), potatoes, legumes, and cruciferous vegetables. Meat products were consumed relatively rarely. An important element of activities aimed at securing food and prolonging its shelf-life was fermentation. That is why pickled mushrooms, cabbage (sauerkraut), and other similar products were very popular in the region. It should be noted, however, that native products were often supplemented with ingredients rarely found in other areas of Poland, which is a result of many trade routes leading through easily accessible passes of the LowerBeskid. The culinary heritage of Lemkovyna as a part of Małopolska and Podkarpacie, however, can be also considered as two mutually pervasive culinary traditions that contribute to what can be considered modern products and regional dishes. In addition to poor (peasant) cuisine, there was also court cuisine. Both these traditions undoubtedly influence the contemporary eating habits of the region's inhabitants.

**Keywords** Culinary traditions · Lemkovyna · Carpathian mountains · Małopolska

## 8.1 Introduction

Lemkivshchyna (lem. Лемковина, pol. Łemkowszczyzna, ukr. Лемківщин) is an area within one of the lowest mountain ranges of the Carpathian Mountains—the Low Beskids, inhabited by an ethnic group of Lemkos. Geographically speaking, it is a narrow strip of area stretching along the ridge of the Carpathians starting from the upper course of the Dunajec River at the west. The end of the region, i.e., the eastern border is basically marked by the upper courses of the Osława and San rivers. According to that it may be stated that the Lemkivshchyna is placed latitudinally along the Carpathians. In terms of geography origin, the region covers the south slopes of the Eastern Polish Carpathians (by reference to the current borders of Poland) including the Low Beskids and a small part of Beskid Sądecki as well as very western part of Bieszczady Mountains. The north border of the region is designated by the basins of Kotlina Sądecka and Doły Jasielsko Sanockie on the west and the belt of Pogórze foothills on the east (Földvary 1988). This geographical location naturally defined the multicultural tradition of the populations living in the Lemko region, where the Ruthenian, Polish, Slovak and later Jewish, Hungarian and Roma elements met together.

## 8.2 Lemko History

Contacts and clashes of these cultures in the Low Beskids started in the early Middle Ages with the (Vladimir the Great expedition in 981 CE). However, the distinct division of land into Polish and Ruthenian terrain took place in the fourteenth century. Removing the political barrier, as a result of the incorporation of Halychian and Vladimirian Rus' into the Polish kingdom in the middle of fourteenth century, opened up the possibilities of free movement of ethnic elements in both directions. In the northern lowland zone, the migration was directed from west to east, while in the southern, mountainous region from east to west starting from the Dukla Pass. This last region since the nineteenth century was referred to as the western and central Lemko region (Nabywaniec 1995). The issues of modern Ruthenian settlement in Lemkivshchyna, and thus the origin of the Lemkos, have prompted attempts at explanation using many scientific concepts, as well as indigenous ones. At present, however, the migration hypothesis is best documented. According to this, the migration wave reached the Lemko region from today's Romania (Reinfuss 1990). Wallachian colonization was strongly associated with herding and farming. Its hiking trail led along the Carpathian Mountains in a north and west direction. Until the end of the sixteenth century, the Wallachian settlers conquered the Bieszczady, Low Beskids, Beskid Sądecki, and part of Pogórze. They continued their settlements up to Moravia (in Czechia). The migration stream was created mainly by Romanian Balkan people with some Slavic elements. During the fifteenth century, the Ruthenian element prevailed over the Wallachian one and caused first partial and then complete Ruthenization of the Wallachians in the Eastern and Central Carpathians. The Ruthenization process was proportional to the tendency to start a stable, sedentary lifestyle. The dominance of the element of Ruthenian culture remained in the upper parts of the Wallachian settlement, while clusters of the predominance of the Polish element began to appear in the San River basin and in the Pogórze region. It seems, however, that the term "Wallachian-Ruthenian settlement" is defined not by means of nationality, but also economic and socio-religious status. This is supported, among other considerations, by the fact that there is a lack of cultural unity of the Bieszczady population (Boykos), with the culture living on the west (Lemkos); as well as by the historical links between the political and economic situation in Hungary in parallel times to migration traces (migration from Hungary to the north) (Nabywaniec 1995).

The nineteenth century was a key period in the history of many countries and nations due to the intensive process of awakening national consciousness and the resulting friction between local communities at the meeting point of cultures. The situation was similar in the Lemko region, where Polish, Ukrainian, Czech, Slovak, Hungarian, and Lemko influences clashed. Religion was the main arena of these cultural transformations. Despite the mixed Polish/Ukrainian cultural domination in the region, the Lemkos avoided Latinisation (Polonization) on the one hand and Ukrainisation on the other. The effect of this phenomenon was a tendency toward affiliation with the Moscow Orthodox Church (in a religious sense) and sympathy with Russia in general. Such tendencies were also a consequence of the presence of Russian troops in the Lemkos region during the century (the Bar Confederation, Revolution of 1848, etc.) (Nabywaniec 1995). During World War I, Lemkos' Rusophilism became the cause of Austrian repression in the face of the approaching Russian front. Fearful of the subversive activity of Russophile leaders, the Austrians interned in the Talerhof concentration camp near Graz nearly 5,000 Greek Catholic clergy, teachers and peasants who favored or were suspected to favor Russia. There were also many other victims among the civilian population (Piecuch et al. 2012). Also, there were considerable losses because the war front lasted for a long time here. After the withdrawal of the Russians, a wave of terror fell on the Lemkos, who were regarded recognized as supporters of Russia. These experiences meant that Lemkos began to orientate themselves politically increasingly by Ukrainian slogans, and to act accordingly. When the Habsburg Empire collapsed in November 1918, the Lemkos in the Sanok poviat, referring to President Wilson's message on self-determination of nations, the Lemkos in established the Ukrainian Poviat National Council. At the same time in the western part of the Lemko region where there were not many supporters of the Ukrainian option, local politicians with a Moscow-based attitude established in Gładyszów the Lemko-Rusyn Council, which decided to include the Lemko region in Great Russia. At the end of 1918, formation of the Lemko-Rusyn People's Republic (Лемко:Руска Народна Република Лемків /Ruska Narodna Respublika Lemkiv) was proclaimed in Florynka. Existing Councils became part of the Republic as part of the formation of the Lemko Supreme Council (Verkhovny Lemkivskyi Soyuz). Similar councils have also been formed on the Slovak side, with various political options: pro-Czech, pro-Hungarian, and pro-Ukrainian. When both the Great Russian and Czech plans failed, the supporters of the Russian option decided to merge with Soviet Russia (Magocsi 2015). At that time, however, the situation was already under the control of Poland. The interwar period was, in turn, a time of Ukrainian-Lemko-Polish friction called a "religious war" (Barwinski 1999). During World War II and the German occupation, the process of expelling Lemkos from the Carpathian Mountains began. First, under the German-Soviet agreement, some of the Lemko inhabitants were displaced to the Soviet Union (about 5,000 people). At the end of the war under the Polish-Soviet agreement about 70,000 Lemkos (60–70% of the Lemkos population) were relocated by force into Ukraine. The remaining Lemkos were affected by Polish deportation actions (Operation Vistula) in 1947 (Magocsi 2015). The displaced people were then sent into the western territory of the newly reconstituted Polish state from which the ethnic Germans had previously been expelled (European Charter for Regional or Minority Languages 2011). The turbulent history of the Lemko region, including especially recent history, resulted in the once overpopulated areas of the Low Beskids remaining basically deserted in all and deprived of their native inhabitants for nearly 50 years (Magocsi 2015).

## 8.3 The Origin of Lemko Culinary Traditions

Historical factors and broadly understood evolution of culture being a derivative of the social history of the region are key culture-forming factors in the field of culinary traditions. As it is known, these traditions are also the result of natural conditions, i.e., the place where food is produced, as well as the method of its preparation. Natural and geographical conditions are the starting point in considering the composition and shape of dishes and their nutritional value. These include climate, water resources, soil type, topography and the flora and fauna that depend on them. The variety of food, as well as its quantity, increases with the development of agriculture and trade. Sometimes passed from generation to generation, recipes and food preparation technologies are equally important. The traditional diet of the inhabitants includes normal (everyday) food consumed during the hunger times, as well as holiday meals on the occasion of annual holidays or family occasions (Kowalska 2012).

In the case of Lemkos, Ruthenian highlanders, who lived until 1947 in the Low Beskid range, constituted for the most part an agricultural population. The Low Beskid range mostly belongs to the montane level and partly to a foothill elevation level. The climate in the Low Beskid is transitional between the Atlantic and continental one. In spring, winter and autumn, warm and dry southern winds blow here, causing frequent changes in weather. The average annual temperature is 4–8 °C and the highest rainfall occurs in June and July. Such terrain and climate strongly influenced local Lemko agriculture, food production, and the quality of food of the inhabitants (Warszyńska 1995). Accordingly, to that Lemkos engaged in farming, animal husbandry, and lived on what they themselves were able to produce or gather in the forest.

Lemko cuisine is a typical cuisine of not very wealthy highlanders, in which dishes based on flour, potatoes and gifts of the forest, i.e., mushrooms, herbs or berries were the most important. Meat was rarely eaten by Lemkos, and if so, it was mostly mutton. Poverty, especially in the pre-harvest period, forced people to eat any food giving energy and helping to avoid starvation. Accordingly, to that methods for food preservation and storage were crucial and the most important food products were those that were easy to store throughout the winter: cabbage, swede, turnip, kohlrabi, groats, beans, and peas (Zielinski 2007). Cabbages and dishes made of them were the basics of meals not only in Lemkivshchyna, but also in Polish villages around the whole country. In fact, cabbage as the life support system for millions of poor people was not only a specifically Polish phenomenon, but can be found in all of Europe, from western France to Russia.

The everyday Lemko diet was supplemented with vegetable fats this being the only spread usable during fasting periods. Mainly linseed, rapeseed, and hemp oils were in use. Despite the fact that meat was not eaten too often, animal fats were frequently consumed including salted or smoked bacon or tallow (Zielinski 2012).

## 8.4 Cereals and Cereals-Based Products

Cereal dishes were prepared from ground cereal, i.e., flour. The most common meals were various soups obtaining by mixing and cooking vegetables or meat stocks with a small amount of flour. When larger amounts of flour were used, mash related meals were prepared, i.e., semi-liquid dishes much more concentrated than soups. Such kinds of dishes were eaten as meals themselves, but they were also used as the base for production of flatbreads. The most famous Lemko dish in this group was *mastyło,* i.e., the thick pudding made of flour, milk, egg, and salt. All the liquid and semi-liquid farinaceous meals listed above were generally mixed with other ingredients. Milk, whey, or quark were added to them (Wacławski 1965). The most popular cereal dish, eaten daily by Lemkos was called *czyr* or *paciara*, i.e., coarse wholemeal rye, oat, (or mixed) flour, cooked in salted water (Zielinski 2014). This emblematic meal was eaten as a warm meal for breakfast. In some cases, that were less common, *czyr* was made with sweet or sour milk and sprinkled with fat and greaves. Rural poor Lemkos prepared *czyr* based on cabbage stock (water obtained after cabbage cooking)—*war*, or with soup made from fruits (fresh or dried). Another example is *kaluga*—a dish made of buckwheat flour cooked in hot water with the addition of some fruits dried (apples) or fresh (berries). Noodles were also commonly prepared from flour. Noodles were made mainly from oat flour, rye flour and less often from wheat. The dish was buttered with bacon or cheese and served with milk (usually eaten for dinner). During the Christmas Eve supper, noodles with honey or sugar were served, sometimes poppy seeds were added as well. Another variety of flour dish were small noodles (*zacierki*) which were usually made from wheat, rye, oat flour, water and a little salt, and cooked in boiling water or milk. Such kinds of noodles were commonly consumed for breakfast (Bańkosz 2007).

The very specific cereal-based meal was sour soup (*kiselycia*) that was prepared by fermentation of wholemeal oat or rye flour with garlic and some spices (bay leaves, allspice). The fermented base was then poured into boiling water and served with potatoes. The soup was also greased with bacon and eaten, broad beans, peas, rutabaga and sometimes with cream or milk. Fasting sour soup with mushrooms was always served on Christmas Eve (Wacławski 1965).

Groats were mashed by hand using home-made wooden equipment. It was made mainly of barley. It was a thick pearl barley or used smaller groats. Millet groat, cooked in milk, considered a delicacy, was eaten on holidays in wealthy homes. From the 1880s, when a railway line from Hungary was opened buyers massively imported Hungarian corn to Western Galicia (Magocsi 2015). Corn groats spread quite quickly among Lemkos (as well as in the Pogórze), compensating for the shortage of other cereals in the cropless years. Thick *mamalyga*, commonly eaten with milk, was cooked from yellow maize flour or groat. The dish in fact is common in the whole range of the southern and eastern Carpathians including the territory of Romania, Moldova, Ukraine, and Poland (Marchenay et al. 2004). Other groats were also cooked mainly in water, whey or hot milk and served with milk or greaves, or with dried plums, mushrooms, cottage cheese, or *bryndza* (Zielinski 2014). The emblematic Lemko dish was *pamula*—thick wheat groats with dry plums, apples, or pears. However, the buckwheat was not so popular in the eastern Carpathians (as compared with other Slavic cultures); Lemkos culinary tradition was particularly associated with some dishes made of them including *hrechanyki,* i.e., cutlets of groats mixed with minced meat, eggs, garlic and spices or *knysze,* i.e., deep-fried dumplings stuffed with buckwheat cottage cheese and onion (Zielinski 2014). Dumplings are a well-known dish presented in many cultures including the Slavic ones. Lemko dumplings called *pirohy* were most often based on a filling of potatoes and cottage cheese or *bryndza.* Dumplings with cabbage were also served. A specific variety of dumplings were small dumplings filled with meat—*zlepience* that were often added to *war,* the soup made of cabbage stock (Zielinski 2014).

Bread is an international symbol of prosperity and satiety. It has always played a very important role in peasant food of any culture. In this meaning bread was considered as a luxury and festive product. In Lemko culture other cereal-based products played a role of everyday food. Bread was usually baked once a week; the dough was prepared from flour mixed with sourdough. Lemkos bread was usually baked from rye flour, sometimes wheat, often mixed rye flour with barley or oat flour. Yeast as a conditioner was used beginning with the starting the end of World War I, mainly in the northern part of the region where bread was a kind of an independent meal and served with butter, bacon, tallow or cottage cheese together with milk, buttermilk or whey as a drink (Wacławski 1965). Bread was also served as a side dish for various soups including *kiselycia* and *war*. Due to Lemko poverty even a stale bread was eaten, mixed together with hot milk; (in this form it was often given to young children). During hard times poor Lemkos made a bread with a supplement of boiled potatoes. More often than baked bread, oat flatbreads were baked without leaven. Lemkos called them *adzymki*; the dough was made from flour, sometimes with eggs, cream, or potatoes. As a leavening agent baking powder or baking soda were used. The difference between a flatbread and bread comes down to the fact that the first was baked from unleavened dough, the second from sour dough. The flatbreads were sprinkled with poppy seeds, chopped onions, or sliced apples. During periods of increasing hunger, flour was mixed with dried and ground chestnuts, birch bark and even sawdust. Due to the history of the region Ukrainian traditions of preparing meals were present in Lemko culture. *Watruszki* are the best example here. *Watruszki* were sweet, round buns made from yeast dough, stuffed with cottage cheese, potatoes, jam, or fruit (Zielinski 2014).

## 8.5 Potato Based Dishes

Since the nineteenth century potatoes (*kompery, bandurki*) were one of the main ingredients in the Lemko diet and the whole of Lesser Poland (Warszyńska 1995). At the beginning, in the southern part of Lesser Poland potatoes were often unavailable so the main crop was rutabaga or turnip. In the Lemko region, a high level of turnip consumption was observed before potatoes were introduced. Turnip was a typical catch crop for feeding cattle and was eaten raw, cooked or dried, and even pickled (Wacławski 1965).

The easiest way to prepare potatoes was to cook them whole and mixed together with bacon or *mastyło* and served with clabber or buttermilk. It was a popular and well-liked summer dinner dish. Such a meal was often taken as lunch during filed work. Boiled potatoes were also commonly added to *kisełycia* (sour soup) or sauce from overcooked fruit thickened with flour (*kaluga*), as well as to holiday broth with meat. Commonly baked potatoes were also eaten. They were eaten along with clabber, buttermilk, hogweed, cabbage, and other vegetables. In the Lemko region, a specific noodle was prepared from old or spoiled potatoes. They were commonly known as *mordunie* (Rataj 2006). As in the other part of the Carpathian Mountains *haluszki,* a small dumpling was also prepared in Lemkovshina, based on grated potatoes with flour. Another important potato dish was *łewesz,* i.e., potato soup cooked with caraway and peas or beans (Bańkosz 2007). Other typical Lemko product were also *bandurjanki* simple potato/quark pancakes, baked in a bread oven or *tartianyki* (*opalanok* when served with bacon), i.e., potato pancakes prepared on fresh cabbage or maple leaves. Another potato dish, i.e., *stolniki* was also baked in the oven. *Stolniki* were grated potato dumplings with potato stuffing and fried onions (Rataj 2006). Similar to potatoes a commonly grown rutabaga was baked, most preferably on a baking sheet (cut into slices). The rutabaga was also used to make soup, with the addition of carrots and parsley, seasoned with flour and cream (Dulian 2016).

## 8.6 Cabbage

Cabbage was another main Lemko crop and a part of the Lemko menu. This is the result of a cabbage preservation method by fermentation popular in central and eastern Europe. Sauerkraut was the common method of cabbage preparation. The oldest way of cabbage fermentation was to dig an earth pit, which was lined with rutabaga leaves. Whole cabbage heads were thrown into the pit, then water was poured in. The pits were covered with straw and soil. Only after they were removed, the head would be shredded. It was widely believed that this procedure preserved cabbage better than others in order to store sauerkraut longer, up to 7 years. A newer way of cabbage fermentation was processing of shredded vegetables in barrels. Sour apples and several whole heads of cabbage rolls called *kiszeniaki* were added to a barrel in many homes (Krzysztofek 2006). *Kiszeniaki* were cabbage rolls made of whole leaves of sauerkraut stuffed with barley and buckwheat groats or potatoes. The other name for *kiszeniaki* was *sarmale* which is strictly linked with Wallachian culture—*sarmale* is a Romanian word for cabbage or rape leaves rolls.

The simplest dish made of sauerkraut was *war,* i.e., the stock obtained by boiling a liquid obtained after a fermentation process with addition of cracklings and served with potatoes (Zielinski 2014). The fast version of *war,* without spread, was eaten in many homes until noon on Christmas Eve with baked potatoes only.

Cabbage was also commonly cooked with peas, mushrooms, barley, or barley groats (Zielinski 2014). The favorite stuffing for dumplings was cabbage with mushrooms. When lemon balm was added to the stuffing they were called *nupis*. Bacon cooked in cabbage, eaten with bread, was considered a delicacy. The very original Lemko cabbage dish were *fuczki*—a simple pancake made from sauerkraut. In the Lemko villages, a sauerkraut soup *kwaśnica* was also popular. *Kwaśnica* was boiled in many ways including with wild mushrooms (Zielinski 2012, 2014).

## 8.7 Fruits and Mushrooms

The everyday food of Lemkos was supplemented with both wild (forest) and homegrown fruits (from orchards). Forest berries (raspberries, blueberries, wild strawberries, blackberries, blackthorns) were sometimes eaten raw with sugar or mixed with dumplings and noodles. Berries, pears, and apples were very often dried. Most often, the fruits were dried in an oven. During drying, the fruits were also smoked. Dried fruit was used to prepare many dishes, mainly for fasting. Such fruits were also used in a spread for rutabaga, potatoes, or noodles (Warszyńska 1995).

Forest mushrooms were most often dried, preserved by fermentation (e.g., fermented red pine mushrooms) or salting. The fermentation process in this case is similar to those observed for fermentation of cabbage. The only ingredients needed are salt and optionally onion. The process takes about three weeks. The product obtained may be additionally preserved by pasteurization (Zielinski 2014). Besides red pine mushrooms other type of mushrooms were popular. Among them porcini, bay bolete, leccinum, and golden chanterelle were the most emblematic (Szelągowska 2017). Drying was one of the most common methods of preserving mushrooms. Mushrooms threaded on pine branches previously stripped off needles or spread on wicker frames, dried in the sun or near the kitchen were very often seen in Lemko villages. Dried mushrooms were stored in cloth bags in airy, dry places—next to the stove or kitchen, in the attic by the chimney or in the pantry. Dried mushrooms served primarily as an addition to dishes, most often with soups including sauerkraut soup, broth or dishes made of sauerkraut, peas or beans, groats or noodles. Mushrooms were often eaten also right after mushrooming. They were baked over a bonfire, strung on sticks, sometimes sliced with bacon or, as shepherds used to do, spread out on stones heated by the fire. Soups were prepared from fresh mushrooms with the addition of various types of groats, noodles, cabbage (fresh and sour), and potatoes. They were often whitened with milk or cream and served with sauerkraut acid or sour rye soup. For the rural population, mushrooms were not only an important ingredient of food, but also generated the possibility of additional income or exchange for other food products (Szelągowska 2017).

## 8.8 Supplemental Plant Food Sources

The quality of diet in rural and foothill regions was particularly influenced by hunger and pre-harvest periods, i.e., short periods of time after stocks run out. Also crop failures caused by long periods of drought or prolonged rains were a cause of the lack of food in the countryside. In the years of famine, the villagers returned to primitive forms of food by collecting wild plants. In this group, among others, nettle (*Urtica dioica* L.) or dwarf nettle (*Urtica urens* L.) should be mentioned. Meals prepared from these herbs were in most case a simple mix of greens with potato or flours (mainly oat flour). Other wild plants were also collected prepared similarly to today's spinach, e.g., lamb's quarters. Wild plants were sometimes converted into groats or flour. The best examples are water manna grass (*Glyceria fluitans* R. BR), couch grass, or beech nuts (Bylak and Mazik 1965). Among other herbs collected and used in the Lemko region peppermint was extraordinarily popular. Peppermint was used fresh or dried for seasoning of quark (quark balls covered by herbs—*homiłki*) or for preparation of drinks. In some cases, lemon balm was also collected and used (Anonim 2011).

## 8.9 Animal Products

Dairy products were a significant part of traditional Lemko everyday food. However, most of the milk was processed into cheese and butter, which, together with cream and milk, was intended for sale. The rest of the milk, often diluted with water, whey and buttermilk was served as a morning and evening meal with potatoes, rutabaga, groats. Milk, buttermilk, and whey were used as spreads for thick dishes such as peas with cabbage, thick porridge. Quarks were salted and then dried in the sun; such cheeses could be stored in a ventilated place for a longer period of time (Oleśniewicz 2009).

In the case of Lemkos, natural conditions and the culture-shaping role of Wallachian colonization meant that sheep breeding, associated meat consumption, and the production and consumption of rennet sheep's cheese was and still is very important for the region's culinary traditions. Since the wave of Wallachian settlement affected the entire Carpathian Mountains, cheese production was the same in this whole area. Lemkos produced plenty of types of sheep cheeses including *bunc* cheese, *bryndza*, hard (Lemko) cheese, and cottage cheese. Natural and smoked Lemko *bunc* was an everyday cheese made in almost every Lemko country cottage. *Bunc* is a rennet unripened cheese obtained by dehydration of the sheep's milk curd by slightly heating it, and then separating the cheese mass from the whey in canvas bags. In the smoked version, the cheese is additionally subjected to a brine bath, followed by smoking in cold smoke. It has a fairly short shelf life. It is intended for direct consumption or for frying and preparing of other regional dishes. Due to its light texture and delicate taste, Lemko *bunc* gained popularity in the region (Wacławski 1965). A different type of sheep soft cheese is *bryndza*. Such cheese is very popular in the Carpathian Mountains and known in Romania, Ukraine, Slovakia, Czechia, and Poland. It is produced using sheep curd that is ground and salted, and then matured in wooden molds. The resulting product could be stored for a longer time. Sheep farming has long been the basis of the mountain economy. Hard cheeses were also produced in Lemkivshchyna but, especially in the past, they were treated as a luxury food so, they were consumed during wedding feasts and annual holidays or family celebrations (Trzeszczyńska 2015).

In addition to sheep, goats were an important breeding animal of Lemkos. Until the 1930s, the Carpathian goat was popular in the Carpathian Mountains and well adapted to harsh climatic conditions. These undemanding animals were able to survive where neither cows nor sheep could survive (Sikora and Kawęcka 2015). *Bryndza* made from goat milk was manufactured in the Low Beskids as early as the fifteenth century, and the method of its production has not changed much since then. As in the case of sheep products, the raw material is goat *bunc* that is made directly from milk. Other cheeses were also similarly produced. Goat cheeses were usually soft and have an intense, slightly salty, spicy taste. Sometimes after production they are coated in charcoal ash, sprinkled with spices or herbs (see *homilki*) or wrapped in leaves (Wacławski 1965).

Meat was a side food for Lemkos and did not found a place in the everyday menu. Even the wealthy farmers ate meat only during major celebrations and on Sundays. Sometimes meat was eaten after successful hunting in the forest. But venison was not as popular as pork that was consumed from their own slaughter or even beef bought from butchers in the city. Other animals that was used include rabbits or poultry. Meat was eaten primarily as cooked in broth or processed into sausages. Meat was often salted (or soaked in saline) and stored in barrels or dried near the oven. The assortment of rural butcher's products was limited to sausages, blood brawn, smoked ribs and bacon, bacon salted and dried near the fire, and tallow (Bylak & Mazik 1965).

## 8.10 Conclusion

The Lemkos' culinary tradition largely draws on the multicultural traditions of Galicia and Austria-Hungary in general. In addition to its own elements, it draws on the culinary traditions of Polish Beskid highlanders, Ukrainian settlers, Hungarians, Roma and the Jewish community. Lemko dishes have, of course, their unique character, but often behind the exotic-sounding names of dishes there are hidden dishes present in the entire Carpathian cuisine (Hornsby 2015).

Lemko cultural and culinary heritage as a part of Slavic heritage built and still builds the very specific and unique world of a multicultural environment that was influenced by sacral and secular life customs, functioning as part of ceremonies as well as aspects of food.

Culinary traditions of the Lemko region, despite the difficult history of the Lemko nation and the geographical conditions of the Low Beskids, are one of the key factors of the renaissance of Lemko culture, not only in Local social terms, but also in economic and tourism terms (Trzeszczyńska 2015). In the era of widespread interest in culinary tourism, this tradition is an important factor in the development of Lemkivshchyna. One of the types of initiatives that are parts of this trend are so-called culinary tourist routes. The best example of this is the Podkarpackie Smaki Culinary Trail created in 2013. The trail includes inns, taverns, and restaurants. In 2017, there were several dozen facilities on the trail, including over 40 restaurants, vineyards, confectioneries, and agritourism farms (Anonim 2019). The goal of the trail is to promote the region and its culinary offerings. The trail is divided into three separate sections, two of which offer contact with the Lemko culture and Lemko culinary traditions. In addition, events promoting the region, regional products, and regional food are regularly organized on the trail (Zieliński 2012).

These activities show that Lemkivshchyna can be rebuilt, newly arising from the homeland once thought lost due to deportations and erosion of social bonds; there can be once more a fully vital region with a unique culture and a high potential for development.

**Acknowledgements** Special thanks to Orianna Katarzyna Wędrowska and Piotr Bazylewicz for great support and a lot of helpful information about the traditions, cuisine, and history of Lemkos.

## References

Anonim (2011) Ugotujmy łemkowskie potrawy [Let's cook Lemko dishes], Fundacja Viribus Unitis

Anonim Retrieved at 10.08.2019 from http://podkarpackiesmaki.pl/pl/kulinaria-podkarpackie/aktualnosci/art816,szlak-kulinarny-podkarpackie-smaki.html

Bańkosz R (2007) Podkarpackie zaprasza smakiem. 13, nr 0(1)

Barwinski M (1999) Acta Universitatis Lodziensis. Folia Geographica Socio-Oeconomica Nr 02:53

Bylak M, Mazik J (1965) Zwyczaje rodzinne Łemków [Lemkos' family habits] in Nad rzeką Ropą, zarys kultury ludowej powiatu gorlickiego [On the river Ropa, an outline of folk culture of the Gorlice poviat], Ed. R Reinfuss, Wydawnictwo Literackie

Dulian D (2016) Ginący świat. Kultura materialna Łemków [A dying world. Material culture of Lemkos], 100%

European Charter for Regional or Minority Languages. Application of the charter in Poland. Strasbourg: Council of Europe, 2011. Retrieved at 10.08.2019 http://www.coe.int/t/dg4/education/minlang/Report/EvaluationReports/PolandECRML1_en.pdf

Földvary GZ (1988) Geology of the carpathian region, World Scientific

Hornsby M (2015) Constructing a Lemko identity: tactics of belonging. Int J Multiling 12(1):1–12

Kowalska S (2012) Cultural heritage in poland – the background, opportunities and dangers, UAM

Krzysztofek K (2006) Kultura Wspołczesna 86(1):47

Marchenay P, Barrau J, Bérard L (2004) JATBA. Revue D'ethnobiologie 65:42

Magocsi PR (2015) With their backs to the mountains. Central European University Press, A History of Carpathian Rus' and Carpatho-Rusyns

Nabywaniec S (1995) Łemkowszczyzna i Łemkowie w świetle badań i opinii [Polish Lemko and Lemkos in the light of research and opinions], Resovia Sacra. Studia Teologiczno-Filozoficzne Diecezji Rzeszowskiej 2:85-114

Oleśniewicz M (2009) Dola Łemka – wspomnienia [The fate of Lemko - memories], Wydawnyctwa Lemkiwskoho ansamblu piśni i tancia "Kyczera", Legnica

Piecuch A, Harkawy A, Janowska – Harkawy M (2012) Opuszczone wsie ziemi gorlickiej [Abandoned villages of Golrlice land], Gondwana
Rataj A (2006) Zwyczaje żywieniowe i sztuka kulinarna a dziedzictwo kulturowe narodów [Food habits and culinary arts versus the cultural heritage of nations]. Conference materials; Tajemnice smaku produktów regionalnych i tradycyjnych [Secrets of taste of regional and traditional products], Kraków
Reinfuss R (1990) Śladami Łemków [In the footsteps of Lemkos] PTTK Kraj
Sikora J, Kawęcka A (2015) Hodowla kóz rasy karpackiej w Polsce [Breeding of Carpathian breed goats in Poland], Instytut Zootechniki
Szelągowska G (2017) Studia i Materiały Ośrodka Kultury Leśnej 213:16
Trzeszczyńska P (2015) The Lemko land remembered: about the research on the memory of the Lemkos in Poland and Ukraine. Anthropol East Europe Rev 33(2):1–25
Wacławski A (1965) Pożywienie ludowe [Folk food] in Nad rzeką Ropą, zarys kultury ludowej powiatu gorlickiego [On the river Ropa, an outline of folk culture of the Gorlice poviat], Ed. R Reinfuss, Wydawnictwo Literackie
Warszyńska J (1995) Karpaty Polskie: przyroda, człowiek i jego działalność [Polish Carpathians: nature, man and his activities]; Uniwersytet Jagielloński. Kraków
Zielinski K (2007) Podkarpackie zaprasza smakiem, 10, nr 0(1)
Zieliński K (2012) Kulinarne opowieści, czyli wędrówki po smakach województwa Podkarpackiego [Culinary stories, or wandering through the flavors of the province Podkarpackie], Urząd Marszałkowski Województwa Podkarpackiego, Rzeszów
Zieliński K (2014) Jadło: przepisy kulinarne pogranicza polsko-słowackiego [Food: recipes of the Polish-Slovak borderland], Stowarzyszenie na Rzecz Rozwoju i Promocji Podkarpacia "Pro Carpatia", Krosno

# Chapter 9
# Genetic Uniqueness of Local Cattle Populations as Part of Homeland Heritage

**Radovan Kasarda and Nina Moravčíková**

**Abstract** The Slovak Pinzgau breed is characterized by many excellent features, and that is why it has spread from Austria across the whole world. In spite of the fact that the Pinzgau cattle are now an endangered population in Europe, many farmers are still interested in its preservation. The breeding aim is primarily oriented to increase the purebred population of the Pinzgau cattle. Its development can serve as a basis for a strategy of further improvement and distribution under wider European conditions. The presence of purebred local individuals has become rare and thus highlights the need to implement a national conservation strategy. There is clearly a race between the characterization of genetic resources and their loss. In a similar way, it is planned that the development of genomic tools will allow optimizing breeding strategies for ensuring the improvement of performance together with the preservation of genetic diversity. Since it is important to keep Pinzgau cattle in the original phenotype, thus pursuing a dual purpose, the objective was to appropriately distinguish between phenotypically similar populations. The approach used for populations' structural assessment is characterized as unsupervised learning methods with specific computation algorithms. We were able to separate even closely related breeds of Slovak and Austrian Pinzgau since they incorporated an admixture with breeds involved in historical development as well as inbreeding, selection signatures and migration. Moreover, the selection tests that lack the direction of selection should be implemented within the breed types and analyses performed using breeds across the breed types should be interpreted as demographic rather than selection signatures. Results show the possibility to classify even unknown samples according to genetic data. Genetic diversity written in genetic data provides useful information for the identification of the origin of individuals. An established methodology for the distinction of genealogically close populations with high-throughput molecular information based on Slovak and Austrian Pinzgau can be proposed as a general method for the analysis

R. Kasarda · N. Moravčíková (✉)
Faculty of Agrobiology and Food Resources, Department of Animal Genetics and Breeding Biology, Slovak University, Tr. A. Hlinku 2, Nitra 949 76, Slovakia
e-mail: nina.moravcikova@uniag.sk

R. Kasarda
e-mail: radovan.kasarda@uniag.sk

© Springer Nature Switzerland AG 2022 
J. Hernik et al. (eds.), *Cultural Heritage—Possibilities for Land-Centered Societal Development*, Environmental History 13,
https://doi.org/10.1007/978-3-030-58092-6_9

of differences in all highly related breeds. Several methods have been used to assess the evidence of positive selection including the analysis of population differences. Identification of genomic regions affected by positive or natural selection associated with economically and biologically important traits is possible and improves our understanding of the connections between changes as a result of interaction with the local environment in which animals live.

**Keywords** Diversity · Genetic admixture · High-density SNP data · Lnkage disequilibrium · Selection footprints

## 9.1 Introduction

There is increasing discussion on the loss of biodiversity, especially when we consider small homelands and the diversity of plant and animal species there, whose number is declining, which is well documented (Thomas 2013). These plant and animal populations represent part of the cultural and historical heritage of the particular area and are connected with the history of human development. We have responsibility for proper monitoring of changes in the living environment, and for developing specific metrics to recognize gains and losses. Such monitoring is needed to develop evidence-based strategies for the coexistence of nature and people (Vellend et al. 2013). This can be well documented with the example of dual-purpose cattle breeding in Slovakia, which has a historically long tradition not only in Central and Eastern, but also in Western Europe.

Breeding of dual-purpose cattle has a long tradition in Slovakia. Both Slovak Spotted and Slovak Pinzgau breeds originated from the mating of autochthonous primitive Carpathian red (brown) and Carpathian grey cattle in the seventeenth and eighteenth century with Swiss Simmental and Austrian Pinzgau cattle. In the nineteenth century, both breeds were very common in the Austro-Hungarian Empire and currently are recognized as part of the cultural and national heritage in Slovakia.

## 9.2 Pinzgau Cattle

The Pinzgau breed belongs to the group of mountain dual-purpose cattle. The origin of the breed is in the alpine area of Austria, settled by Celtics who began to breed their cattle early after 500 AD. The breeding was influenced by the availability of red cattle of those German tribes, who settled around Salzburg alongside the rivers Enns and Mur during migration. Later in the period 1690–1740, Pinzgau cattle were influenced by the Tux-Zillertal breed from which the Pinzgau inherited their typical coat colour. The name "Pinzgau" was given to the breed in the first half of the eighteenth century referring to the Pinzgau valley in the area of Zell am See, where even earlier one strain of cattle had this name. The Pinzgau breed is characterized by many excellent features, and that is why it has spread from Austria across the whole world.

Pinzgau cattle have spread overseas also and not only in temperate climate zones, but even in extreme conditions of subtropical, tropical and steppe areas around the world. Distribution of Pinzgau cattle to those areas took place in the 60's and 70's of the last century. To Canada from Austria, 88 dams and 23 sires of Pinzgau cattle were imported in the period 1972–1976. Between 1980–1990, another 2 bulls, 1 cow and A. I. doses of 6 bulls were imported. To USA, Pinzgau cattle were imported in 1976 from Austria through Canada. In South Africa, the Pinzgau cattle breeders association was established in 1962 (at its founding with 30 farms and almost 2000 animals registered).

In all countries mentioned, the Pinzgau breed is appreciated for its practicality and outstanding features such as fertility, earliness, adaptability and hardiness, resistance to diseases, walkability, performance, tractability, feed efficiency, carcass traits, quality, tenderness and taste of meat. Pinzgau cattle are considered mostly as a specialized (single-purpose) beef breed. In USA, it is recognized as a breed which along with its middle size frame has high feed conversion, adaptable and resistant (Pinzgau is a breed present in every state) with very good character and temperament.

Pinzgau dams are appreciated for excellent fertility, because usually calving occurs at the age of 2 years (2.5 years in USA). Cows give birth regularly, fertility on family farms is over 90%, about 80% on commercial farms. A typical trait is the birth of twins. Breeders appreciate a high pregnancy rate during the suckling period. Generally, mothering traits are highly appreciated, especially good milk production (due to dual-purpose genetic background) and good shape of lactation curve, which ensures calves with a rearing weight around 250 kg and more.

Due to good lactation performance, cows are successfully integrated into hybridization schemes. In experimental trials, Pinzgau cattle steers reached up to 2530 g average daily gain during a 103-days fattening period. The average carcass weight was 297 kg with a dressing percentage over 55.7%. Comparable results have been achieved in all countries mentioned. In sensory evaluation, Pinzgau meat was rated in the second position in tenderness (after Black Angus) and took first place in taste. Excellent results were achieved in other parameters as well according to the USDA Meat Research Centre.

In spite of the fact that Pinzgau cattle are now an endangered population in Europe, many farmers are still interested in their preservation. The breeding aim is primarily oriented to preserve the purebred population of the Pinzgau with its characteristic phenotype. Its development can serve as a basis for a strategy of further improvement and distribution under wider European conditions.

## 9.3 Local Population of Pinzgau Cattle in Slovakia

The Slovak Pinzgau cattle have been bred in Slovakia for almost 200 years. The history of the breed has been described by Pšenica (1990). It has been defined as a dual-purpose type of cattle used in milk and meat production, known economically in lowland areas with good feed basis as well as good conformation adaptability to harsh conditions throughout mountain areas (Kasarda et al. 2014). The production system of mountain areas in Slovakia is characterized by a high proportion of slope pastures on shallow soils, extreme climatic conditions and a long winter-feeding period. Both, the recent and ancient history of breeding confirm the suitability of this breed in the given production conditions, where its performance traits could be evaluated only in relation to the environment, which is limited, especially in nutritional resources.

In past, the breed was multipurpose, intended for workability in pulling (carriages, ploughs), and production of milk and beef. Workability retreated due to the technical development of agriculture, and later, the main attention was focused on performance traits. Therefore, the Slovak Pinzgau breed is at present dual-purpose, bred for milk and meat purposes, with preserved high modularity allowing for the future possibility of specialization, especially in beef production (Pšenica et al. 1998). Thanks to unique traits such as longevity, fertility, health and good pasture conversion was suitable for mountain areas in north Slovakia (Pavlík et al. 2014). But, due to structural changes and significant decrease in population size, Slovak Pinzgau cattle have become endangered (Kadlečík et al. 2008).

At the beginning of Pinzgau breeding in Slovakia, only purebred animals were used. In the 60's and 70's of the last century, grading-up hybridization with other breeds took place to increase milk production besides preservation of the dual-purpose character. Jersey cattle were used experimentally as the phylogenetically closest breed, but due to the low birth weight of calves and increased need for treatment, it did not succeed. The Ayrshire, MRI (Meuse-Rhine-Issel) and Red Holstein succeeded in hybridization trials, and Red Holstein is also used nowadays in the grading-up process.

At present, sires from Slovakia are used in insemination and natural servicing, with some imports from Austria; thus both populations are genealogically closely connected. In 2005, Pinzgau A.I. doses of sires from the Canadian population of beef type were experimentally imported to the Slovak population. The genetic nucleus of dual-purpose type is anyway stored in populations registered in Herd books of Austria and Slovakia. The genetic reserve of Slovak Pinzgau cattle is farmed under an intensive production system used in sub-mountainous areas, contrary to Austria where the majority (65%) of cows is bred under an extensive farming system of suckling cows (Kasarda et al. 2016).

The presence of purebred local individuals has become rarer and thus highlights the need to implement a national conservation strategy. Breeders in Austria and Slovakia are interested in preserving the breed and its development in the future whether through genetic evaluation of growth intensity (Bulla et al. 2013), longevity (Mészáros et al. 2008) or traits, which are included in selection indices at present. Moreover, the availability of genomic tools at present allows for more detailed insight into the genome of the Slovak Pinzgau cattle and the development of a strategic proposal with emphasis on sustainable genetic progress in the future (Kasarda et al. 2018a).

The trend of Slovak Pinzgau population size from a long-term perspective is negative and it is recognized by FAO UN (1998) as endangered by extinction and classified as a genetic resource (Kadlečík et al. 2008). Highest population size was

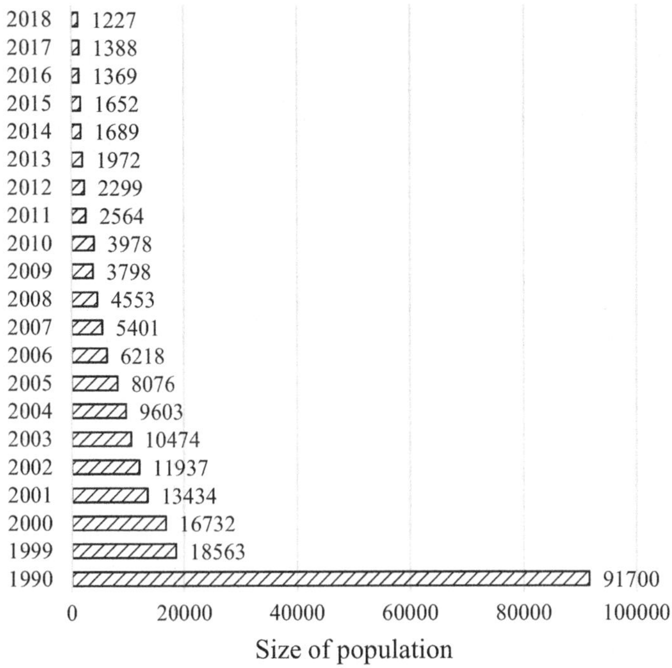


**Fig. 9.1** Trend of Slovak Pinzgau cattle population size in the period 1990–2018

reached in 1978, when in Slovakia were over 160 thousand animals. However, after 1990, due to the transformation to a market economy and essential changes in Slovak agriculture and widespread holsteinization, population size significantly decreased (Kadlečík et al. 2013). The decrease of population size from 2000 is more than 10% annually, where under milk recording was in 2018 registered 659 purebred cows (1227 in total) and 1101 purebred suckling cows (1251 in total) (Fig. 9.1). According to information from Central Farm Animal Registry, only 9068 Slovak Pinzgau animals are currently registered as purebred (all sex and age categories). Because of this, the future development of Slovak Pinzgau breed will depend on several decisions, but above all will be important, if farmers and breeders will be able to support and enhance genetic diversity from short as well as long-term perspective. There is clearly a race between the characterization of genetic resources and their loss.

## 9.4 Genetic Diversity of Local Populations

Biodiversity is generally described as differences between species, breeds, lines or families. Similarly, it could reflect phenotypic differences between individuals, e.g. performance, reproduction, health status or fitness of animals. Those parameters are easily recognizable and measurable. The genetic diversity covers parameters on the gene pool level, which are not visible, but equally important for the management of particular populations. These parameters include inbreeding (intensity of relatives mating), effective population size, e.g. animals which act in the production of the next generation, and genetic distances expressing mutual genetic similarity of individuals.

As proven by several complex studies in cattle, genetic diversity decreased globally in the last decades. This genetic erosion was caused mainly by specialization of production, preferring worldwide recognized high-producing commercial breeds. It is important to highlight that the intensive selection of best bulls for insemination of a high number of cows continuously results in a decrease in the genetic diversity of particular breeds. Genetic diversity monitoring is, therefore, an important task in population management, especially in case of local populations.

Globally, the local cattle populations are out of main interest, especially in connection to the use of modern tools of genetic/genomic selection. From this perspective, if the future generation of farmers would like to use those breeds, it will be essential to develop preconditions for breeding programs.

The genetic diversity is characterized in many papers mainly with respect to fitness traits and survival of populations. Biologists were the first, who studied genetic changes in populations of wild animals and pointed out difficult genetic processes, running in every population. If the management of genetic resources doesn´t act at a particular level, population diversity and genetic resources will be lost. Therefore, conservation genetics was developed. The process of genetic management in small populations begins by definition of diversity. Many breeds become endangered due to intensive breeding work. "Genetically valuable" animals are sold better, have more progeny, which could at the end result in the mating of relatives. Close relatives are produced with higher homozygosity and negatively affect the degree of genetic diversity and decrease the ability of the population to react to the changes in the environmental condition in which they are bred (Kasarda et al. 2018b).

Genetic diversity within the breeds is usually quantified as expected heterozygosity (Nei 1987), but could be estimated by alternative methods based on allelic diversity (Caballero et al. 2010). It was suggested that through maximizing heterozygosity, it is possible to significantly increase the diversity of alleles, and then through the diversity of alleles minimizing of inbreeding level. In this context, Toro et al. (2009) proposed an alternative system for the preservation of diversity considering to some extent diversity of alleles. The higher number of alleles, the higher is potential diversity of subpopulations and since the maximum diversity is achieved when the alleles have the same frequency. A possible way is to find contributions based on maximum diversity available when all alleles in subpopulations have equal frequencies. Estimation of gene diversity in this case match, at least particularly, to diversity of alleles in population according to the potential of each subpopulation according the number and type of alleles containing.

Expected heterozygosity or number of alleles within breeds expresses also the effect of drift on breeds' diversity, while decreased diversity is usually connected with increased genetic drift. Moreover, the difference between expected and observed heterozygosity, as well as deviation from Hardy–Weinberg equilibrium, indicate preferential mating and population fragmentation, i.e. existence of subpopulations.

Systematic animal breeding requires performance recording and genealogical information. This information is used in population management to increase animals' performance in several ways. Genealogical information is also used for genetic diversity evaluation and its value is increasing simultaneously with precision and completeness of pedigrees (Cassell et al. 2003). Estimation of population parameters based on genealogical data is derived from key papers of Sewall Wright (Wright 1920, 1921—inbreeding; 1931, 1938—effective population size; 1951—genetic structure of the population).

Classically, the calculation of genetic diversity parameters is based on genealogical data. However, even if the pedigrees are easily available, they have some disadvantages compared to genomic information. One of them is mainly the fact that pedigree data are not fully complete, e.g. in case of cattle, usually from 3 to 5 generations of ancestors are available, where founders are considered as unrelated animals without known pedigree information. This assumption underestimates the level of inbreeding and overestimates effective population size that is in case of genealogical analysis derived from the level of inbreeding in population.

The development of genomic tools allows optimizing the breeding strategies for ensuring the improvement of performance together with the preservation of genetic diversity. Currently, the SNP (single nucleotide polymorphism) loci are the most typically used genetic markers in conservation genetics. The utilization of SNP markers allows us to reliably evaluate the manifestation of traits, e.g. if the trait will develop or if the individual will be the carrier of a particular trait. The SNP loci belong to the group called biallelic markers, which means that in populations three groups of individuals (genotypes) are expected; homozygous dominant, homozygous recessive and heterozygous animals. If a given population is in equilibrium without loss of heterozygosity, each group of animals has an equal proportion in the population (25:25:50). The loss of heterozygosity usually occurs in case of f.e. inbreeding increase when proportion of homozygous animals is increasing, and natural variability is decreasing.

Globally, the SNP arrays are preferred to obtain genome-wide data. Usually, there are arrays with low density (min. 3000–6000 SNP markers), high density (min. 50,000–777,000 SNP markers) and whole genome scans (2 mil. and more bases). The SNP arrays with min. 50,000 SNP markers cover a substantial proportion of the genome because of the unique selection of the position of those markers over the particular chromosomes. Such SNP arrays exist for almost all farm animal species (cattle, horses, sheep, goat, pig, chicken). These markers express SNPs, e.g. variation of single loci resp. bases on one position in genome between individuals or chromosome pairs. The majority of SNP markers are neutral from a point of trait variability, only a minor proportion of markers is localized in genes or regulatory sequences (Hayes and Goddard 2010).

High-density SNP panels allow for more accurate calculation of summary statistics and genetic distances between individuals, as well as very accurate estimation of species/breeds phylogeny, using shared haplotypes for quantification of relationships on intra- and inter-population levels. Moreover, genetic mapping and genotyping arrays open ways to the identification of mutations causing variation in adaptation (Lenstra et al. 2012), which could allow for rational evaluation of the effect of breeds´ conservation, and thus fulfil the primary expectation of genetic diversity studies (Toro et al. 2009). The SNP arrays as a powerful technology of genotyping were used in many studies oriented on, e.g. estimation of genomic inbreeding (Ferenčaković 2013, Ferenčaković et al. 2013, Marras et al. 2015; Zavarez et al. 2015), level of linkage disequilibrium (Beghain et al. 2012, Pérez O'Brien et al. 2014), historical effective population size (Flury et al. 2010; Uimari and Tapio 2011), genetic diversity, population structure, level of admixture or genomic selection (Mastrangelo et al. 2014, Pérez O'Brien et al. 2015).

Besides genomic data used for the estimation of genomic breeding values, the great advantage of SNP arrays from our point of view is mainly due to the possibility of detailed analysis of allelic diversity in the Pinzgau genome, as well as the adjustment of selection to preserve the majority of the diversity available. Genomic data could be used also for the study of the inheritance of various production, reproduction and fitness-related traits (Kasarda et al. 2016). Even more this technology allows append of genotyping information about high number of cattle populations (McTavish et al. 2013, Decker et al. 2014) and following storage in public deposits as Dryad Digital Repository.

Results of genetic diversity analyses in the population of Slovak Pinzgau cattle were presented in several papers, e.g. Kadlečík et al. (2007, 2008, 2011, 2017), Hazuchová et al. (2010), Pavlík et al. (2012, 2013, 2014) and Kasarda et al. (2015).

## 9.5 Effective Population Size and Inbreeding

Pedigree analyses resulted in the group of genetic diversity parameters devoted from a common ancestor. Coefficient of inbreeding ($F$), increase of inbreeding per generation ($\Delta F$), average relationship coefficient ($AR$) and its increase per generation ($\Delta R$) belong to this group. Another important parameter reflecting the degree of genetic diversity is effective population size ($N_e$). The effective population size is described as the size of an idealized population, which would lose heterozygosity at a rate equal to that of the observed population. There are many factors affecting this parameter. The effective population size is always lower than real population size ($N$), because considering fluctuation in number of animals, sex ratio, family size, migration etc. Intensity of inbreeding and effective population size are closely related and could be mathematically defined as $N_e = 1/2\ \Delta F$. The estimation of $N_e$ can help in the prediction of loss of genetic variation and rate of increase in inbreeding and also provides useful information about the evolutionary history of populations (Leroy et al. 2013; Curik et al. 2014). In addition, the knowledge about the trend of $N_e$ in particular populations provides relevant information for the monitoring of genetic diversity and helps to explain the observed extent of genetic variation in population from a retrospective point of view (Flury et al. 2010). The pattern of historical $N_e$ in local populations can increase our understanding of the impact of selective breeding strategies on the genetic variation within the framework of population genetics (Shin et al. 2013).

Selection in small populations is restricted and has to account for strict control of the intensity of inbreeding as the main criterion in the evaluation of endangerment degree of a particular population (Gandini et al. 2014). The generally accepted rule is that the population becomes endangered when the increase of inbreeding intensity is more than 1% per generation (Kasarda and Kadlečík 2007). Inbreeding could be defined as the mating of individuals related through common ancestors, or as the probability that a particular individual inherits two identical alleles from one ancestor (Gutiérrez and Goyache 2005).

In classical breeding strategies, the inbreeding intensity is derived from the degree of relationship of parents based on the genealogical data from pedigrees and herd-books. Pedigree inbreeding is based on Mendelian inheritance theory, e.g. progeny inherits the same proportion of genetic information from both parents, i.e. the intensity of inbreeding in full sibs will be always the same. Modern breeding programs of cattle are using accurate methods of genetic evaluation and broad use of reproduction technologies. Even those programs are leading to fast genetic progress, lead as well to accumulation of inbreeding due to higher impact of the lower number of selected individuals or families (higher risk of co-selection of relatives). Therefore, inbreeding increases in almost all cases and results in economic losses.

Inbreeding depression affects production traits like growth, milk production, health, fertility or longevity. Various studies oriented to the protection of endangered breeds and preservation of diversity in purebred livestock populations have been implemented in practises. But, the problem is larger. It was showed that inbreeding is a function of selection intensity and effective population size (ratio of male and females in reproduction), and therefore, a population high in size could present signs of depression as well. It´s off-putting that even commercial populations with several millions of animals could have a low effective population size (Weigel 2001). The use of genealogical data alone for the assessment of inbreeding intensity usually leads to underestimation of reality, especially if we consider the development of the population which is not available from historical sources (Zhang et al. 2015; Forutan et al. 2018).

The inbreeding coefficient expressed based on the genomic information is more accurate for the determination of genome-wide autozygosity and ancient inbreeding compared to genealogical data (Peripolli et al. 2017). The genomic inbreeding ($F_{ROH}$) can be derived from the genome-wide proportion of continuous homozygous segments referred to as runs of homozygosity (ROH). When an individual's parents share a relatively recent common ancestor, they will share large parts of their genomes that are identity by descent (IBD). If both the parents transmit the same segment to the offspring, the offspring will be homozygous for that segment, thus creating runs of homozygosity (ROH) (Kim et al. 2013; Mastrangelo et al. 2018). The frequency, size and distribution of ROH segments in the genome are affected by many factors such as intensity of selection, recombination rate, linkage disequilibrium, population structure or mutation rate (Peripolli et al. 2017). In cattle, the selection of superior animals was recognized as the most important factor affecting the ROH patterns in the particular local population. The intensive selection for specific traits of interest leads to an increase in the frequency of homozygous genotypes around the target loci that are involved in the genetic control of given phenotypic traits. Thus, ROH segments can be also defined as genomic regions with reduced diversity and, consequently, high homozygosity around the selected locus that might harbour targets of positive selection and are under strong selective pressure (Pemberton et al. 2012). Assuming that the frequency of certain alleles increases due to positive selection, such regions can be a valuable resource for the mapping of causative mutations.

## 9.6 Uniqueness of Slovak Pinzgau Cattle

### 9.6.1 *Genomic Loss of Pinzgau Cattle Demonstrated by Inbreeding*

In the genome of Slovak Pinzgau cattle, the ROH segments covered an average of 4.72% (117.86 Mb) of the autosomal genome (Kasarda et al. 2018b). The obtained average number of ROHs per animal (44.88) is comparable with previous studies on dual-purpose cattle breeds (Marras et al. 2015; Szmatoła et al. 2016). On the other hand, the Slovak Pinzgau cattle show significantly lower genome autozygosity compared to those reported for dairy cattle. For example, Kim et al. (2013) and Mastrangelo et al. (2016) identified dairy cattle as one of the most homozygous animals among various cattle breeds. Such high genome autozygosity in dairy cattle could be a consequence of intensive artificial selection as well as repeated use of superior and proven sires in breeding practises (Table 9.1).

**Table 9.1** The number (in brackets) and length (Mb) of detected ROHs by categories (Kasarda et al. 2018a, b)

| Category | Mean ± SD | Lower 95% CI | Upper 95% CI | Range | Genome coverage % |
|---|---|---|---|---|---|
| >1 Mb | 117.86 ± 57.67 (44.88 ± 11.23) | 108.58 (43.07) | 127.13 (46.69) | 18.83–398.71 (13.00–70.00) | 4.72 |
| >2 Mb | 76.58 ± 53.98 (13.44 ± 6.52) | 67.89 (12.39) | 85.26 (14.49) | 2.16–350.75 (1.00–34.00) | 3.07 |
| >4 Mb | 58.96 ± 51.28 (6.26 ± 4.21) | 50.71 (5.59) | 67.20 (6.94) | 0.00–339.89 (0.00–19.00) | 2.36 |
| >8 Mb | 38.50 ± 43.76 (2.54 ± 2.39) | 31.47 (2.16) | 45.54 (2.93) | 0.00–333.68 (0.00–16.00) | 1.54 |
| >16 Mb | 21.95 ± 35.06 (0.88 ± 1.22) | 16.31 (0.69) | 27.58 (1.08) | 0.00–264.49 (0.00–7.00) | 0.88 |

CI—confidence interval

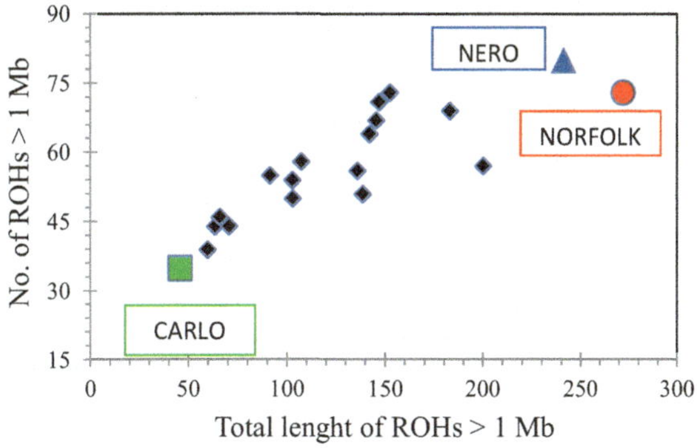


**Fig. 9.2** Relationship between number and total length of ROH segments >1 Mb in active population of Slovak Pinzgau bulls (Šidlová et al. 2014)

The comparison of pedigree-based and genomic inbreeding in Slovak Pinzgau cattle showed contradictory results. Kasarda (2019) revealed that the level of pedigree-based inbreeding, as well as average relatedness, could be significantly underestimated due to the lower quality and completeness of pedigree information. For example, the genealogical analysis showed that only four sires from active breeding bulls of Slovak Pinzgau population were inbred, whereas based on the $F_{ROH>16\,Mb}$ were inbred 6 and based on $F_{ROH>8\,Mb}$ even 13 from the total 19 bulls under consideration (Kukučková et al. 2017; Moravčíková et al. 2017).

In the active population of Slovak Pinzgau sires, the highest level of inbreeding, as well as longest homozygous DNA segments, were found in the genome of Slovak bull Norfolk (Fig. 9.2, red circle), whose father comes from Austria. Even if the pedigree analysis indicated for this bull inbreeding at a level of 2.15%, the molecular analysis revealed up to 8%. A similar level of inbreeding showed also the Austrian bull Nero (father of Norfolk, Fig. 9.2, blue triangle), which confirms the fact that highly inbreed animals carry homozygous stretches to their offspring, and its frequent use in the population can negatively affect the genetic diversity. Even if these bulls are highly appreciated and frequently used by farmers, they may endanger the diversity of the entire population by inappropriate mating plans. On the other hand, the Slovak bull Carlo (Fig. 9.2, green square), born after a father from Canada, appears to be an outbred individual based on both, pedigree and genomic analysis.

Performing analyses with ROH of different lengths allows estimation of the distance of the current population from the base population, hence provides information on the age of inbreeding. Different ROH inbreeding coefficients are expected to have different remote common ancestors. If the genome of an individual contains segments as short as 1 Mb, then the individual´s autozygosity originated from common ancestors was up to 50 generations in the past. Very long runs represent recent inbreeding (>16 Mb segments are expected in average after 3 ≈ generations), so part of autozygosity that is due to a more distant common ancestor is not covered with them (Howrigan et al. 2011; Curik et al. 2014). In comparison with other length categories, ROH > 1 Mb had clearly overestimated results (Ferenčaković 2013). Kukučková et al. (2017) reported for Slovak Pinzgau cattle the total average value

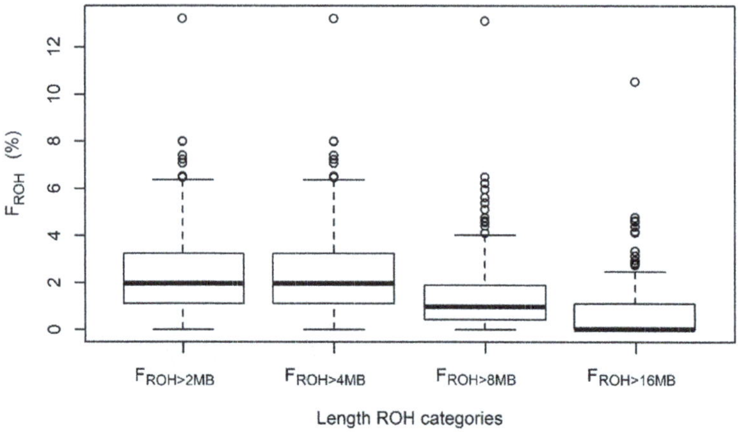


**Fig. 9.3** Boxplot distribution of $F_{ROH}$ within four length categories

of $F_{ROH}$ at level 0.0218 ± 0.0224 (falling within the category between > 2 and > 4 Mb). This confirms also the study of Moravčíková et al. (2017) and Kasarda et al. (2018b), which showed that ROH segments greater than 4 Mb ($F_{ROH>4\,Mb}$) cover in average 2.22% of the genome, whereas inbreeding estimates greater than 16 Mb achieved 0.81% that signalized recent inbreeding in analyzed population (Fig. 9.3).

## 9.7 Genomic Loss by Effective Population Size

The ancient and recent effective population size was estimated based on the linkage disequilibrium among the SNP markers. Compared to genomic estimates, based on the genealogical data, it is possible to analyze the trend of effective population size only a few generation ago, depending on the depth of pedigree. The pedigree analysis indicated the maximum $N_e$ at level 296.11, while the historical effective population size computed based on the genomic data was significantly higher (2697). Figure 9.4 shows estimated ancestral and recent effective population size, where *Ne1* was calculated according to Sved (1971) assuming no selection or mutation and *Ne2* was computed following Weir and Hill (1980) formula taking into account the restricted size of the population. Even if the total number of animals in the genomic analysis is relatively low, the estimation of historical effective population size ($N_{eT}$) takes into account a huge number of chromosomal regions derived from a much larger number of ancestors, thus is less sensitive to population size. The $N_{eT}$ derived from 150 generations ago showed a clearly linear decrease of 4.11 animals per generation (Kukučková et al. 2017).

The trend of recent effective population size was estimated separately for sires, dams of sires and dams of dams (Fig. 9.5). In general, the current $N_e$ is mainly

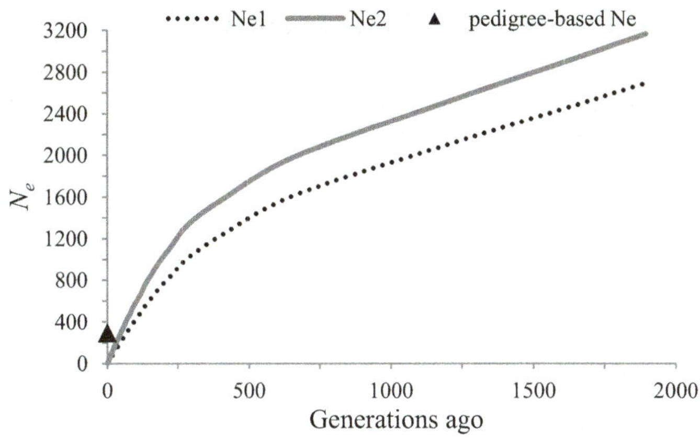


**Fig. 9.4** Estimated ancestral and recent effective population size based on the squared correlation coefficient ($r^2$) between SNPs

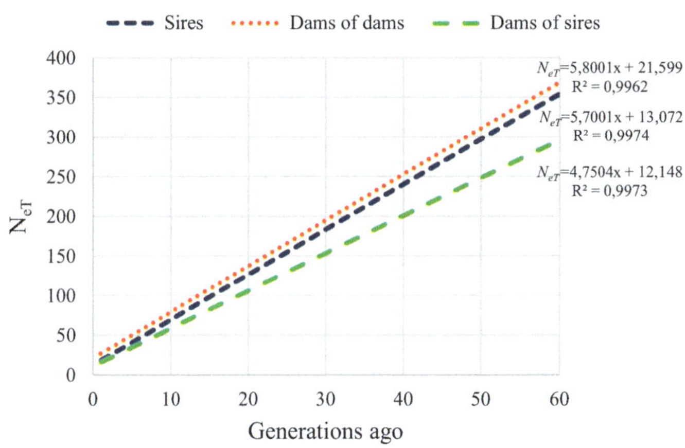


**Fig. 9.5** The trend of genomic-based estimate of current effective population size (Moravčíková et al. 2017)

reflected by the larger distances in the genome, while shorter segments express the trend of $N_e$ in past generations. As can be seen in Fig. 9.5, the effective population size decreased linearly across all of analyzed groups. The current effective population size, estimated using linear regression over generation 10 to 60, was in average 30.29 animals (at 95% confidence interval from 28.95 to 33.46). A decrease of approximately 7.81 animals per generation was found ($R^2 = 0.996$). Compared to the study of Pavlík et al. (2014), that reported for dairy Slovak Pinzgau $N_e = 188.94$ and for beef Slovak Pinzgau $N_e = 809$, the obtained estimates of effective population size were significantly lower. One of the reasons can be mainly the overestimation of $N_e$ due to incomplete pedigree data.

## 9.8 Genetic Uniqueness of Local Populations

Kukučková et al. (2017), Kasarda et al. (2018a, b) and Moravčíková et al. (2018) showed that even if all breeds included in the grading-up process of Slovak Pinzgau cattle left indelible traces in its genome, the Slovak Pinzgau breed can be regarded as genetically unique local population. From this point of view, the genetic distances, Bayesian cluster analysis and discriminant analysis of principal components (DAPC) proved to be the most successful methods for fine-scale differentiation between genetically closely connected breeds.

The previous studies of Kasarda et al. (2015, 2018b) revealed that the majority part of autosome residing in ROH is in Slovak Pinzgau cattle located on BTA4 and BTA6, very close to genes responsible for milk and beef production as well as coat colour pattern (Fig. 9.7b). Moreover, the comparison of ROH distribution in the genome of Slovak Pinzgau and Slovak Spotted cattle revealed genomic areas, in particular single mutations, which are unique for the genome of each local population.

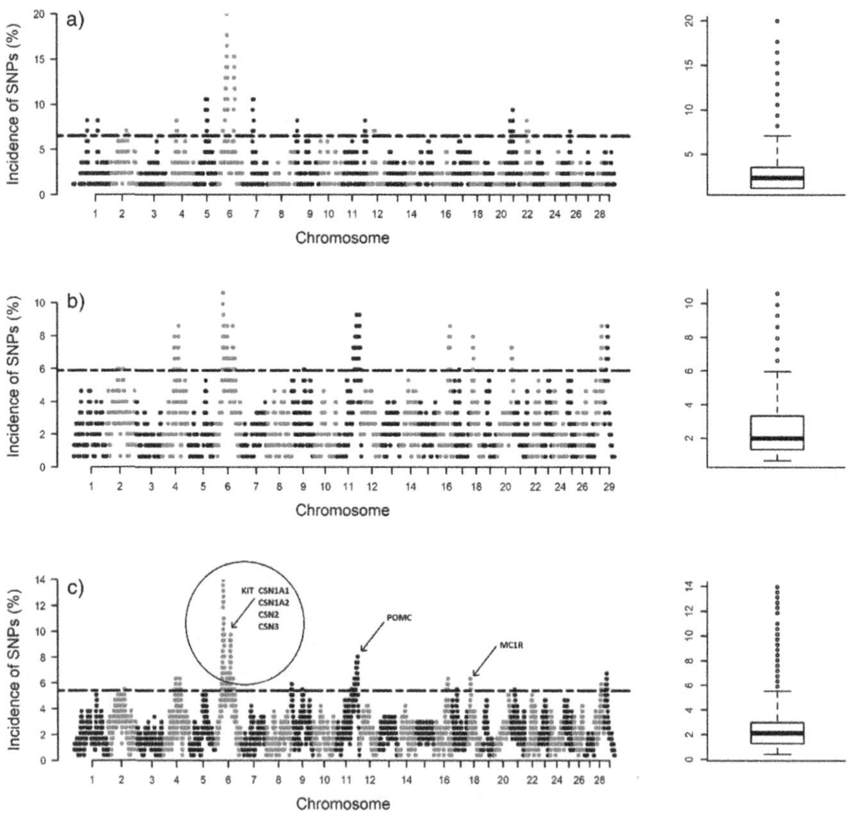


**Fig. 9.7** Genome-wide occurrence of SNPs in ROH for Slovak Spotted (**a**) and Slovak Pinzgau (**b**) cattle and incidence of runs for each SNP across both breeds (Kasarda et al. 2018b)

Based on the scan of autozygosity islands in the genome of Slovak Pinzgau cattle, totally 21 genomic regions with extreme ROH frequencies on BTA2, BTA4, BTA6, BTA9, BTA11, BTA16, BTA18, BTA20, BTA28 and BTA 29 were detected. The strongest ROH pattern was identified on BTA4 in the region from 65,400,608 bp to 80,706,119 bp that included QTLs affecting teat length (Ashwell et al. 2001), longissimus muscle area and marbling score (Mizoshita et al. 2004). The subset of biologically and economically most important quantitative trait loci located in each region is summarized. Besides them, we identified in target regions of selection some of the genes involved in multiple signalling and signal transduction pathways in a wide variety of biological processes, including the genetic control of milk production and reproduction (LCT, CSN1S1, CSN1S2, CSN2, CSN3, BMPR1B), body conformation and meat quality (GHRHR, POMC, MYO1G), coat colour (MCR1, KIT) and immunity response (IGFBP, IGJ, MR1, TLR10, TLR6).

The regions reflecting the genetic differentiation between Slovak and Austrian Pinzgau cattle were decipher based on the estimation of genome-wide $F_{ST}$ distribution. The origin of the analyzed cattle populations and especially the breeding strategy of both countries can lead to increases in allele frequencies of SNPs, due to positive selection of the populations. The higher allele frequencies of these SNPs are representative of differences in selection, neutrality or other processes which were used in breeding programs. The SNPs under selection are expected to show an allele frequency that deviates from neutral loci, leading to an increased level of genetic differentiation (Mancini et al. 2014). The signal reflecting differences caused by the recent selection in Austrian and Slovak Pinzgau populations was defined as genomic regions with the $F_{ST}$ frequency based on the outliers according to the obtained boxplot distribution (Fig. 9.8). The threshold value identifying SNPs as outliers were set to 0.12.

The analysis of differences in genome-wide linkage disequilibrium patterns between Slovak Spotted and Slovak Pinzgau populations revealed overall 10 genomic regions distributed across autosomes BTA4, BTA5, BTA6, BTA7, BTA11, BTA12, BTA20, BTA22 and BTA23 (Moravčíková et al. 2019) (Fig. 9.10). The longest region

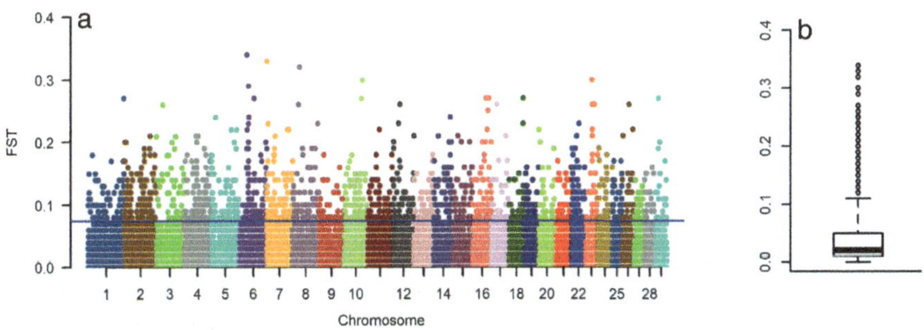


**Fig. 9.8** Genome-wide distribution of $F_{ST}$ values as signals of recent selection reflecting breeding programs of both Slovak and Austrian Pinzgau cattle

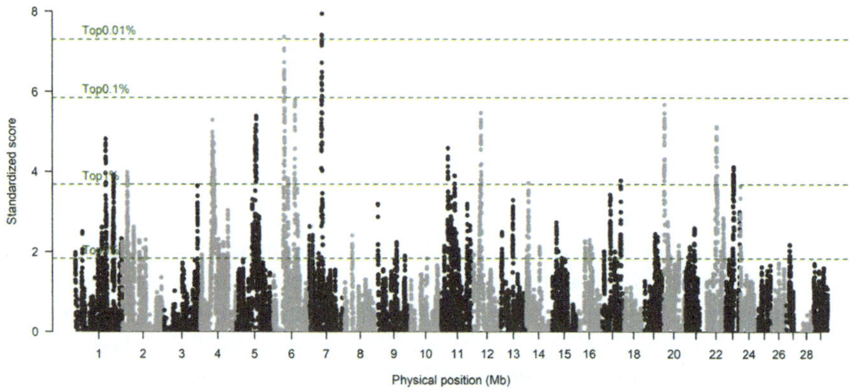


**Fig. 9.10** Differences in the genome-wide LD patterns (upper line is showing the genome-wide significance threshold) (Moravčíková et al. 2019)

found on BTA5 included QTLs for longissimus muscle area (Casas et al. 2003), backfat EBV (Li et al. 2004) and follicle-stimulating hormone (Casas et al. 2004).

The shortest region was detected on BTA11 near QTLs for yearling and weaning weight. Besides them, we identified in target selection regions, similarly as in the previous approach, some of the genes involved in the genetic control of milk production (casein family, HAL, IGF1, ABCG2, SPP1), muscle formation and body composition (MYBPC1, MYH9, PACRGL), reproduction (AMDHD1), temperament (SNRPF), and coat colour (KIT, KDR), what confirms that the regions displaying selection signatures in the Slovak Spotted and Slovak Pinzgau cattle are linked mostly to milk production and muscle development, thus selection for dual-purpose performance.

## 9.9 Conclusion

As it was shown, genome-wide analysis is more accurate in comparison to pedigree analysis because it is free of historical and missing records. Genomic analyses are suitable also in extreme cases when no historical or present records are available. Analysis of homozygous segments distribution along the entire genome of animals gives clear insight into the level of diversity of the local population. Genome-based effective population size is another parameter describing the state of genetic diversity. Compared to other parameters it has one advantage, because has "simple" explanation and is easy to understand. In contrary to the pedigree-based estimates. it gives not only a single value, but allows to present historical development. Moreover, the genome-based estimates of effective population size are automatically accounting for all effects influencing frequency of alleles as migration, selection, mutation, drift and recombination.

Genome-based methodology presented here is unique and allows for differentiation of closely (historically or phylogenetically) related populations as presented in case of Slovak and Austrian Pinzgau. Detailed view on the estimated parameters allows not only to distinguish between Slovak dual purpose Pinzgau and Spotted cattle, but also between closely connected Slovak and Austrian Pinzgau. The differences are traceable along the entire genome, not only on the population level. Different selection goals applied in both populations could be seen as unique genotypes, in particular allelic combinations, which differ between populations. The presented methodology allows for application in every population of choice and provides clear arguments for the identification of local populations as part of homeland heritage.

## References

Ashwell MS et al (2001) A genome scan to identify quantitative trait loci affecting economically important traits in a US Holstein population. J Dairy Sci 84(11):2535–2542

Beghain J et al (2012) Genome wide linkage disequilibrium in the Blonde d'Aquitaine cattle breed. J Anim Breed Genet 130(4):294–302

Bulla J et al (2013) Pinzgauer cattle in Slovakia. Slovak J Anim Sci 46(4):151–114

Caballero A et al (2010) Management of genetic diversity of subdivided populations in conservation programmes. Conserv Genet 11(2):409–419

Casas E et al (2003) Detection of quantitative trait loci for growth and carcass composition in cattle. J Anim Sci 81(12):2976–2983

Casas E et al (2004) Quantitative trait loci for male reproductive traits in beef cattle. Anim Genet 35(6):451–453

Cassell BG et al (2003) Effect of incomplete pedigrees on estimates of inbreeding and inbreeding depresion for days to first service and summit milk yield in Holsteins and Jeeseys. J Dairy Sci 86:2967–2976

Curik I et al (2014) Inbreeding and runs of homozygosity: a possible solution to an old problem. Livest Sci 166:26–34

Decker JE et al (2014) Worldwide patterns of ancestry, divergence, and admixture in domesticated cattle. PloS Genet 10(3):e1004254

FAO (1998) Secondary guidelines for development of national farm genetic resources management plans. Management of small populations at risk. FAO, Rome

Ferenčaković M (2013) Estimates of autozygosity derived from runs of homozygosity: empirical evidence from selected cattle populations. J Anim Breed Genet 130:286–293

Ferenčaković M et al (2013) Estimating autozygosity from high-throughput information: effects of SNP density and genotyping errors. Genet Sel Evol 45(1):42

Flury C et al (2010) Effective population size of an indigenous Swiss cattle breed estimated from linkage disequilibrium. J Anim Breed Genet 127:339–347

Forutan M et al (2018) Inbreeding and runs of homozygosity before and after genomic selection in North American holstein cattle. BMC Genomics 19:98

Gandini G et al (2014) Selection with inbreeding control in simulated young bull schemes for local dairy cattle breeds. J Dairy Sci 97:1790–1798

Gutiérrez JP, Goyache F (2005) A note on ENDOG: a computer program for analysing pedigree information. J Anim Breed Genet 122:172–176

Hayes BJ, Goddard ME (2010) Genome-wide association and genomic selection in animal breeding. Genome 53(11):876–883

Hazuchová E et al (2010) Pedigree analysis of the Slovak Pinzgau bulls. Acta Fytotechn Zootechn 13:62–64
Howrigan DP et al (2011) Detecting autozygosity through runs of homozygosity: a comparison of three autozygosity detection algorithms. BMC Genomics 12(1):460
Kadlečík O et al (2007) Pedigree analysis of purebred Pinzgau population. Acta Fytotechn Zootechn 10:29–32
Kadlečík O et al (2008) Inbreeding in purebred Slovak Pinzgau dual – purpose cattle population. Arch Zootech 11(2):21–28
Kadlečík O et al (2011) Pedigree analysis of Slovak Pinzgau breed. Agric Conspec Sci 76(3):165–168
Kadlečík O et al (2013) Diversity of cattle breeds in Slovakia. Slovak J Anim Sci 46(4):145–150
Kadlečik O et al (2017) Inbreeding and genetic diversity loss in Slovak Pinzgau breed. Agric Conspec Sci 82(3):259–262
Kasarda R (2019) Genomic and pedigree-based inbreeding in Slovak spotted cattle. AGROFOR 4(1):102–110
Kasarda R et al (2014) Influence of mating systems and selection intensity on the extent of inbreeding and genetic gain in the Slovak Pinzgau cattle. Czech J Anim Sci 59(5):219–226
Kasarda R et al (2015) Genome-wide selection signatures in Pinzgau cattle. Slovak Journal of Food Science 9:268–274
Kasarda R et al (2016) Inheritance of coat colour in Slovak Pinzgau cattle. J Cent Eur Agric 17(1):48–55
Kasarda R et al (2018a) Genome-wide scan for loci under selection in local populations of Pinzgau cattle. In: Proceedings of the World Congress on Genetics Applied to Livestock Production, Auckland, 11–16 February 2018
Kasarda et al (2018b) The impact of artificial selection on runs of homozygosity in Slovak Spotted and Pinzgau cattle. Slovak J Anim Sci 51(3):91–103
Kasarda R, Kadlečík O (2007) An economic impact of inbreeding in the purebred population of Pinzgau cattle in Slovakia on milk production traits. Czech J Anim Sci 52:7–11
Kim ES et al (2013) Effect of artificial selection on runs of homozygosity in U.S. Holstein cattle. PLoS One 8:e80813
Kukučková V et al (2017) Genomic characterization of Pinzgau cattle: genetic conservation and breeding perspectives. Conserv Genet 18(4):893–910
Lenstra JA et al (2012) Molecular tools and analytical approaches for the characterization of farm animal genetic diversity. Anim Genet 43(5):483–502
Leroy G et al (2013) Methods to estimate effective population size using pedigree data: examples in dog, sheep, cattle and horse. Genet Sel Evol 45(1):1–10
Li C et al (2004) Identification and fine mapping of quantitative trait loci for backfat on bovine chromosomes 2, 5, 6, 19, 21, and 23 in a commercial line of Bos taurus. J Anim Sci 82(4):967–972
Mancini G et al (2014) Signatures of selection in five Italian cattle breeds detected by a 54K SNP panel. Mol Biol Rep 41:957–965
Marras G et al (2015) Analysis of runs of homozygosity and their relationship with inbreeding in five cattle breeds farmed in Italy. Anim Genet 46:110–121
Mastrangelo S et al (2014) The genome-wide structure of two economically important indigenous Sicilian cattle breeds. J Anim Sci 92:4833–4842
Mastrangelo S et al (2016) Genomic inbreeding estimation in small populations: evaluation of runs of homozygosity in three local dairy cattle breeds. Animal 10:746–754
Mastrangelo S et al (2018) Genome-wide identification of runs of homozygosity islands and associated genes in local dairy cattle breeds. Animal 26:1–9
McTavish EJ et al (2013) New World cattle show ancestry from multiple independent domestication events. PNAS 110(15):E1398–E1406
Mészáros G et al (2008) Genetic evaluation for length of productive life in Slovak Pinzgau cattle. Arch Tierz 51:438–448

Mizoshita K et al (2004) Quantitative trait loci analysis for growth and carcass traits in a half-sib family of purebred Japanese Black (Wagyu) cattle. J Anim Sci 82(12):3415–3420
Moravčíková N et al (2017) Effective population size and genomic inbreeding in Slovak Pinzgau cattle. Agric Conspec Sci 82(2):97–100
Moravčíková N et al (2018) Genomic response to natural selection within alpine cattle breeds. Czech J Anim Sci 63(4):136–143
Moravčíková N et al (2019) Genomic signatures of selection in cattle through variation of allele frequencies and linkage disequilibrium. J Cent Eur Agric 20(2):576–580
Nei M (1987) Molecular evolutionary genetics. Columbia University Press, New York
Pavlík I et al (2012) Pedigree analysis of Slovak Holstein cattle. In: Book of Abstracts of the 63rd Annual Meeting of the European Federation of Animal Science, Bratislava, 27–31 August 2012
Pavlík I et al (2013) Comparison of genetic diversity in dual-purpose and beef Pinzgau populations. Acta Fytotechn Zootechn 16(3):69–73
Pavlík I et al (2014) Joint genealogical analysis as a tool for diversity evaluation in Pinzgau cattle populations. Arch Tierz 57(14):1–12
Pemberton T et al (2012) Genomic patterns of homozygosity in worldwide human populations. Am J Hum Genet 91:275–292
Pérez O´Brian AM et al (2014) Linkage disequilibrium levels in Bos indicus and Bos taurus cattle using medium and high density SNP chip data and different minor allele frequency distributions. Livest Sci 166:121–132
Pérez O´Brian AM et al (2015) Low levels of taurine introgression in the current Brazilian Nelore and Gir indicine cattle populations. Genet Sel Evol 47:31
Peripolli E et al (2017) Runs of homozygosity: current knowledge and applications in livestock. Anim Genet 48(3):255–271
Pšenica J (1990) Pinzgau cattle in Slovakia. Príroda, Bratislava
Pšenica J et al (1998) Slovak Pinzgau cattle. Slovenský chov 16
Shin DH et al (2013) Accurate estimation of effective population size in the Korean dairy cattle based on linkage disequilibrium corrected by genomic relationship matrix. Asian-Australas J Anim Sci 26:1672–1679
Šidlová V et al (2014) Inbreeding coefficient derived from runs of homozygosity estimation in Pinzgau cattle using bovineSNP50 BadChip. In: Book of abstract International scientific genetic conference, XXVI. Genetic Days", Praha, 3–4 September 2014
Sved JA (1971) Linkage disequilibrium and homozygosity of chromosome segments in finite populations. Theor Popul Biol 2:125–141
Szmatoła T et al (2016) Characteristics of runs of homozygosity in selected cattle breeds maintained in Poland. Livest Sci 188:72–80
Thomas ChD (2013) Local diversity stays about the same, regional diversity increases, and global diversity declines. Proc Natl Acad Sci U S A 110(48):19187–19188
Toro MA et al (2009) Molecular characterization of breeds and its use in conservation. Livest Sci 120(3):174–195
Uimari P, Tapio M (2011) Extent of linkage disequilibrium and effective population size in Finnish Landrace and Finnish Yorkshire pig breeds. J Anim Sci 89:609–614
Vellend M et al (2013) Global meta-analysis reveals no net change in local-scale plant biodiversity over time. Proc Natl Acad Sci USA 110:19456–19459
Weigel KA (2001) Controlling inbreeding in modern breeding programs. J Dairy Sci 84:E177–E184
Weir BS, Hill WG (1980) Effect of mating structure on variation in linkage disequilibrium. Genetics 95:447–488
Wright S (1920) Principles of livestock breeding. US Dept Agric Bull 905
Wright S (1921) Systems of mating. Genetics 6:111–178
Zavarez LB et al (2015) Assessment of autozygosity in Nellore cows (Bos indicus) through high-density SNP genotypes. Front Genet 6:1–8
Zhang Q et al (2015) Runs of homozygosity and distribution of functional variants in the cattle genome. BMC Genomics 16:542

# Chapter 10
# Objectification of Reliability of Selected Methods of Identification and Quantification of Meat and Its Substitutes

**Jozef Golian, Zuzana Drdolová, and Lucia Benešová**

**Abstract** A current problem in the food and agricultural industry is the detection of animal species substitution. Emphasis is placed on the selection of simple, reliable, reproducible and repeatable methods for identifying and quantifying food ingredients. For this reason, it is necessary to know about effective systems to ensure effective control, correct identification, and quantification of meat and its possible replacement. This chapter focuses on the design of a system for identifying and quantifying the DNA of the animal species from which the meat originated and possible meat substitutes. Thermal treatment of the products and their mechanical processing was also considered in order to obtain relevant and repeatable results for several samples simultaneously so that several aspects are examined in one analysis at a time. The Meat 5.0 LCD-Array kit was verified in this chapter first. To eliminate cross-reaction and to evaluate the specificity of the method, pure samples were tested for each animal species included in the analysis. Tissue samples from 21 species of animals, including domestic livestock, wildlife and avian species, were used as biological material. Specific analyses were focussed on pig meat. Furthermore, 11 blends of a combination of three animal species were analyzed to verify functionality, adding two adulterants to the main blend constituent of pork in three different ratios. The animal species under investigation were identified by their specific probes with 100% specificity. Tests were also performed on samples after various heat treatments, with no observed changes, which means that the kit could also be used to test cooked or sterilized foods. It was not possible to quantify the DNA using the Meat 5.0 LCD-Array. Also, the LCD-Array kit has been evaluated as successful in identifying non-meat animal proteins. The innuDETECT Assay was verified as a second. The combination of 6 binary mixtures of meats in eighteen different concentration orders of the addition was analyzed. The addition concentration range was from 100 to 0.1%. All analyzed animal species included in the experiment were equally correctly identified after reaching the 5% addition limit. In the first binary mixture where pork was added to the beef, the pig DNA was detected at 0.5% of the addition. In the

J. Golian (✉) · Z. Drdolová · L. Benešová
Faculty of Biotechnology and Food Sciences, Department of Food Hygiene and Safety, Slovak University of Agriculture, Tr. A. Hlinku 2, Nitra 949 76, Slovakia
e-mail: Jozef.golian@uniag.sk

© Springer Nature Switzerland AG 2022 
J. Hernik et al. (eds.), *Cultural Heritage—Possibilities for Land-Centered Societal Development*, Environmental History 13,
https://doi.org/10.1007/978-3-030-58092-6_10

second mixture, the chicken was identified only to a level of 5% of the addition of chicken to pork. In the third binary mixture, the turkey DNA was detected to a level of 1% of the addition of turkey to pork. In the fourth binary mixture where sheep was added to pork, the sheep DNA was captured to a level of 1% addition. In the fifth combination of goat and pork meat, goat DNA was captured at 1% of the addition. In the sixth binary mixture combination of beef and pork was able to capture the bovine DNA to a level of 0.5% addition. The results were identical for both studied mixtures before and after the heat treatment. This means that the heat treatment has no effect on the performance of the detection kits. These validated kits for the identification of animal species in food have a high potential in the field of laboratory food analysis and food control. These compared techniques are highly specific and allow easy identification of animal species, sufficiently sensitive and provide repeatable results.

**Keywords** Identification · Quantification · DNA extraction · Detection kits · Meat products · Animal species

## 10.1 Introduction

Reliable, accurate and efficient methods for identifying animal species in food are the basis for proper and objective analysis of food. They provide a framework for ensuring health security, maintaining fair trade and enabling detection or prevention of economic fraud and consumer deception. Due to their high economic value, meat and meat products are the subjects of unfair commercial practices and counterfeiting. In this chapter, two techniques were evaluated for the identification of animal species in food, possibilities of quantification by selected techniques and real application in practice in identifying and quantifying the proportion of animal species contained in meat and meat products.

A variety of DNA isolation techniques are used to isolate DNA from foods that generally provide more or less sufficient amounts of isolated DNA together with the reduction of potential inhibitors. Most of these methods are time-consuming and technically demanding. Components of extraction mixtures, known as PCR inhibitors, include: chelating agents such as ethylenediaminetetraacetic acid (EDTA), which can complex the $Mg^{2+}$ ions required for polymerase activity, sodium hydroxide (NaOH), which causes DNA degradation and polymerase denaturation phenol, which also causes denaturation of the polymerase, because it binds to the enzyme molecule by hydrogen bonds and ethanol, isopropanol, which causes DNA precipitation (Bar et al. 2012). In dairy and meat products, calcium ions have been identified as a source of PCR inhibition (Bickley et al. 1996). DNA molecules are thermally more stable than proteins and this is one of the reasons why DNA-based methods have been of great interest in recent years (Cottenet et al. 2011; Sakaridis et al. 2013). The growing number of DNA-based food studies is due to the very good specification and low detection limits that can be obtained by various PCR-based techniques (Darwish et al. 2009; Drummond et al. 2013). DNA for these methods is usually extracted from cells that are present in products in large but very variable numbers (Sharma et al. 2011). Requirements to be met for successful PCR include extracting as much pure DNA from the sample as possible and optimizing the reaction conditions. As there are several possibilities for extracting DNA from food, research protocols and commercial kits are included. Extraction of high-quality DNA with high yield is a limiting factor in genetic analysis. DNA quality from each line should be consistent to allow a proper genetic analysis of multiple individuals. Over the past two decades, PCR-based methods have been used in different food areas, whether to determine different contaminants and food ingredients such as pathogens and toxin-producing organisms, food ingredients, allergens, or to detect adulteration. Effective DNA extraction from food is one of the parameters that influence the successful implementation of these methods (Volk et al. 2014).

## 10.2 Meat 5.0 LCD-Array Validation

### *10.2.1 Biological Material*

Tissue samples from 21 animal species were used: pig (*Sus scrofa*), cattle (*Bos taurus, Bos bison*), sheep (*Ovis aries*), equine (*Equus caballus, E. asinus*), goat (*Capra hircus*), hare (*Lepus europaeus*), rabbit (*Oryctolagus cuniculus*), roe deer (*Capreolus capreolus*), red deer (*Cervus elaphus*), fallow deer (*Dama dama*), chicken (*Gallus gallus*), turkey (*Meleagris gallopavo*), goose (*Anser* sp.), mallard duck (*Anas platyrhynchos*), Muscovy duck (*Cairina moschata*), pheasant (*Phasianus* sp.), ostrich (*Struthio camelus*), dog (*Canis* sp.), cat (*Felis silvestris*), water buffalo (*Bubalus bubalis*) and reindeer (*Rangifer tarandus*). Authentic tissue samples of the mentioned animal species included in the experiment were obtained from regional sources within the territory of Slovakia. The samples were cut into smaller pieces using a sterile scalpel and placed in sterile containers in which they were stored. All samples were stored at −18 °C.

### *10.2.2 Tested Mixtures*

Mixtures were prepared using a blender (Blender 8008, Connecticut, USA) to a final weight of 100 g (Table 10.1). To avoid any cross-contamination, each blend was treated separately using a different blender container. The mixtures were immediately transferred to sterile tubes and stored at −18 °C until DNA extraction. Series of mixtures were prepared in duplicates. One variant was analyzed in the raw state and the other after heat treatment at 100 °C for 60 min.

**Table 10.1** Scheme of preparation and combination of tested species

| Mixture | Major component [%] | Minor component 1 [%] | Minor component 2 [%] |
|---|---|---|---|
| | Pork | Goat | Sheep |
| A1 | 98.0 | 1.0 | 1.0 |
| A2 | 99.0 | 0.5 | 0.5 |
| A3 | 99.8 | 0.1 | 0.1 |
| | Pork | Beef | Equines |
| B1 | 98.0 | 1.0 | 1.0 |
| B2 | 99.0 | 0.5 | 0.5 |
| B3 | 99.8 | 0.1 | 0.1 |
| | Pork | Hare | Rabbit |
| C1 | 98.0 | 1.0 | 1.0 |
| C2 | 99.0 | 0.5 | 0.5 |
| C3 | 99.8 | 1.0 | 1.0 |
| | Pork | Red deer | Fallow deer |
| D1 | 98.0 | 1.0 | 1.0 |
| D2 | 99.0 | 0.5 | 0.5 |
| D3 | 99.8 | 0.1 | 0.1 |
| | Pork | Turkey | Pheasant |
| E1 | 98.0 | 1.0 | 1.0 |
| E2 | 99.0 | 0.5 | 0.5 |
| E3 | 99.8 | 0.1 | 0.1 |
| | Pork | Chicken | Goose |
| F1 | 98.0 | 1.0 | 1.0 |
| F2 | 99.0 | 0.5 | 0.5 |
| F3 | 99.8 | 1.0 | 1.0 |
| | Pork | Mallard duck | Muscovy duck |
| G1 | 98.0 | 1.0 | 1.0 |
| G2 | 99.0 | 0.5 | 0.5 |
| G3 | 99.8 | 0.1 | 0.1 |
| | Pork | Roe deer | Fallow deer |
| H1 | 98.0 | 1.0 | 1.0 |
| H2 | 99.0 | 0.5 | 0.5 |
| H3 | 99.8 | 0.1 | 0.1 |
| | Pork | Cat | Dog |
| I1 | 98.0 | 1.0 | 1.0 |
| I2 | 99.0 | 0.5 | 0.5 |
| I3 | 99.8 | 1.0 | 1.0 |
| | Pork | Buffalo | Reindeer |
| J1 | 98.0 | 1.0 | 1.0 |
| J2 | 99.0 | 0.5 | 0.5 |
| J3 | 99.8 | 0.1 | 0.1 |

(continued)

**Table 10.1** (continued)

| Mixture | Major component [%] | Minor component 1 [%] | Minor component 2 [%] |
|---|---|---|---|
| | Pork | Turkey | Ostrich |
| K1 | 98.0 | 1.0 | 1.0 |
| K2 | 99.0 | 0.5 | 0.5 |
| K3 | 99.8 | 0.1 | 0.1 |
| | Pork | Hare | Rabbit |
| C1 | 98.0 | 1.0 | 1.0 |
| C2 | 99.0 | 0.5 | 0.5 |
| C3 | 99.8 | 1.0 | 1.0 |
| | Pork | Red deer | Fallow deer |
| D1 | 98.0 | 1.0 | 1.0 |
| D2 | 99.0 | 0.5 | 0.5 |
| D3 | 99.8 | 0.1 | 0.1 |
| | Pork | Turkey | Pheasant |
| E1 | 98.0 | 1.0 | 1.0 |
| E2 | 99.0 | 0.5 | 0.5 |
| E3 | 99.8 | 0.1 | 0.1 |
| | Pork | Chicken | Goose |
| F1 | 98.0 | 1.0 | 1.0 |
| F2 | 99.0 | 0.5 | 0.5 |
| F3 | 99.8 | 1.0 | 1.0 |
| | Pork | Mallard duck | Muscovy duck |
| G1 | 98.0 | 1.0 | 1.0 |
| G2 | 99.0 | 0.5 | 0.5 |
| G3 | 99.8 | 0.1 | 0.1 |
| | Pork | Roe deer | Fallow deer |
| H1 | 98.0 | 1.0 | 1.0 |
| H2 | 99.0 | 0.5 | 0.5 |
| H3 | 99.8 | 0.1 | 0.1 |
| | Pork | Cat | Dog |
| I1 | 98.0 | 1.0 | 1.0 |
| I2 | 99.0 | 0.5 | 0.5 |
| I3 | 99.8 | 1.0 | 1.0 |
| | Pork | Buffalo | Reindeer |
| J1 | 98.0 | 1.0 | 1.0 |
| J2 | 99.0 | 0.5 | 0.5 |
| J3 | 99.8 | 0.1 | 0.1 |
| | Pork | Turkey | Ostrich |
| K1 | 98.0 | 1.0 | 1.0 |
| K2 | 99.0 | 0.5 | 0.5 |
| K3 | 99.8 | 0.1 | 0.1 |

*Note* Composition is expressed in weight percent

**Table 10.2** PCR product preparation

| Title | 1 sample | 8 samples |
|---|---|---|
| 2 × Master Mix | 12.5 μl | 112.5 μl |
| Primer Mix 'MEAT' | 1.5 μl | 13.5 μl |
| PCR grade water | 6.0 μl | 54.0 μl |
| DNA samples | 5 μl | 5 μl |
| Total volume | 25 μl | |

### *10.2.3 DNA Isolation*

DNA was isolated from each species individually using the Maxwell® 16 Tissue DNA Purification Kit (Promega, Madison, USA) and the Maxwell® 16 System (Promega, Madison, USA), innuPREP DNA Mini Kit (Analytik Jena, Germany) and the NucleoSpin Food Isolation Kit (Macherey–Nagel, Düren, Germany) according to the manufacturer's instructions. DNA was isolated from a 50 mg sample. The amount of DNA in each sample was quantified using a Quantus fluorimeter (Promega, Madison, USA). The DNA was stored at −18 °C until use. For each analysis, two parallel samples were analyzed in two replicates.

### *10.2.4 PCR Products Preparation*

The PCR was performed in a TOptical Gradient 96 thermocycler (Biometra, Göttingen, Germany). Mix for preparation of PCR product (for one sample): 2 × Master Mix – 12.5 μl, Primer Mix 'MEAT' – 1.5 μl, PCR grade water – 6.0 μl. Into the prepared mixture with a total volume of 20 μl, 5 μl of the isolated DNA from the analyzed sample were added (Table 10.2).

### *10.2.5 Cycler Settings*

Thermocycler setup is presented in Table 10.3.

### *10.2.6 Agarose Gel Electrophoresis*

Analysis of PCR amplicons was performed using agarose gel electrophoresis with a Consort Maxi Series EV243 (Cleaver Scientific, Rugby, United Kingdom). Agarose gel (2%) was prepared by dissolving the appropriate quantities of agarose (SERVA Electrophoresis, Heidelberg, Germany) in 1 × Tris–Acetate-EDTA buffer (Focus

**Table 10.3** Cycler mode setting

| Step | Duration | Temperature (°C) | Temperature gradient (°C/s) |
|---|---|---|---|
| 1. Initial denaturation | 5 min | 95 | |
| 2. 35 repetitions: | | | |
| Denaturation | 30 s | 94 | 3 |
| Annealing | 45 s | 57 | 3 |
| Elongation | 45 s | 72 | 3 |
| 3. End | 2 min | 72 | |

Bioscience, Queensland, Australia) (pH 8.0) in a ProLine SM117 microwave oven (ProLine, Schiphol, Netherlands). Ethidium bromide GelRed 10.000 × (Biotium, Fremont, California, USA) stock solution was added directly to molten agarose at a concentration of 1 $\mu g \cdot ml^{-1}$, before pouring the gel. The electrophoretic samples were mixed with 5 × Green GoTaq Flexi Buffer (Promega) before loading into the gel. After electrophoresis, DNA fragments in the agarose gel were visualized with an EB-20 UV transilluminator (Ultra Lum, Claremont, California, USA).

### *10.2.7 LCD-Array Hybridization and Detection*

Each LCD-chip contains eight identical matrices in rectangular reaction chambers, which can be used individually. Functional chip controls for hybridization, secondary labelling and staining are located in the three corners of the matrix. The reaction field of the Meat 5.0 LCD-Array kit consists of an 8 × 8 point pattern with an average point diameter of 350 μm. Dots contain species-specific capture probes positioned as vertical duplicates (Table 10.4).

Labelled PCR fragments, coupled to a hybridization buffer, were bound to a positive array and were hybridized. During hybridization, labelled PCR fragments bind to specific immobilized capture probes as spots at the bottom of each array field. These spots were visualized using a PF3650u LCD-array scanner (PacificImage Electronics, Torrance, USA) and evaluated using the SlideReader V12 software (Chipron, Berlin, Germany). The default detection threshold was 2000 pv (Chipron 2014).

### *10.2.8 Hybridization*

The chip and the associated components were tempered to room temperature. The thermostat was warmed to 35 °C and the hybridization mixture was prepared according to Table 10.5.

**Table 10.4** List of species studied with numerical code sequence

| No | Name | Species | No | Name | Species |
|---|---|---|---|---|---|
| 1 | Hyb-Ctrl | | | | |
| 2 | Cattle | *Bos taurus, Bos bison* | 14 | Red deer | *Cervus elaphus* |
| 3 | Sheep | *Ovis aries* | 15 | Fallow deer | *Dama dama* |
| 4 | Equine | *Equus caballus, E. asinus* | 16 | Springbok | *Antidorcas marsupialis* |
| 5 | Goat | *Capra hircus* | 17 | Canine/dog | *Canis sp.* |
| 6 | Camels | *Camelus* sp. | 18 | Cat | *Felis silvestris* |
| 7 | Buffalo | *Bubalus bubalis* | 19 | Chicken | *Gallus gallus* |
| 8 | Pork | *Sus scrofa* | 20 | Turkey | *Meleagris gallopavo* |
| 9 | Kangaroo | *Macropus rufus/ giganteus* | 21 | Goose | *Anser* sp. |
| 10 | Hare | *Lepus europaeus* | 22 | Ostrich | *Struthio camelus* |
| 11 | Rabbit | *Oryctolagus cuniculus* | 23 | Mallard duck | *Anas platyrhyncos* |
| 12 | Reindeer | *Rangifer tarandus* | 24 | Muscovy duck | *Cairina moschata* |
| 13 | Roe deer | *Capreolus capreolus* | 25 | Pheasant | *Phasianus* sp. |

**Table 10.5** Hybridization mixture

| Title | 1 sample (μl) | 8 samples (μl) |
|---|---|---|
| Hybridization buffer | 22 | 198 |
| Modulator | 2 | 18 |
| Total volume | 24 | |

10 μl of each PCR product was pipetted into the prepared hybridization mixtures. Chip was placed in a cover for incubation with the possibility of closing. Using a pipette, 28 μl of the resulting mixture of PCR product and hybridization mixture was transferred into the reaction chambers. The sealed chip cover was placed in a thermostat and incubated at 35 °C for 30 min. 150 ml wash bottles with wash solution were prepared by adding 5 ml Wash Buffer Concentrate to 995 ml deionized water. After incubation, the chip was washed in a series of wash flasks and then dried, rotating for 10 s in the CHIP Spin FVL2400N (Chipron 2014).

28 μl of the labelling mixture (Table 10.6) was applied to each array of the chip, the chip was incubated at room temperature for 5 min and repeatedly washed and dried.

In the final staining phase, 28 μl of staining solution was used. A determined volume was added to each chip array and the mixture was allowed to incubate for 5 min at room temperature. The staining process was stopped by flushing in the beaker for 10 s. It is necessary to dry the chip again for scanning.

**Table 10.6** Labelling mixture

| Title | 1 sample (μl) | 8 samples (μl) |
|---|---|---|
| Dilution buffer | 27 | 270 |
| Modulator | 3 | 30 |
| Label | 0.2 | 2 |
| Total volume | 30.2 | |

### *10.2.9 Statistical analysis*

The results of the analysis were evaluated using the MS Excel (2017) statistical package.

## 10.3 Results

The amount of DNA isolated from the analyzed animal species ranged from 5.9–30 ng.$\mu l^{-1}$, taken from a non-heat-treated sample and prepared raw mixtures. In the case of heat-treated samples and mixtures, 0.93–21 ng.$\mu l^{-1}$ were measured.

In order to evaluate the specificity of the chosen method and to verify the absence of a potential cross-reaction, pure samples of all animal species included in the analysis were first tested (Table 10.7).

By comparing the amount of isolated DNA, it was found that lower values were observed in the samples after heat treatment. Differences in the range of values between mixtures containing different concentrations of contaminants (0.1%, 0.5%, 1%) were also observed. We believe that this may be caused by heating the samples during homogenization in an electric mixer. When preparing the mixtures, great emphasis was placed on thorough mixing of the mixed samples. In addition, the isolation of DNA from different animal species containing different amounts of contaminating species required several additional steps during the preparation and isolation process that could affect the extraction power. When working with the Meat 5.0 LCD-Array kit, different DNA concentrations did not affect the sensitivity of the technique or the repeatability of the result.

The first part is focussed on the evaluation of the specificity of Meat 5.0 LCD-Array probes as a modern approach to animal species identification. With a focus on pork, species-specific analyses were carried out on 20 animal species combined in a mixture of three species together. Pork was used as an essential part of each mixture, also focussing on the specificity of the pig identification probe, which contains multiple sequences for detecting polymorphisms of individual nucleotides in the target region.

The kit manufacturer declares a detection limit of <0.5% depending on the level of sample processing (Chipron 2014). To determine the potential cross-reactivity or competitiveness of individual probes, raw meat and meat after heat treatment were

**Table 10.7** Evaluation of the specificity of hybridization probes encoding animal species involved in the study

| Pork | Sheep | Equine |
|---|---|---|
| 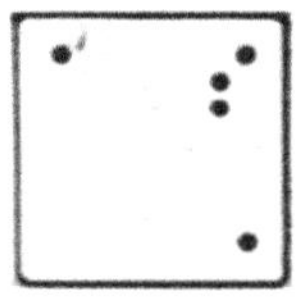 | 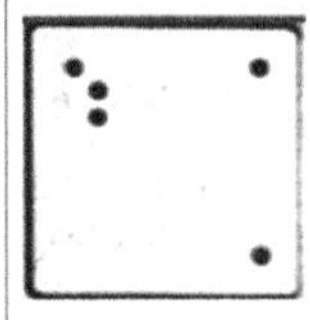 | 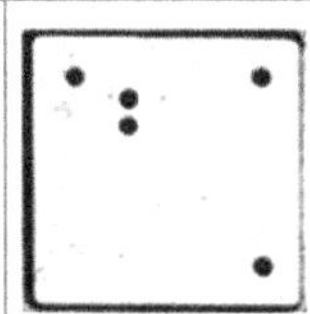 |
| Goat | Cattle | Red deer |
| 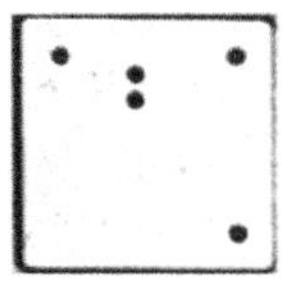 | 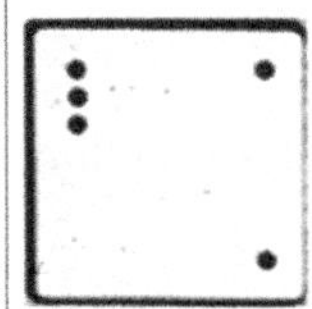 | 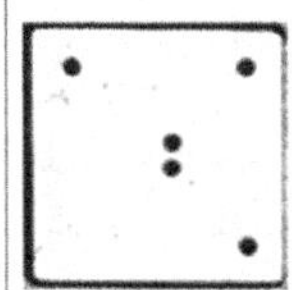 |
| Goose | Buffalo | Hare |
| 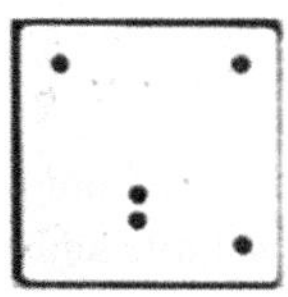 | 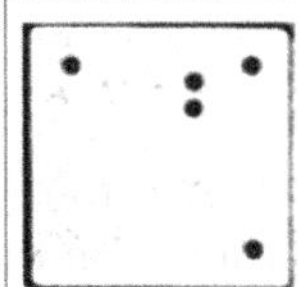 | 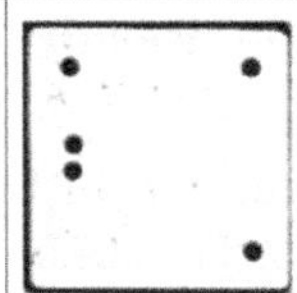 |
| Rabbit | Reindeer | Pheasant |
| 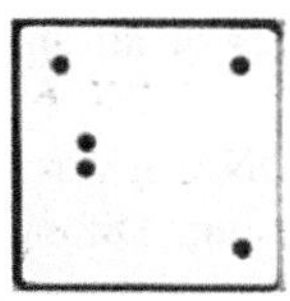 | 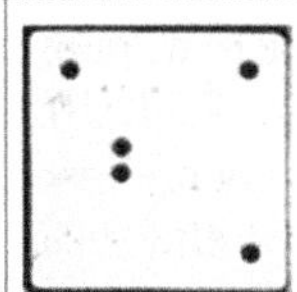 | 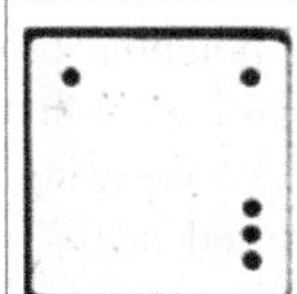 |
| Roe deer | Canine/dog | Turkey |
| 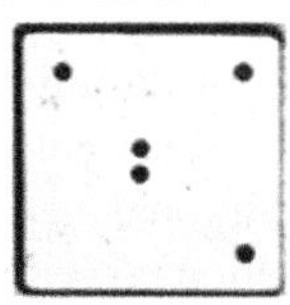 | 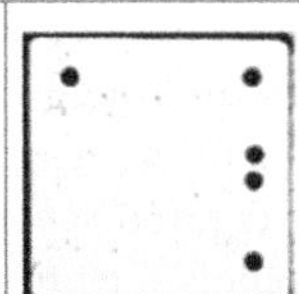 | 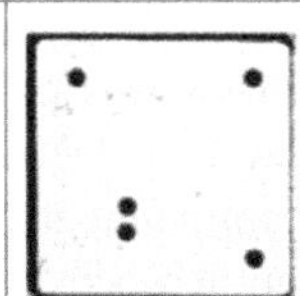 |
| Muscovy duck | Chicken | Cat |
| 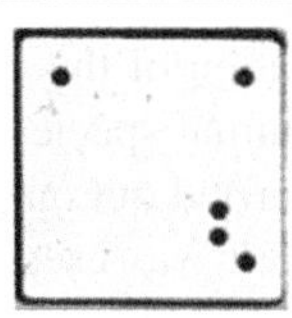 | 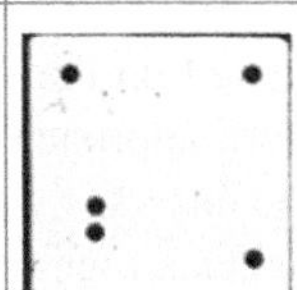 | 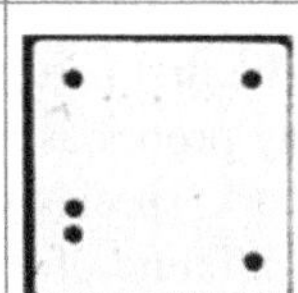 |

(continued)

**Table 10.7** (continued)

| Mallard duck | Ostrich | Fallow deer |
|---|---|---|
| 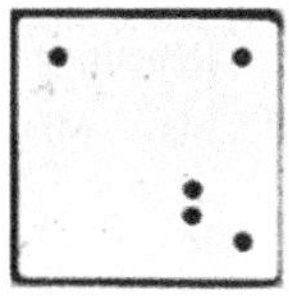 | 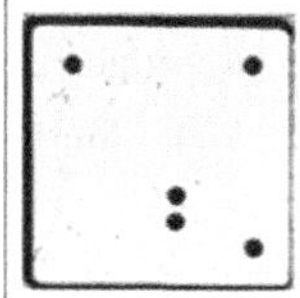 | 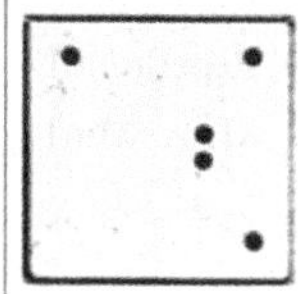 |

tested in two replicates of a mixture of pork containing two types of meat in equal proportions (Table 10.1). In general, 100% species specificity of all probes tested was observed, with no false positive or false negative results. All animal species provided hybrid signals on their respective species-specific probes. Examined animal species were identified by their specific probes with 100% specificity in addition to the analyzed sample containing reindeer, where cross-reactivity was noted, but this was stated in the kit user manual.

Raw and heat-treated samples and mixtures were analyzed at three different addition concentrations. Each analysis was repeated twice. All animal species were correctly identified only after reaching the detection limit of 0.5% in each sample and mixtures. In samples containing 0.1% adulterant, the kit detected the DNA of the adulterants, but the repeatability was lower because there was no comparable result for the two replicates that were performed in each analysis.

For comparison with other authors, the repeatability of the result obtained by Kohen kappa (κ) calculation was evaluated based on data obtained from three different addition concentrations (1%, 0.5%, 0.1%). $\kappa > 0.63$ was the average of the results of all mixtures analyzed. This value was considered a threshold for a good level of accuracy. Kohen kappa was calculated separately for the results of the analysis of the addition of 1 and 0.5% adulterants. In this case, $\kappa = 1$ because of the high repeatability rate. There was no significant difference between the raw and heat-treated samples, suggesting that heat treatment does not affect kit accuracy (Table 10.8).

The sensitivity of the Meat 5.0 LCD-Array was determined to be 0.5% for the detection of contaminating species present in the samples, both raw and boiled, although the amount of DNA extracted from the raw samples was higher.

This higher level of sensitivity was obtained from a single analysis for each meat mix at different levels of contamination. In the validation process described in this study, each mixture was analyzed twice to better evaluate the sensitivity of the method and its repeatability. For meat, samples containing 0.1% adulterants were not recorded in both replicates. This lower repeatability of the meat samples may be due to either a lower sensitivity of the probe or the difficulty of obtaining a perfectly homogeneous sample.

**Table 10.8** Evaluation of identification of combinations tested animal species

| | Before heat treatment | | | | | | After heat treatment | | | | | |
|---|---|---|---|---|---|---|---|---|---|---|---|---|
| | Repetition 1 | | | Repetition 2 | | | Repetition 1 | | | Repetition 2 | | |
| | Maj | Min1 | Min2 | Maj | Min1 | Min2 | Maj | Min1 | Min2 | Maj | Min1 | Min2 |
| *Pork + sheep + goat* | | | | | | | | | | | | |
| A1 | + | + | + | + | + | + | + | + | + | + | + | + |
| A2 | + | + | + | + | + | + | + | + | + | + | + | + |
| A3 | + | − | − | + | + | − | + | + | + | + | + | + |
| *Pork + cattle + equine* | | | | | | | | | | | | |
| B1 | + | + | + | + | + | + | + | + | + | + | + | + |
| B2 | + | + | + | + | + | + | + | + | + | + | + | + |
| B3 | + | + | + | + | + | − | + | + | + | + | + | + |
| *Pork + hare + rabbit* | | | | | | | | | | | | |
| C1 | + | + | + | + | + | + | + | + | + | + | + | + |
| C2 | + | + | + | + | + | + | + | + | + | + | + | + |
| C3 | + | − | − | + | + | + | + | + | + | + | + | + |
| *Pork + red deer + fallow deer* | | | | | | | | | | | | |
| D1 | + | + * | + | + | + | + | + | + | + | + | + | + |
| D2 | + | + * | + | + | + | + | + | + | + | + | + | + |
| D3 | + | − * | + | + | − | − | + | + | + | + | + | − |
| *Pork + turkey + pheasant* | | | | | | | | | | | | |
| E1 | + | + | + | + | + | + | + | + | + | + | + | + |
| E2 | + | + | + | + | + | + | + | + | + | + | + | + |

(continued)

Beltramo et al. (2017), in their study, worked with the Meat 5.0 LCD-Array detection kit with DNA levels ranging from 6.6 to 25 ng.μl$^{-1}$ for raw meat mixtures containing different concentrations of contaminating species and from 0.60 to 13 ng.μl$^{-1}$ for heat-treated samples.

Cottenet et al. (2016) in addition to animal testing have used plant species commonly found in culinary products such as wheat, corn and rice, where their identity has been previously verified. As expected, all tested meat and fish samples led to 16S rRNA PCR amplification of 115–125 bp (base pairs), while plant samples were randomly led to nonspecific amplicons. When PCR products from meat, fish and plant were further hybridized to Meat LCD-Array, they provided hybridization signals to their respective species-specific capture probes, only meat species, while fish and plant samples did not lead to any hybridization points (Cottenet et al. 2013).

Fajardo et al. (2008) found that red deer and reindeer had low cross-reactivity on deer probes. Such weak cross-reactions between deer species have already been observed using the real-time PCR approach.

**Table 10.8** (continued)

| | Before heat treatment | | | | | | After heat treatment | | | | | |
|---|---|---|---|---|---|---|---|---|---|---|---|---|
| | Repetition 1 | | | Repetition 2 | | | Repetition 1 | | | Repetition 2 | | |
| | Maj | Min1 | Min2 | Maj | Min1 | Min2 | Maj | Min1 | Min2 | Maj | Min1 | Min2 |
| E3 | + | + | + | + | + | + | + | + | + | + | + | + |
| *Pork + chicken + goose* | | | | | | | | | | | | |
| F1 | + | + | + | + | + | + | + | + | + | + | + | + |
| F2 | + | + | + | + | + | + | + | + | + | + | + | + |
| F3 | + | + | + | + | + | – | + | + | + | + | + | + |
| *Pork + mallard duck + Muscovy duck* | | | | | | | | | | | | |
| G1 | + | + | + | + | + | + | + | + | + | + | + | + |
| G2 | + | + | + | + | + | + | + | + | + | + | + | + |
| G3 | + | + | – | + | – | + | + | + | + | + | + | – |
| *Pork + roe deer + fallow deer* | | | | | | | | | | | | |
| H1 | + | + | + | + | + | + | + | + | + | + | + | + |
| H2 | + | + | + | + | + | + | + | + | + | + | + | + |
| H3 | + | + | – | + | + | – | + | + | + | + | + | + |
| *Pork + cat + dog* | | | | | | | | | | | | |
| I1 | + | + | + | + | + | + | + | + | + | + | + | + |
| I2 | + | + | + | + | + | + | + | + | + | + | + | + |
| I3 | + | + | + | + | + | + | + | + | - | + | + | - |

(continued)

**Table 10.8** (continued)

| | Before heat treatment | | | | | | After heat treatment | | | | | |
|---|---|---|---|---|---|---|---|---|---|---|---|---|
| | Repetition 1 | | | Repetition 2 | | | Repetition 1 | | | Repetition 2 | | |
| | Maj | Min1 | Min2 | Maj | Min1 | Min2 | Maj | Min1 | Min2 | Maj | Min1 | Min2 |
| *Pork + buffalo + reindeer* | | | | | | | | | | | | |
| J1 | + | + | + | + | + | + | + | + | + | + | + | + |
| J2 | + | + | + | + | + | + | + | + | + | + | + | + |
| J3 | + | + | + | + | + | + | + | + | + | + | + | + |
| *Pork + turkey + ostrich* | | | | | | | | | | | | |
| K1 | + | + | + | + | + | + | + | + | + | + | + | + |
| K2 | + | + | + | + | + | + | + | + | + | + | + | + |
| K3 | + | + | – | + | + | + | + | – | – | + | – | – |

Maj—Major component; Min—Minor component, + detected; - undetected; * cross reaction

As already mentioned, *Cervidae* species exhibit a very high genetic similarity in their mitochondrial DNA, making it difficult to devise methods for identifying specific deer species (Hoffmann et al. 2014; Cottenet et al. 2016).

Genus *Bison* led to strong signals on bison and cattle probes, which were designed to target genus *Bison* and genus *Bos*. Although *Bison* corresponds to the whole genus, some taxonomists would place the bison in their very close *Bos* family (Wilson and Ruff 1999).

Except for weak interactions, no false positive results were observed. Genus *Aves* did not show any cross-reactions, even among poultry species. Probes for chicken, turkey, goose, pheasant and two duck species were species specific. Furthermore, considering all experiments performed, no hybridization signals were observed in the negative controls. These data show that the species-specific probes found on the Meat LCD-Array provide specific signals when testing only the corresponding meat species (Cottenet et al. 2016).

## 10.4 Validation of the innuDETECT Assay Kit

### *10.4.1 Biological Material*

Tissue samples from 6 animal species were used: pig (*Sus scrofa*), cattle (*Bos taurus, Bos bison*), sheep (*Ovis aries*), goat (*Capra hircus*), chicken (*Gallus gallus*) and turkey (*Meleagris gallopavo*). Authentic tissue samples of the mentioned animal species were obtained from regional sources within the territory of Slovakia. Meat samples were cut into smaller pieces using a sterile scalpel and placed in sterile containers in which they were stored. All samples were stored at −18 °C.

### *10.4.2 Tested Mixtures*

Meat mixtures were prepared using a blender (Blender 8008, Connecticut, USA) to a final weight of 100 g. Six combinations of binary mixtures of meat in eighteen different concentration levels of addition were created (Table 10.9). To avoid cross-contamination, each blend was treated separately using a different blender container. The meat mixtures were transferred to sterile tubes and stored at −18 °C until DNA extraction. Series of mixtures were prepared in duplicate. One variant was analyzed in the raw state and the other after heat treatment at 100 °C for 60 min.

**Table 10.9** Scheme of testing and mixtures preparation

| Tested meat mixtures | | | | | | | | | | | | |
|---|---|---|---|---|---|---|---|---|---|---|---|---|
| Tested mixtures | A | | B | | C | | D | | E | | F | |
| | beef/ pork meat [%] | | pork/ chicken meat [%] | | pork/ turkey meat [%] | | pork/ sheep meat [%] | | pork/ goat meat [%] | | pork/ beef meat [%] | |
| 1 | 100 | 0 | 100 | 0 | 100 | 0 | 100 | 0 | 100 | 0 | 100 | 0 |
| 2 | 0 | 100 | 0 | 100 | 0 | 100 | 0 | 100 | 0 | 100 | 0 | 100 |
| 3 | 10 | 90 | 10 | 90 | 10 | 90 | 10 | 90 | 10 | 90 | 10 | 90 |
| 4 | 20 | 80 | 20 | 80 | 20 | 80 | 20 | 80 | 20 | 80 | 20 | 80 |
| 5 | 30 | 70 | 30 | 70 | 30 | 70 | 30 | 70 | 30 | 70 | 30 | 70 |
| 6 | 40 | 60 | 40 | 60 | 40 | 60 | 40 | 60 | 40 | 60 | 40 | 60 |
| 7 | 50 | 50 | 50 | 50 | 50 | 50 | 50 | 50 | 50 | 50 | 50 | 50 |
| 8 | 55 | 45 | 55 | 45 | 55 | 45 | 55 | 45 | 55 | 45 | 55 | 45 |
| 9 | 60 | 40 | 60 | 40 | 60 | 40 | 60 | 40 | 60 | 40 | 60 | 40 |
| 10 | 65 | 35 | 65 | 35 | 65 | 35 | 65 | 35 | 65 | 35 | 65 | 35 |
| 11 | 70 | 30 | 70 | 30 | 70 | 30 | 70 | 30 | 70 | 30 | 70 | 30 |
| 12 | 75 | 25 | 75 | 25 | 75 | 25 | 75 | 25 | 75 | 25 | 75 | 25 |
| 13 | 80 | 20 | 80 | 20 | 80 | 20 | 80 | 20 | 80 | 20 | 80 | 20 |
| 14 | 85 | 15 | 85 | 15 | 85 | 15 | 85 | 15 | 85 | 15 | 85 | 15 |
| 15 | 90 | 10 | 90 | 10 | 90 | 10 | 90 | 10 | 90 | 10 | 90 | 10 |
| 16 | 95 | 5 | 95 | 5 | 95 | 5 | 95 | 5 | 95 | 5 | 95 | 5 |
| 17 | 99 | 1 | 99 | 1 | 99 | 1 | 99 | 1 | 99 | 1 | 99 | 1 |
| 18 | 99.5 | 0.5 | 99.5 | 0.5 | 99.5 | 0.5 | 99.5 | 0.5 | 99.5 | 0.5 | 99.5 | 0.5 |
| 19 | 99.9 | 0.1 | 99.9 | 0.1 | 99.9 | 0.1 | 99.9 | 0.1 | 99.9 | 0.1 | 99.9 | 0.1 |

### *10.4.3 DNA Isolation*

DNA was isolated from prepared blends of admixtures using the innuPREP DNA Mini Kit (Analytik Jena, Germany), Maxwell® 16 System (Promega, Madison, USA) and Maxwell® 16 Tissue DNA Purification Kit (Promega, Madison, USA) and NucleoSpin Food kit (Macherey–Nagel, Düren, Germany) according to the manufacturer's instructions. For each analysis, three parallel samples were obtained, which then were analyzed in five replicates.

### *10.4.4 RT-PCR*

Detection was performed using innuDETECT kit. Positive and negative controls were performed. All solutions and materials from the kit were allowed to reach the

**Table 10.10** PCR preparation

| Reagent | Addition volume (μl) |
|---|---|
| 2 × Master Mix | 10 |
| Primer/Probe Mix x IC | 3 |
| IC | 1 |
| Sample | 5 |
| PCR grade $H_2O$ | 1 |

**Table 10.11** Cycler mode setting

| Step | Duration (s) | Temperature (°C) |
|---|---|---|
| 1. Initial denaturation | 120 | 95 |
| 2. 35 repetitions: | | |
| Denaturation | 10 | 95 |
| Anealing | 45 | 62 |
| Elongation | 45 | 62 |

required temperature before use. Preparation of PCR was with reagents provided in Table 10.10.

The tubes were tightly capped and placed in a LightCycler® 2.0 (Roche Diagnostics, Mannheim, Germany). Cycler mode setting is provided in Table 10.11. Analysis was performed using LightCycler® 2.0 Software 4.1 (Roche Diagnostics, Mannheim, Germany).

### 10.4.5 Statistical Analysis

The results of the analysis were evaluated using the MS Excel (2017) statistical package.

### 10.4.6 Results

In the validation experiment of the innuDETECT Assay kits, meat samples from 6 animal species that are applied extensively in the production of meat products, were used. Six combinations of binary mixtures of eighteen produced in eighteen different concentration levels of addition were analyzed.

Each species was also tested individually with positive and negative controls included in each analysis. The concentration range of the addition was from 100 to 0.1%. Individual animal species were detected in the raw state and after heat treatment. The amount of DNA isolated from the animal species analyzed ranged from 4.6–26 ng.$\mu l^{-1}$ for uncooked samples and prepared mixtures of raw meat.

The concentration of 0.83–19.8 $ng.\mu l^{-1}$ was measured in heat-treated samples and mixtures of meat.

All analyzed animal species included in the experiment were identically correctly identified after reaching the 5% addition limit. Identical results were also observed in the detection of heat-treated samples and binary mixtures of tested animal species.

In the first binary mixture (A), a combination of pork and beef where pork was added in a descending concentration order, pig DNA was detected in both 1 and 0.5% addition. The addition rate of 0.1% pork to beef has not been identified by the detection kit. The second binary mixture (B), a combination of chicken and pork where the chicken was added in descending concentration order, the chicken was identified only to a level of 5% chicken addition. Detection kit not detected 1, 0.5 and 0.1% addition of chicken meat mixed with pork.

The binary mixture (C), a combination of turkey and pork where the turkey meat was added in a descending concentration order, was able to detect the DNA of the turkey with the detection kit to the level of 1% addition. The addition of 0.5 and 0.1% of turkey meat to the pork was not detected by the detection kit. The binary mixture (D), a combination of sheep and pork meat where the sheep was added in a descending concentration order, was able to detect sheep DNA by the kit up to the level of 1% addition. The addition of 0.5 and 0.1% of sheep meat mixed with pork was not captured by the detection kit.

Mixture (E), a combination of goat and pork where goat meat was added in a descending concentration order, detection kit was able to capture goat DNA at a level of 1% addition. The addition of 0.5 and 0.1% of goat meat mixed with pork was not detected by the detection kit. The binary mixture (F), a combination of beef and pork where beef was added in a descending concentration order, detection kit was able to capture the beef DNA up to 0.5% addition. The addition of 0.1% of beef meat mixed with pork was not detected by the detection kit.

The results presented in (Table 10.13) are identical to the results of the analyses performed on the mixtures before heat treatment (Table 10.12).

Al-Kahtani et al. (2017) detected pig DNA in meat mixtures using both conventional and real-time PCR (RT-PCR). They analyzed thirty meat mixtures containing beef, chicken, camel, rabbit, goat and sheep meat with varying percentages of pork (0%, 1%, 5%, 10% and 20%).

Tanabe et al. (2007) concluded in their study that the LOD (Limit Of Detection) for porcine DNA using RT-PCR and conventional PCR was 10 $fg.ul^{-1}$ and 1 $pg.ml^{-1}$. These statements are consistent with those of Kesmen et al. (2012), which found that RT-PCR can detect chicken and turkey DNA at 0.0001 $ng.\mu l^{-1}$.

These results are consistent with those of Al-Kahtani et al. (2017), which detected the presence of pork DNA in meat mixtures using both conventional PCR and RT-PCR. They also used a standard curve and Ct (Cycle threshold) values for indicative addition quantification. Analysis revealed pork in all meat mixtures with 1 to 20% with one exception of 1% addition of pork in chicken meat mixture where 1% of pork has not been identified.

Tested mixtures of pork with beef were positive for pork by RT-PCR and recorded Ct values of 19.09, 22.78, 24.26 and 23.90 for beef containing 20, 10, 5 and 1%

**Table 10.12** Success of innuDETECT Assay kits in detecting animal species before heat treatment

| Tested mixtures | A | | B | | C | | D | | E | | F | |
|---|---|---|---|---|---|---|---|---|---|---|---|---|
| Kit | innuDETECT Pork Assay | innuDETECT Beef Assay | innuDETECT Chicken Assay | innuDETECT Pork Assay | innuDETECT Turkey Assay | innuDETECT Pork Assay | innuDETECT Sheep Assay | innuDETECT Pork Assay | innuDETECT Goat Assay | innuDETECT Pork Assay | innuDETECT Beef Assay | innuDETECT Pork Assay |
| *Detection success before heat treatment* | | | | | | | | | | | | |
| 1 | + | − | + | − | + | − | + | − | + | − | + | − |
| 2 | − | + | − | + | − | + | − | + | − | + | − | + |
| 3 | + | + | + | + | + | + | + | + | + | + | + | + |
| 4 | + | + | + | + | + | + | + | + | + | + | + | + |
| 5 | + | + | + | + | + | + | + | + | + | + | + | + |
| 6 | + | + | + | + | + | + | + | + | + | + | + | + |
| 7 | + | + | + | + | + | + | + | + | + | + | + | + |
| 8 | + | + | + | + | + | + | + | + | + | + | + | + |
| 9 | + | + | + | + | + | + | + | + | + | + | + | + |
| 10 | + | + | + | + | + | + | + | + | + | + | + | + |
| 11 | + | + | + | + | + | + | + | + | + | + | + | + |
| 12 | + | + | + | + | + | + | + | + | + | + | + | + |
| 13 | + | + | + | + | + | + | + | + | + | + | + | + |
| 14 | + | + | + | + | + | + | + | + | + | + | + | + |
| 15 | + | + | + | + | + | + | + | + | + | + | + | + |
| 16 | + | + | + | + | + | + | + | + | + | + | + | + |
| 17 | + | + | − | + | + | + | + | + | + | + | + | + |
| 18 | + | + | − | + | − | + | − | + | − | + | + | + |
| 19 | − | + | − | + | − | + | − | + | − | + | − | + |

+ detected; − undetected

**Table 10.13** Success of innuDETECT Assay kits in detecting animal species after heat treatment

| Tested mixtures | A | | B | | C | | D | | E | | F | |
|---|---|---|---|---|---|---|---|---|---|---|---|---|
| Kit | innuDETECT Pork Assay | innuDETECT Beef Assay | innuDETECT Chicken Assay | innuDETECT Pork Assay | innuDETECT Turkey Assay | innuDETECT Pork Assay | innuDETECT Sheep Assay | innuDETECT Pork Assay | innuDETECT Goat Assay | innuDETECT Pork Assay | innuDETECT Beef Assay | innuDETECT Pork Assay |
| *Detection success after heat treatment* | | | | | | | | | | | | |
| **1** | + | - | + | - | + | - | + | - | + | - | + | - |
| **2** | - | + | - | + | - | + | - | + | - | + | - | + |
| 3 | + | + | + | + | + | + | + | + | + | + | + | + |
| 4 | + | + | + | + | + | + | + | + | + | + | + | + |
| 5 | + | + | + | + | + | + | + | + | + | + | + | + |
| 6 | + | + | + | + | + | + | + | + | + | + | + | + |
| 7 | + | + | + | + | + | + | + | + | + | + | + | + |
| 8 | + | + | + | + | + | + | + | + | + | + | + | + |
| 9 | + | + | + | + | + | + | + | + | + | + | + | + |
| 10 | + | + | + | + | + | + | + | + | + | + | + | + |
| 11 | + | + | + | + | + | + | + | + | + | + | + | + |
| 12 | + | + | + | + | + | + | + | + | + | + | + | + |
| 13 | + | + | + | + | + | + | + | + | + | + | + | + |
| 14 | + | + | + | + | + | + | + | + | + | + | + | + |
| 15 | + | + | + | + | + | + | + | + | + | + | + | + |
| 16 | + | + | + | + | + | + | + | + | + | + | + | + |
| 17 | + | + | - | + | + | + | + | + | + | + | + | + |
| 18 | + | + | - | + | - | + | - | + | - | + | + | + |
| 19 | - | + | - | + | - | + | - | + | - | + | - | + |
| 14 | + | + | + | + | + | + | + | + | + | + | + | + |
| 15 | + | + | + | + | + | + | + | + | + | + | + | + |

+ detected; - undetected

of pork meat. The results of this test confirm the reliability of the pork detection assay at different levels. The limits of detection of pig DNA in meat mixtures were 0.025 ng.$\mu l^{-1}$ and in the case of combinations of pork and beef in the second variant of the mixture of pork and chicken, the reaction was unsatisfactory at 0.911 ng.μl-1 (Al-Kahtani et al. 2017).

Ulca et al. (2013) in a survey of 42 samples of meat products (Turkey) found four samples positive for pig DNA. Ali et al. (2014) identified chicken nuggets containing pork. Demirhan et al. (2011) found in two of the 11 retail products obtained from Germany the content of animal gelatine with Ct values of 30.04 and 43. They also tested 32 samples from Turkey and one of these products (gelatin-coated cake) was positive for the presence of pig DNA with Ct 36.3. Sahilah et al. (2012) found that 37.2% of the pharmaceutical capsules they tested contained pig DNA. Ct values ranged from 21.40 to 31.07 and signal values were from 2045 to 53,812. Kesmen et al. (2012) reported that Ct values greater than 30 indicate an LOD of 0.0001 ng in samples of cooked and raw meat mixtures. Such small amounts of pig DNA in commercial food products are likely to consist in cross-contamination of the production line rather than deliberate adulteration of food products with pork (Al-Kahtani et al. 2017). There are no acceptable levels of pork contamination specified by any Halal food regulators; Ulca et al. (2013) proposed 0.1% as the permissible limit.

A study by Al-Kahtani et al. (2017) shows that detection of adulteration or contamination of commercial food products with pork additions by RT-PCR is more accurate, more reliable than conventional PCR methods and more sensitive. Quality control laboratories could, therefore, use RT-PCR for sensitive products including Halal. The presence of pork less than 0.1% in meat products could be considered as cross-contamination in a commercial production line.

## 10.5 Conclusion

Based on the results, we can conclude that the Meat 5.0 LCD-Array kit gave correct and repeatable results when analyzing all of the species tested. No difference in efficacy was observed from DNA obtained from samples subjected to different treatments, suggesting that the kit could also be used for testing cooked or sterilized foods. Using the innuDETECT Assay detection kit, all analyzed animal species included in the experiment were identically correctly identified after reaching the 5% addition limit. Identical results were also observed in the detection of heat-treated samples and binary mixtures of tested animal species. Based on the findings, heat treatment has no effect on the performance of this detection kit. There is a possible implementation of these methods for finding adulteration of other meats than those originating from traditionally fed animals of native breeds origin (with firstly done mapping of standards of native animal breeds). They could also be useful for possible adulterations concerning raw materials price advantage and assuring allergenicity clearance.

**Acknowledgements** This work was supported by the Slovak Research and Development Agency under the contract no. APVV-17-0508.

## References

Ali ME, Razzak MA, Abd Hamid SB (2014) Multiplex PCR in species authentication: probability and prospected review. Food Anal Methods 7(10):1933–1949. https://doi.org/10.1007/s12161-014-9844-4

Al-Kahtani H, Ismail EA, Ahmed MA (2017) Pork detection in binary meat mixtures and some commercial food products using conventional and real-time PCR techniques. Food Chem 219:54–60. https://doi.org/10.1016/j.foodchem.2016.09.108

Bar T, Kubista M, Tichopa A (2012) Validation of kinetics similarity in qPCR. Nucleic Acids Res 40:1395–1406. https://doi.org/10.1093/nar/gkr778

Beltramo C, Riina MV, Colussi S, Campia V, Maniaci MG, Biolatti C, Trisorio S, Modesto P, Peletto S, Acutis LP (2017) Validation of a DNA biochip for species identification in food forensic science. Food Control 78:366–373. https://doi.org/10.1016/j.foodcont.2017.03.006

Bickley J, Short JK, Mcdowell DG, Parkes HC (1996) Polymerase chain reaction (PCR) detection of Listeria monocytogenes in diluted milk and reversal of PCR inhibition caused by calcium ions. Lett Appl Microbiol 22(2):153–158. https://doi.org/10.1111/j.1996.tb01131.x

Chipron (2014) User manual LCD-Array Kit MEAT 5.0 DNA – based identification of animal species. Berlin, Germany: Chipron GmbH. 22 p. http://peramed.com/peramed/docs/Manual%20MEAT%205.0%20Vers.1.12014.pdf

Cottenet G, Blancpain C, Golay PA (2011) Simultaneous detection of cow and buffalo species in milk from China, India, and Pakistan using multiplex real-time PCR. J Dairy Sci 94(8):3787–3793. https://doi.org/10.3168/jds.2011-4195

Cottenet G, Blancpain C, Sonnard V, Chuah PF (2013) Development and validation of a multiplex real-time PCR method to simultaneously detect 47 targets for the identification of genetically modified organisms. Anal Bioanal Chem 405(21):6831–6844. https://doi.org/10.1007/s00216-013-7125-5

Cottenet G, Sonnard V, Blancpain C, Ho HZ, Leong HL, Chuah PF (2016) A DNA macro-array to simultaneously identify 32 meat species in food samples. Food Control 67:135–143. https://doi.org/10.1016/j.foodcont.2016.02.042

Darwish FS, Allam HA, Amin AS (2009) Evaluation of PCR assay for detection of cow's milk in water buffalo's milk. World Appl Sci J 7(4):461–467

Demirhan Y, Ulca P, Senyuva HZ (2011) Detection of porcine DNA in gelatin and gelatin-containing processed food products Halal/Kosher authentication. Meat Sci 90(3):686–689. https://doi.org/10.1016/j.meatsci.2011.10.014

Drummond MG, Brasil BSAF, Dalsecco LS, Brasil RSAF, Teixeira LV, Oliveira DAA (2013) A versatile real-time PCR method to quantify bovine contamination in buffalo products. Food Control 29(1):131–137. https://doi.org/10.1016/j.foodcont.2012.05.051

Fajardo V, Gonzalez I, Martin I, Rojas M, Hernandez PE, Garcia T, Martín R (2008) Real-time PCR for detection and quantification of red deer (*Cervus elaphus*), fallow deer (*Dama dama*), and roe deer (*Capreolus capreolus*) in meat mixtures. Meat Sci 79(2):289 298. https://doi.org/10.1016/j.meatsci.2007.09.013

Hoffmann GS, Johannesen J, Griebeler EM (2014) Species cross- amplification, identification and genetic variation of 17 species of deer (*Cervidae*) with microsatellite and mitochondrial DNA from antlers. Mol Biol Rep 42(6):1059–1067. https://doi.org/10.1007/s11033-014-3845-7

Kesmen Z, Yetiman AE, Sahin F, Yetim H (2012) Detection of chicken and turkey meat in meat mixtures by using real-time PCR assays. J Food Sci 77(2):C167–C173. https://doi.org/10.1111/j.1750-3841.2011.02536.x

Sahilah AM, Fadly ML, Norrakiah AS, Aminah A, Wan AWM, Maaruf AG, Khan MA (2012) Halal market surveillance of soft and hard gel capsules in pharmaceutical products using PCR and southern-hybridization on the biochip analysis. Int Food Res J 19(1):371–375
Sakaridis I, Ganopoulos I, Anagnostis A, Athanasios T (2013) High resolution melting analysis for quantitative detection of bovine milk in pure water buffalo mozzarella and other buffalo dairy products. Int Dairy J 28(1):32–35. https://doi.org/10.1016/j.idairyj.2012.08.006
Sharma N, Singh NK, Bhadwal MS (2011) Relationship of somatic cell count and mastitis: an overview. Asian-Australasian J Ani Sci 24(3):429–438. https://doi.org/10.5713/ajas.2011.10233
Tanabe S, Hase M, Yano T, Sato M, Fujimaru T, Akiyama H (2007) A real-time quantitative PCR detection method for pork, chicken, beef, mutton, and horseflesh in foods. Biosci Biotechnol Biochem 71(12):3131–3135. https://doi.org/10.1271/bbb.70683

Ulca P, Balta H, Çağin İ, Senyuva HZ (2013) Meat species identification and Halal authentication using PCR analysis of raw and cooked traditional Turkish foods. Meat Sci 94(3):280–284. https://doi.org/10.1016/j.meatsci.2013.03.008
Volk H, Piskerik S, Kurinčič M, Klančnik A, Toplak N, Jeršek B (2014) Evaluation of different methods for DNA extraction from milk. Journal of Food and Nutrition Research 53:97–104
Wilson DE, Ruff S (1999) The Smithsonian book of North American mammals. Washington, USA: Smithsonian Institution press. 750 p. ISBN 978–1560988458

# Chapter 11
# Traditional Cheeses from the Malopolska Region

**Dorota Najgebauer-Lejko, Jacek Domagała, and Maria Walczycka**

**Abstract** Recently, there has been an increasing interest in traditional food, produced locally without any artificial additives using old methods and with unique characteristics. Malopolska is a southwestern province of Poland which has distinctively rich local food traditions. Among many food products in this region, cheeses have gained special attention, with *oscypek* cheese being the most popular and recognizable. Oscypek is a scalded-smoked, hard cheese manufactured in the Podhale Region exclusively from unpasteurized ewe's milk or with an additive of up to 40% of cow's milk which must come from the Polish Red cattle breed. This cheese constitutes an important part of the shepherding tradition in the Tatra Mountains as it has been practiced for hundreds of years by local highlanders. "Oscypek" together with "Redykołka" (mini variety of oscypek produced in various fancy shapes, i.e., animals, hearts, spindles) and "Bryndza Podhalańska" (soft rennet cheese) are registered under the European Protected Designation of Origin (PDO) geographical indication. Other cheeses produced traditionally, in the Malopolska Region, included in the Ministry of Agriculture and Rural Development's List of Traditional Products are gazdowski cheese (gołka, pucok, kara), bundz, gomółki kowalowskie, and fresh white cheese (Tvarog). The present chapter focuses on the traditional acid and rennet cheeses produced in the Malopolska Region from different types of milk. Issues related to historical background, manufacturing procedures, product characterization, and authenticity are discussed.

**Keywords** Cheese · Malopolska · Oscypek · Bryndza podhalańska · Redykołka · Tvarog

D. Najgebauer-Lejko (✉) · J. Domagała · M. Walczycka
Department of Animal Product Processing, Faculty of Food Technology, University of Agriculture in Krakow, Balicka 122, 30-149 Krakow, Poland
e-mail: dorota.najgebauer-lejko@urk.edu.pl

J. Domagała
e-mail: jacek.domagala@urk.edu.pl

M. Walczycka
e-mail: maria.walczycka@urk.edu.pl

© Springer Nature Switzerland AG 2022

J. Hernik et al. (eds.), *Cultural Heritage—Possibilities for Land-Centered Societal Development*, Environmental History 13,
https://doi.org/10.1007/978-3-030-58092-6_11

## 11.1 Introduction

Malopolska is a province located in the southwestern part of Poland which has distinctively rich local food traditions. The agrarian structure of the region is characterized by a large number of small-sized farms, often located in the harsh mountainous areas. These circumstances forced farmers to specialize in the production of the niche foodstuffs. Moreover, strong regional identity, well-preserved tradition, and vigorous tourist sector are trigger forces between the valorization of local food (Adamski and Gorlach 2016).

Documented history of cheese production in Poland has a long tradition and dates back to the fifteenth century. The first written evidence of cheese making in Podhale can be found in the charter documents of Ochotnica Village located in the Gorce Mountains and dates from the year 1415. It is connected with the Wallachians (Vlachs), semi-nomadic tribe, speaking Eastern Romanian languages, which moved from the Balkan Peninsula along the Carpathian Range with the sheep flocks to settle as far north as the Polish Podhale Region and also to the western end of Carpathian–żywiecki Region (Silesian Voivodeship). They brought and spread their shepherds' culture, with the specific way of organizing pasturage (transhumance), traditional log hut called *bacówka* and methods of milk processing (Bonczar and Wszołek 2003; Kawęcka and Krupiński 2014; EC 2006a, b, 2009b). Their customs and knowledge of the methods of production of cheeses such as bryndza podhalańska, oscypek, and redykołka were passed down from generation to generation and have been continued to the present day.

Cheese production in the Malopolska Region is strictly linked with an old shepherding tradition and with local breeds of animals, especially Polish Red Cattle and Polish Mountain Sheep. Polish Red is the oldest dual purpose and the only existing native cattle breed in Poland. The herds of this breed are protected by the Genetic Resources Conservation Program. Polish Red cows are relatively small one-colored animals, resistant to hard mountain conditions and to diseases, yielding on average 4,000 L of milk per year, with high concentrations of protein (3.6%) and fat (4–4.5%) and high biological and technological value (Adamczyk et al. 2008). Polish Mountain Sheep (white or colored) is a crossbreed of Podhale Zakel with Frisian rams and Transylvania Zakel. This breed is genetically well adapted to the harsh climate of mountains and provides proper milk yield with properties particularly beneficial for cheese production and also high-quality lambs' meat. Milk from both Polish Mountain Sheep and Polish Red cows is the only allowable source milk for production of regional cheeses certified with European Protected Designation of Origin (PDO) such as bryndza podhalańska, oscypek, and redykołka. The traditional grazing system on the unpolluted mountain pastures and meadows, rich in a wide range of flora species (including endemic and/or medicinal plants), with milk coming from local animal breeds and utilization of traditional production technology affects unique characteristics and high quality of milk products. Therefore, cheeses from Malopolska Region are appreciated at the European, national, and regional levels. Protected Designation of Origin (PDO), Protected Geographical Indication

(PGI), and Traditional Speciality Guaranteed (TSG) are the EU quality certification systems given to promote and protect the names of specific agricultural products and foodstuffs characterized by the unique properties strictly linked to their geographical origin as well as traditional methods of manufacturing. Among these three schemes, products registered as PDO have the strongest links to the specific place as every part of their production, processing, and preparation must take place in the one, strictly defined region. In contrast to PDO, PGI certification means that at least one of the stages of production, processing, or preparation takes place in a specific geographical area and is granted to products with a quality, reputation, or other property which is attributable to this region (European Commission 2019). Among eight food products from Poland registered as PDO, there are three cheeses, namely, Bryndza Podhalańska, Oscypek, and Redykołka and all of them come from Malopolska (two latter also from Silesian Voivodeship). There are also two Polish cheeses awarded with PGI label; however, they do not come from Malopolska ("*Ser koryciński swojski*" and "*Wielkopolski ser smażony*") (EC 2009b). This indicates that the production of cheeses in Poland, especially in the Podhale Region, where it is strictly connected with a rich and centuries-old shepherding tradition, is a very important part of the material and intangible heritage. Agricultural and Food Quality Inspection (AFQI) is the competent authority to maintain and publish a list of producers who own a valid certificate of compliance or quality certificate as regards Protected Designation of Origin (PDO), Protected Geographical Indication (PGI), and Traditional Speciality Guaranteed [TSG) (IJHARS 2019). At this moment, there are nine registered producers of Bryndza podhalańska, 40 producers of Oscypek, and three producers of *Redykołka* (all products with PDO indication). The same inspection body is authorized to conduct official controls, supervision of product certification bodies, and carry out verification of compliance with specification of products with PDO, PGI, and TSG indications (IJHARS 2019).

Cheeses are also valued products at the national level. One category in the national system of food quality is "Quality Tradition" (*Jakość Tradycja*) established by the Polish Chamber of Regional and Local Products to distinguish the products with the long history of production (at least 50-year history of product manufacture while maintaining the traditional recipe, raw materials, and technology should be documented) and with the highest quality arising from higher standards of production, reputation, or other unique characteristics. Among 20 products and 11 producers awarded with the Quality Tradition mark in the Malopolska, there are four cheeses (white fresh acid/curd cheeses) from three producers (Polska Izba Produktu Regionalnego i Lokalnego 2019). The products whose quality and exceptional characteristics and properties result from the usage of the traditional methods (with at least 25-year-old history), which constitute a part of the cultural heritage of the region where they are produced and being a part of the identity of a local community may be registered on the Ministry of Agriculture and Rural Development's List of Traditional Products (Dz. U. 2017 r. poz. 1168). Among 13 dairy products included in the list of traditional products from the Malopolska Region seven are cheeses, i.e., three PDO cheeses as well as gazdowski cheese (gołka, pucok, kara), bundz, gomółki kowalowskie, and fresh white cheeses (Tvarog). Moreover, the products from this list produced in the Malopolska Region may be certified and labeled with the trademark "*Produkt Tradycyjny z Małopolski*" (Traditional Product from Malopolska). Another quality and authenticity-of-origin scheme in the region is "*Produkt Lokalny z Małopolski*" (Local Product from Malopolska) which was implemented in the years 2011–2017 with funding from the Swiss-Polish Co-operation Program. This certification scheme is the basis for a trademark, which distinguishes environmental-friendly (often certified as organic) products directly from producers in the market place. The aim of the project is to promote local entrepreneurship, processing and retail of local agricultural products and thus to support development of the rural areas in Malopolska. In this model, consumers can buy products through the dedicated sales and distribution system with the assurance that the products are local, chemical free, and fresh and the products can be traced directly to their producer. Among certified producers, there are also farms engaged in the production of cheeses (Fundacja Partnerstwo dla Środowiska 2019).

## 11.2 Cheeses Labeled with Protected Designation of Origin (PDO) Indication

### *11.2.1 Bryndza Podhalańska*

Bryndza Podhalańska (from Romanian "brinze" and "Podhale" Region where it is produced) is a soft, spreadable, rennet cheese and thanks to Kazimierz Furczoń senior shepherd (*baca*) from Leśnica was the first Polish product registered with PDO indication in 2007 (EC 2007). It is produced solely from ewe's or from a mixture of ewe's and cow's milk by a traditional technology (Scheme 11.1). Milk must come from animals of the specified, native to Poland breeds of sheep and cows: Polish Mountain Sheep and Polish Red cows, respectively. Both breeds have a long history of rearing on the territory of Poland. Animals are grazed on green pastures from the end of April to the beginning of the October, out of the vegetation season, animals are fed hay derived from the pastures and meadows of the specific region. The characteristic of bryndza podhalańska is shown in Table 11.1.

Sheep and cows are milked two or three times a day. The production of Bryndza Podhalańska (Scheme 11.1) takes place in a shepherd hut using original, usually wooden tools and vessels. Milk is heated in the special, often copper, kettle (*kotlicek*) then poured into wooden tapered downward barrel called *puciera*. The powdered rennet (1.1–2.7 g per 100 L of milk) diluted in a small amount of warm water is added to the milk. After mixing for a short time, milk is left at the temperature of 28–39 °C to coagulate. To maintain the constant temperature, a wooden cover is put on the *puciera*. Milk coagulates from 30 to 60 min, depending on the milk temperature. After whey separation on the surface of the curd is observed, the senior shepherd (*baca*) starts to cut the curd. In the beginning, it is gently cut crosswise and just after a few minutes vigorously using a wooden tool, shaped as a fork,

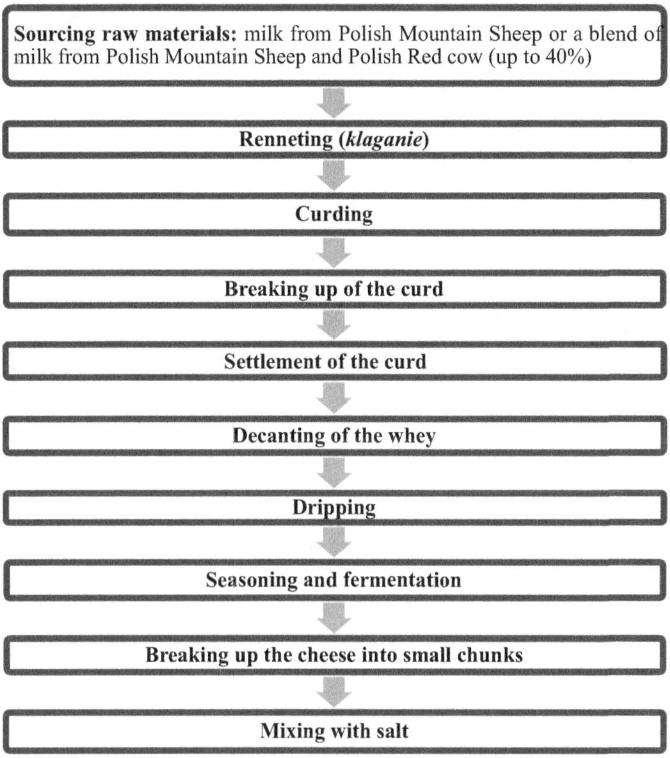


**Scheme 11.1** Production procedure of the Bryndza podhalańska (EC 2006a)

called a *ferula*. The process is performed at the temperature of 20–35 °C, which may vary among producers. After about 3 min, pieces of cheese of cereal grain size are obtained. During the process of breaking up the curd, hot water is added to raise the temperature to 35 °C to facilitate aggregation of the cheese grains into a lump. After that, the content of the *puciera* is left to settle the cheese lumps on the bottom of the barrel and then 30–50% of whey is removed. Removed whey is used during cheese processing for production of Żętyca (or Żentyca, kind of whey beverage) or to feed animals. Cheese mass is picked, kneaded, and put into the special cloth which is then hung out to drip for over 12 h. As a result, cheese mass takes shape of a loaf, which in the highlander dialect is called *bundz, bunc, gruda,* or *udój*. Cheese is then left for maturation for 4–12 days depending on the temperature. During that time lactic fermentation takes place, which results in the slightly sour taste and a large amount of small eyes inside the cheese. After maturation, cheese is broken up into small chunks. The method of breaking up the cheese is specific to each producer and different techniques may be applied according to the production technology practiced by the senior shepherd. Broken "*bunc*" is mixed and kneaded with salt in a kneading-trough (in the highlander dialect called: *dzieżka, dziczka, dzierżka*) in such a manner so as to avoid air bubbles inside. The amount of added salt depends on

**Table 11.1** Characteristics of the Bryndza Podhalańska (EC 2006a; IJHARS 2019)

| Product specification | | |
|---|---|---|
| Name | *Bryndza Podhalańska* | – |
| Number of registered producers (PDO) | 9[a] | – |
| Raw materials | Ewe's milk<br>Cow's milk | Breed: Polish Mountain Sheep (at least 60%)<br>Breed: Polish Red (up to 40%) |
| Season of production | From May to September | – |
| Place of production (geographical area) | Malopolska Voivodeship:<br>Silesian Voivodeship | – The entire Nowotarski and Tatrzański Districts<br>– Municipalities in the Sucha District: Zawoja, Bystra Sidzina<br>– Municipalities in the Żywiecki District: Milówka, Węgierska Górka, Rajcza, Ujsoły, Jeleśnia, and Koszarawa |
| Physical characteristic | Color | White, creamy-white, or with a willow-green shade |
| | Taste | Strong, salty, or slightly salty, sometimes slightly sour |
| Chemical composition | Water | Up to 60% (dry mass not less than 40%) |
| | Fat content | Not less than 38% in dry mass |

[a]Situation as of 23.02.2019 (IJHARS 2019)

the taste and preference of the senior shepherd. Sometimes, for preservation, melted butter is poured onto the surface of the final product—Bryndza Podhalańska (EC 2006a; Regionalny Związek Hodowców Owiec i Kóz 2006).

Besides the addition of salt, the type and composition of microbiota, i.e., bacteria (*Lactococcus* ssp., *Lactobacillus* ssp., *Streptococcus* ssp., *Enterococcus* ssp.) and fungi (including *Kluyveromyces marxianus, Galactomyces candidus/Geotrichum candidum, Yarrowia lipolytica*) present in Bryndza cheese influence its specific taste and aroma. The highest quality Bryndza is believed to be produced at the beginning of the production season in May, which is probably connected with grazing sheep on the spring pastures (Pangallo et al. 2014).

## *11.2.2 Oscypek*

Not only is the Oscypek cheese the most popular cheese of the Malopolska Region but also it has become a flagship Polish product and the symbol of Tatra Mountains. Its name, characteristic shape, weight, and ingredients from which it is made and production procedure have been protected with UE PDO certificate since 2008 (Adamski and Gorlach 2016; EC 2008; Fonte 2008). The characteristic of the Oscypek (Fig. 11.1) according to EU standard is given in Table 11.2.

It is hard scalded and smoked cheese produced from May to September from ewe's or mixed ewe's and cow's milk. The name "*Oscypek*" is etymologically connected with the word "*oszczypywać*"—meaning to pinch rapidly, to beat like during one of the production stages or from "*oszczep*", meaning "javelin," in reference to the shape of this cheese. The name of the cheese is also related with the word "*oscypiarka*"—the curved wooden molding form used to give the final product characteristic patterns (Fig. 11.2).

Oscypek is produced using traditional hand-processing technology, in shepherds' huts mainly from unpasteurized milk of sheep herds pastured on high mountain pastures. According to EU standard, it can be also made from a blend of ovine and bovine milks, with the exception that the last comes from the indigenous Polish Red cow breed and is not added in the amount of more than 40% of the whole milk for cheese production. PDO label guarantees authenticity of Oscypek connected with the strictly defined geographical area where it is produced, with ingredients and traditional production procedures (Scheme 11.2). These factors ensure high quality of the final products which distinguish them from *oscypek*-like cheeses produced from cheaper and more easily available ingredients and/or using industrialized technology (Majcher et al. 2015).

The chemical composition of the Oscypek cheese results from the composition of sheep milk or also cow's milk, if the mixed Oscypek type is produced. The average chemical composition of the Oscypek cheese is shown in Table 11.3. The presented data indicates that there is no difference in the basic composition between the cheeses produced exclusively from ewe's milk and those obtained from ewe's milk blended with cow's milk.

Kawęcka and Sosin-Bzducha (2014) revealed that month of the grazing season has a significant effect on the fatty acid (FA) profile of the Oscypek cheese as cheeses produced in May, as compared to those produced from the Summer ewe's milk (August), contained less saturated FA (namely, C14:0, C16:0) and more monounsaturated (C18:1) and polyunsaturated FA (CLA, including c-9,t-11 isomer), and were characterized by a lower atherogenicity index. Sheep cheeses, such as Oscypek, Bundz, and Bryndza, are characterized by the higher content of short-chain and

**Fig. 11.1** Oscypek from the shepherd's hut in Ochotnica Górna

**Table 11.2** Characteristics of the Oscypek cheese (EC 2006b; IJHARS 2019)

| Product specification | | |
|---|---|---|
| Name | *Oscypek* or *Oszczypek* | – |
| Number of registered producers | 40[a] | |
| Raw materials | Ewe's milk<br>Cow's milk | Breed: Polish Mountain Sheep (at least 60%)<br>Breed: Polish Red (up to 40%) |
| Place of production (geographical area) | Malopolska Voivodeship:<br>Silesian Voivodeship | – The entire Nowy Targ and Tatrzański Districts,<br>– Municipalities in the Sucha District: Zawoja, Bystra Sidzina<br>– Municipalities in the Limanowa District: Niedźwiedź and part of the Kamienica municipality situated within the boundary of the Gorczański National Park or south of the Kamienica River and the following *sołectwa* (municipal sub-divisions) in Mszana Dolna municipality: Olszówka, Raba Niżna, Łostówka, Łętowe, and Lubomierz<br>– Municipalities in the Nowy Sącz District: Piwniczna, Muszyna, and Krynica<br>– Municipalities (*gminy*) in the Cieszyn District (*powiat*): Istebna<br>– Municipalities in the Żywiec District: Milówka, Węgierska Górka, Rajcza, Ujsoły, Jeleśnia, and Koszarawa |
| Physical characteristic: | Shape | double cone or spindle |
| | Weight | 0.6–0.8 kg |
| | Length | 17–23 cm |
| | Diameter | 6–10 cm |
| | Color on cross Section | Light cream, darker at the rind, shade closer to white is permissible |
| | Rind | Straw-colored gleam, light brown, soft shine |
| Chemical composition | Water | Up to 44% (dry mass not less than 56%) |
| | Fat content | At least 38% in dry mass |

[a]Situation as of 23.02.2019 (IJHARS 2019)

**Fig. 11.2** Wooden ring form for the Oscypek with characteristic carved patterns

medium-chain FA (C4:0-C12:0), especially capric acid, and a lower share of palmitic acid (C16:0), stearic (C18:0), and palmotoleic (C16:1) than cheeses from cow's milk (Gołka) and, therefore, determination of the FA profile can be an useful tool in detection of the sheep cheese adulteration (Kudełka 2014).

Flavor is a key element of the characteristic sensory profile of the Oscypek. The unique, perceived as slightly sour, piquant, salty, and smoked aroma of this cheese is shaped by many factors, from which the most important are the milk flavor influenced by the animal breed and applied feeding system (fresh, unpolluted pastures rich in herbs and other plants often endemic to the specific region of Podhale), specific microflora coming from unpasteurized milk, traditional hand-made production procedures performed in huts with the use of mainly wooden utensils and vessels (*puciera, ferula, czerpak*), and natural preservation methods including brine salting and slow, cold smoking (Majcher and Jeleń 2011). Chemical evaluation conducted by Majcher and Jeleń (2011) using gas chromatography-olfactometry in conjunction with aroma extract dilution analysis revealed that there are 11 compounds contributing to the greatest manner in the overall aroma of the Oscypek cheese, with guaiacol being the most important odorant (Table 11.4).

Generally, the most important flavors found in the Oscypek were described as smoked, pungent, butyric, and buttery, and the other detected odors were toasted, milky, rennet, and brine. Flavors in the Oscypek cheese come mainly from the smoking stage of production (phenolic compounds, aldehydes, alcohols, furans/furanones) and to the lower extent from microbial action (methyl esters, ketones, aldehydes, alcohols, sulfur compounds) or native milk enzymes and/or rennet activity (e.g., free fatty acids) during curding, scalding, and brining as well as from raw milk (e.g., terpenes, methyl esters, alcohols) (Majcher et al. 2011). Moreover, the volatile profile of the Oscypek turned out to be specific enough to distinguish the original products that met the EU requirements for PDO from the *oscypek*-like cheeses produced from cow's milk and/or using industrial methods such as the solid-phase microextraction-mass spectrometry method (SPME-MS) combined with chemometric detectors (Majcher et al. 2015). Oscypek when compared to its imitations, especially those made by industrial methods from pasteurized milk, is

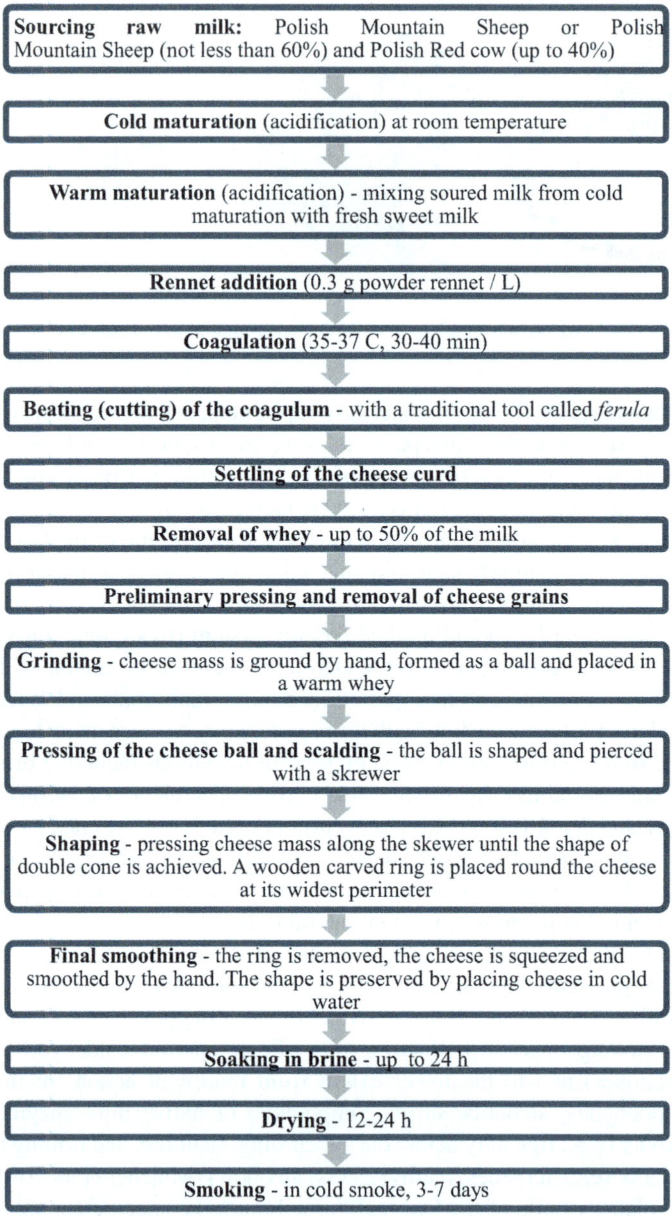


**Scheme 11.2** Production procedure of the Oscypek cheese (EC 2006b; Majcher et al. 2010)

**Table 11.3** Average energy value, content of basic chemical components, selected minerals and vitamins in the Oscypek cheese (Kokotkiewicz et al. 2018)

| Type of *oscypek* | Energy value (kcal/100 g) | Basic chemical composition (g/100 g) | | | | | Selected minerals and vitamins (mg/100 g) | | | |
|---|---|---|---|---|---|---|---|---|---|---|
| | | Dry matter | Protein | Fat | Carbohydrates | Ash | Ca | Na | Thiamine | Niacin |
| Ewe's | 371 | 66.0 | 29.6 | 27.1 | 2.7 | 6.6 | 866 | 1619 | 0.050 | 0.18 |
| Mixed | 371 | 66.1 | 29.7 | 27.0 | 2.9 | 6.5 | 918 | 1285 | 0.038 | 0.14 |

**Table 11.4** The list of potential key aroma compounds in the Oscypek (Majcher and Jeleń 2011)

| Group | Name | Odor | Origin |
|---|---|---|---|
| Phenolic compounds | Guaiacol (2-methoxyphenol) | Smoked | Wood smoking—thermal degradation and depolymerization/oxidation of lignin |
| | 4-methylphenol | Burnt, phenolic, fecal | |
| | 2-methoxy-4-methylphenol | Smoked | |
| | 3-ethylphenol | Soap, plastic | |
| | 2,6-dimethylphenol | Burnt, smoked | |
| | 2,4-dimethylphenol | Plastic, phenolic | |
| Ketones | Diacetyl (2,3-butanedione) | Butter | Fermentation process—biochemical reactions during curding and ripening between milk lipids, amino acids, and carbohydrates |
| Carboxylic acids | 3-methylbutanoic acid | Sweaty, rancid | |
| Aldehydes | 3-(methylthio)propanal | Cooked potatoes | |
| Free fatty acids | Acetic acid | Sour | |
| | Butanoic acid | Sweaty | |

characterized by the most developed flavor bouquet composed of the outstandingly larger number of volatile compounds (Majcher et al. 2010).

Dominant bacteria in sheep milk are mesophilic lactic acid bacteria (LAB) present at the average amount of 2–6 log cfu in 1 mL of milk. The main LAB genera found in this milk type comprise *Lactococcus, Lactobacillus, Leuconostoc, Streptococcus,* and *Enterococcus*. Milk can be also contaminated with yeast and molds and psychrotrophic bacteria, or rarely with other microorganisms such as *Staphylococcus aureus, Salmonella* spp., *Escherichia coli, Clostridium perfringens, Bacillus cereus, Listeria* spp., *Helicobacter pylori,* pathogenic streptococci, etc. (Alexandraki et al. 2016). Oscypek is produced from unpasteurized milk, so unlike the commercial cheeses, the microorganisms responsible for lactic acid fermentation come from milk not from starter culture added during processing. Some other microorganisms may also enter from the environment, e.g., shepherd's hut (yeasts). Therefore, it has been observed that there is great microbiological biodiversity between cheeses coming from different producers and even different batches. Generally, the predominant microorganisms determined in Oscypek belong to the genera *Streptococcus, Lactobacillus, Lactococcus, Leuconostoc,* and *Enterococcus*. These bacteria have favorable conditions for growth but only until the scalding stage of production. The subsequent scalding, brining, and smoking have an opposite effect due to high temperatures involved (scalding 50–70 °C), low water activity (brining) as well as bacteriostatic or bactericidal properties of phenolic compounds formed (during cold or warm smoking). As a result, the number of bacteria in the final product is considerably reduced (Alegría et al. 2012; Majcher et al. 2011). Wszołek and Bonczar (2003) reported that the total number of microorganisms in the fresh smoked Oscypek cheese produced traditionally in shepherds' huts, located in the mountainous regions of Southern Poland, was in the range of $10^8$–$10^9$ cfu/g and LAB constituted the main group of the inherent microflora. Fresh cheeses contained also coliforms (10–10,000 cfu/g), yeasts, and in some samples molds. The number of enumerated microorganisms in the samples after smoking was lower than the respective level before smoking. Moreover, the microorganism count decreased and the number of coliforms diminished after 60 days of storage. *Listeria* and *Salmonella* cells were not present in the samples as confirmed by the studies performed by Surówka et al. (2016). However, some cheeses contained *Staphylococcus aureus* in the count higher than the maximum allowable level (Council Directive 1992). Further studies, conducted in accordance to legal regulations concerning products from raw milk, revealed the absence of the staphylococci toxins (Wszołek and Bonczar 2003). The coliforms, *Escherichia coli*, yeasts, and in some cases molds and small numbers of *Clostridium perfringens* spores were also detected in the Oscypek cheeses by the Berthold-Pluta et al. (2011). The presence of the microorganisms from the *Entrobacteriaceae* family indicates some deficiencies in the production hygiene but is quite typical for cheeses produced worldwide from unpasteurized ewe's milk. The authors concluded that hygienic standards during productions are key factors influencing microbiological quality and safety of cheeses made from raw milk.

### *11.2.3 Redykołka*

The name "Redykołka" is derived from the word "*redyk*" which in the tradition of Podhale means the ceremonial practice whereby sheeps are taken up to mountain pastures (in May, Spring *redyk*), kept there and brought back down (in September, Autumn *redyk*) (EC 2009b). Redykołka is a semi-hard, half-fat cheese generally produced in the same way as Oscypek cheese (from cheese mass remaining after Oscypek production), but is much smaller (up to 300 g) and formed in the shape of a bird or other animal, heart, or spindle. Redykołka has been produced under the PDO indication since 2009, but currently there are only three registered producers (two from Malopolska) (EC 2009a; IJHARS 2019).

## 11.3 Cheeses Registered in the Ministry of Agriculture and Rural Development's List of Traditional Products

### *11.3.1 Cheeses Included in the Ministry of Agriculture and Rural Development's List of Traditional Products*

Table 11.5 presents the list of and short characterization of cheeses from Malopolska registered in the list of traditional products maintained by the Ministry of Agriculture and Rural Development of Poland.

**Table 11.5** Cheeses from the Malopolska Region included in the Ministry of Agriculture and Rural Development's List of Traditional Products (MRiRW 2019)

| Cheese name | Date of registration | Type of cheese | Source material | Other certificates |
|---|---|---|---|---|
| Bryndza Podhalańska | 28.09.2015 | Soft rennet | Ovine or mixed ovine/bovine (at least 60/40 ratio) | PDO |
| Bundz/Bunc | 10.10.2005 | Rennet | Ovine | – |
| Gomółki Kowalowskie | 19.08.2013 | Acid/curd, dried | Cow's milk, mint, or caraway | – |
| Oscypek/Oszczypek | 28.09.2005 | Smoked, scalded, rennet | Ovine or ovine/bovine (at least 60/40 ratio) | PDO |
| Redykołka | 10.10.2005 | Smoked, scalded, rennet | Ovine or mixed ovine/bovine (at least 60/40 ratio) | PDO |
| Gazdowski cheese—Gołka (Pucok, Kara) | 03.06.2018 | Semi-hard rennet | Unpasteurized cow's milk | – |
| Fresh curd cheeses | 22.12.2014 | Acid/curd, fresh | Cow's milk (including milk from PR breed) | – |

PR—Polish Red cattle breed

#### 11.3.1.1 Bundz (Bunc) Cheese

Bundz is a soft, rennet, lump cheese produced from raw ewe's milk in a shape of a loaf. The production steps of this cheese are the same as first stages of the Bryndza production and were described before. Fresh Bundz has a sweet taste which after few days becomes slightly acidic due to the lactic fermentation process. The consistency should be smooth with rare pea-sized eyes. The color on the cross section is white or white with a willow-green tint. The rind is thin, white-cream, and flexible. The mean basic composition of this cheese is as follows: dry matter 45%, protein 18%, and fat 22%. From 2.5 L of ewe's milk, about 2.5 kg of Bundz can be obtained, which may be served fresh, after few days of ripening, smoked, or un-smoked (Bonczar and Wszołek 2003; Danków and Pikul 2011).

#### 11.3.1.2 Gazdowski Cheese—Gołka (Other Names: Kara, Pucok)

Gazdowski Cheese ("gazda" means mountain farmer) was added on the Ministry of Agriculture and Rural Development's List of Traditional Products in 2008 in the Malopolska Voivodeship and in the Silesian Voivodeship. The technology of production of the gazdowski cheese is similar to the Oscypek. The main difference is the milk type. Generally, Gołka is made from cow's milk and can be produced

**Fig. 11.3** Gazdowski Cheese (Gołka)

throughout the whole year. It is also produced in different shape than Oscypek, in the form of 300–600 g cylinder 6–15 cm long but also with characteristic patterns (Fig. 11.3). The color on cross section is uniform, slightly yellowish, darker outside with a light brown and matte rind. The consistency is soft and elastic, and tastes slightly salty with a noticeable aroma of smoke (MRiRW 2019). In comparison to Oscypek, Gazdowski cheese is characterized with higher average moisture and salt content and lower concentration of proteins, whereas the fat content is similar. Both cheese types differ also with regard to sensory properties as Oscypek cheese has a more pronounced, richer taste and aroma and better firmer and more elastic texture (Kędzierska-Matysek et al. 2014). Like Oscypek, Gołka is produced from unpasteurized milk which is subjected to spontaneous fermentation performed by the autochthonous microorganisms coming from milk and from the production environment. The study of Sip et al. (2010) revealed that acidophilic LAB, namely, from the Enterococcus, Lactococcus, Lactobacillus, and Leuconostoc genera, are the most abundant group (about 9 log cfu/g) of microbiota present in cow-ewe's Gołka. Cheeses also contained considerable amount of yeasts (about 6–7 log cfu/g) but were free from pathogenic Salmonella, Listeria, and enterotoxic *S. aureus*. What is more, there is evidence that some LAB strains isolated from gołka and oscypek perform strong antagonistic activity against *Listeria monocytogenes* (Ołdak et al. 2017; Sip et al. 2012).

#### 11.3.1.3 Gomółki Kowalowskie

Gomółki are very old Polish cheeses produced from tvarog: white, acid/curd cheese formed in the shape of a flattened sphere and slowly dried in a shaded drafty place (e.g., an attic) (Bohdziewicz 2008). The main goal of their production process was to preserve cheese for winter supplies. According to another recipe, gomółki can be produced from cow's milk with an addition of butter, eggs, or egg yolks and cream spiced with salt and pepper, formed into small spherical shapes, and subjected to long drying (in the sun in the summer season or in the oven after bread baking). The name "gomółka" was confirmed in the sixteenth-century sources and means hand-made, spindle-shaped, or spherical lump of cheese. Gomółki Kowalowskie are traditional hardened cottage cheeses from Kowalowa Village in the administrative district of *gmina* Ryglice in the Malopolska Voivodeship, registered in the List of

Traditional Products in 2013 (Table 11.5). These products are grounded firmly in the Malopolska cuisine and tradition. They have been served during wedding receptions, harvest festivals, trade fairs, exhibitions, and competitions. Traditionally, they are a part of the Easter Saturday's food baskets. They can be eaten as a snack, with beer or wine.The recipe of the gomółki kowalowskie production has been passed down for generations. It is produced from acid cheese kneaded with fresh milk and herbs (usually mint or less often with caraway) into the shape of small eggs or larger ones mostly spherical or oval (3–8 cm), and dried in the sun or ovens. The color of the final product is cream to yellowish, darker outside, with dark green flecks. The consistency is hard, firm, and even, and the taste is salty, cheesy with mint flavor. For centuries, Gomółki had been made by shepherds from unpasteurized milk but nowadays they are made mainly by housewives for home consumption and for tourists (Frączek 2019; MRiRW 2019; Slow Food Foundation for Biodiversity 2019).

#### 11.3.1.4 Fresh Acid Curd Cheeses (Tvarogs, White Cheeses)

Tvarog, also known as "white cheese" or quark, is a fresh cultured cheese produced by acidic coagulation of milk performed by lactic acid bacteria present in raw milk (spontaneous fermentation) or added as a starter culture. In the past, only hand-processing technology was applied for tvarog manufacturing and it was made almost exclusively from skimmed milk. Nowadays, the traditional production comprises standardization of fat content in milk, pasteurization, cooling, and the addition of starter culture (mesophilic LAB: *Lactococcus* ssp. and *Leuconostoc* ssp.), fermentation (several hours), warming up to 30–40 °C, cutting of the curd, removing of whey, and pressing. This kind of cheese is very popular in Poland with the average annual per capita consumption equal to 5.16 kg (data from 2015) and has an established position on the Polish market (Król et al. 2018). Almost all dairy processing plants in the Malopolska have some types of Tvarog to offer. The popularity of this product results from centuries-long tradition, dietary habits, low prices, and simple production technology for preparation in households. What is more, Tvarog is an ingredient of many traditional dishes of Polish cuisine (e.g., cheese cake, dumplings, etc.). It is known for its high dietetic and nutritional value due to considerable content of highly digestible protein, relatively low energetic value, lactose fermented to lactic acid, and it is also valuated as a source of minerals and vitamins. Tvorog is offered as a full-fat, semi-skimmed, or skimmed product and in different shapes and packages: wedges or cubes, sliced, ground, wrapped in parchment paper or plastic wrap, in cups, buckets or vacuum packed, with or without sour cream added. There are also natural variants or those flavored with herbs or with sweet additives (vanilla, chocolate, fruits). It can be smoked, fried, or steamed. Aside from acidic Tvarog, there are also cheeses produced by an acid-rennet coagulation, such as homogenized-, fromage-, or cottage-type cheeses (Król et al. 2018; Ziarno and Lenart 2016).

Fresh curd cheeses have been registered in the list of traditional products since 2014. According to the requirements, curd cheeses should be uniformly white to light cream, with firm, uniform structure without lumps, slightly granular in case of skimmed cheeses. Taste and smell should be clean, mild, slightly acidic, and characteristic for this kind of cheese. The shape and mass: wedges and cubes 0.3–0.5 kg, blocks—0.5–0.7 kg (MRiRW 2019). Registered cheeses are produced from raw milk coming from cows grazed on the unpolluted mountain and sub-mountain meadows rich in herbs, located in the Limanowa, Wieliczka, and Bochnia Districts. Cows of traditional Polish Red breed are reared in this area. During winter season animals are fed hay, fodder roots, haylage, and concentrated fodders prepared by farmers. These ingredients all affect the high sensory and nutritional quality of milk products including Tvarog (MRiRW 2019).

### *11.3.2 Other Cheeses from Malopolska*

Majority of registered cheeses, such as Oscypek, Redykołka, and Bryndza Podhalańska, are produced locally on a small scale and only during a selected period (from May to September). They are not commonly available in the retail market and can be bought mainly at the place of production, i.e., in log huts, during local events, etc. Therefore, other cheeses, often referred to as mountain or regional cheeses, similar to Oscypek and Gołka are produced from cow's or a blend of cow's and ewe's milk in different shapes. In the Malopolska Region, they are produced in households (also for tourists) or on an industrial scale, in the latter case from pasteurized milk with usage of starter cultures. Cow's rennet scalded-smoked cheeses are produced in the form of strings, sticks, plaits, cylinders, rolls, oval shapes, or balls, and in different sizes (OSM Nowy Sącz 2019; Mlekovita Zakopane 2019; Zakład Przetwórstwa Mleka Dominik 2019).

Korbacze (Korboce) are a type of semi-hard string cheese produced primarily from cow's milk, and are very popular in the Polish Podhale and in the Orava Region in Slovakia (korbáčik), especially among tourists (Fig. 11.4). Cheese strings with slightly salty mild taste, smoked or un-smoked are twisted together or formed in a bunch (Kokotkiewicz et al. 2018; Kruczek and Krauzowicz 2016).

Also industrial dairies in Malopolska produce bryndza cheese from a mixture of cow's and ewe's milk with industrialized techniques, i.e., application of starters (Bryndza Sądecka) (OSM Nowy Sącz 2019).

**Fig. 11.4** Korbacze (Korboce)

Dairy processing plants located in the Malopolska Province meet the consumers' demand for dairy products by production of other cheese types which are not designated as being regional, such as rennet Dutch-type cheeses (e.g., Gouda, Edamski, Salami), Swiss-type cheeses (Ementaler, Starodworski), fromage, cottage, acid-rennet homogenized cheeses, and Tvarog (quark). Moreover, different types of goat's cheeses (fresh, ripened, acid and rennet, plain or with herbs and other additives) are also produced mainly in the agritourism farms in the Malopolska Region.

## 11.4 Conclusion

Today's trends and fashion, including heritage travel and online shopping, (both by inhabitants of a region, here, those living in Malopolska, and by outsiders), encourage consumers to search for food products, including cheeses, characterized by pro-healthy properties, with outstanding characteristics, which can serve both nutritive purposes and the prestige of consumption of these unique foodstuffs. Also, the increasing awareness of ethnic histories, including customs, food recipes, architectural monuments, and unique landscapes, favors the need and the will for protection and preservation and revival of continuing and past cultures. Taking the above into account, governments and legal authorities have prepared a few rules and regulations to be obeyed by all, but only through including the cooperation of each person—each community member—can we hope to protect and preserve cultural heritage. The positive climate to sustain and develop some tools, regulations, and goods in service to inherited values and beliefs is needed to sustain cultural preservation efforts; for it is said, "the nation without a past does not have a future."

## References

Adamczyk K, Felenczak A, Jamrozy J et al (2008) Conservation of polish red cattle. Slovak J Anim Sci 41(2):72–76

Adamski T, Gorlach K (2016) One tradition, many recipes: social networks and local food production—the oscypek cheese case. In: Fonte M, Papadopoulos AG (eds.) Naming food after places: food relocalisation and knowledge dynamics in rural development. Routledge, London & New York, p 173–196

Alegría Á, Szczesny P, Mayo B et al (2012) Biodiversity in Oscypek, a traditional Polish cheese, determined by culture-dependent and-independent approaches. Appl Environ Microbiol 78(6):1890–1898

Alexandraki V, Kazou M, Angelopoulou A et al (2016) The microbiota of non-cow milk and products. In: Tsakalidou E, Papadimitriou K (eds) Non-bovine milk and milk products. Academic Press, London, p 117–159

Berthold-Pluta A, Pluta A, Zaniecka M (2011) Jakość mikrobiologiczna oscypków (Microbiological quality of oscypek cheeses). Med Weter 67(5):335–338

Bohdziewicz K (2008) Twarogi kwasowe—przetwórstwo (Acid tvarogs—processing). Przegląd Mlecz 7:12–15

Bonczar G, Wszołek M (2003) Regionalne produkty mleczarskie w kraju i na świecie (Region-specific milk products in Poland and in other parts of the world). Żywność 36(Supl.):93–102
Council Directive 92/46/EEC of 16 June 1992 with amendments: laying down the health rules for the production and placing on the market of raw milk, heat-treated milk and milk-based products
Danków R, Pikul J (2011) Przydatność technologiczna mleka owczego do przetwórstwa (Technological suitability of sheep milk for processing). Nauka. Przyroda. Technologie 5(2):7
Dz. U. z 2017 r. poz. 1168. Obwieszczenie Marszałka Sejmu Rzeczypospolitej Polskiej z dnia 26 maja 2017 r. w sprawie ogłoszenia jednolitego tekstu ustawy o rejestracji i ochronie nazw i oznaczeń produktów rolnych i środków spożywczych oraz o produktach tradycyjnych (Law Gazette 2017 text 1158. Announcement of the Marshal of the Sejm of 26 May 2017 on the publication of the consolidated text of the act on the Registration and Protection of Denominations and Indications of Agricultural and Food Products and on Traditional Products). Warsaw, 20.06.2017
Dz.U. 2005 nr 10 poz. 68 Ustawa z dnia 17 grudnia 2004 r. o rejestracji i ochronie nazw i oznaczeń produktów rolnych i środków spożywczych oraz o produktach tradycyjnych (Law Gazette 2005 No. 10 text 68. Act of 17 December 2004 on the Registration and Protection of Denominations and Indications of Agricultural and Food Products and on Traditional Products)
EC (2006a) Application for registration pursuant to articles 5 and 17(2) 'Bryndza Podhalańska' EC No: PL/PDO/005/0450/18.02.2005 PDO. Official Journal of the European Union, Brussels, 23.09.2006, C 230/2–4
EC (2006b) Application for registration in accordance with Articles 5 and 17(2) 'Oscypek' No EC: PL/0451/21.02.2005 PDO. Official J Eur Union Brussels, C 180/94–97, 2.8.2006
EC (2007) Commission Regulation No 642/2007 of 11 June 2007 registering a name in the Register of protected designations of origin and protected geographical indications Bryndza Podhalańska (PDO). Official J Eur Union Brussels, 12.06.2007, L 150/4
EC (2008) Commission Regulation No 127/2008 of 13 February 2008 entering a designation in the register of protected designations of origin and protected geographical indications (Oscypek (PDO)). Official J Eur Union Brussels, 14.02.2008, L 40/5
EC (2009a) Commission Regulation (EC) No 1176/2009 of 30 November 2009 entering a name in the register of protected designations of origin and protected geographical indications (Redykołka (PDO)). Official J Eur Union, Brussels, 1.12.2009, L 314/62
EC (2009b) Publication of an application pursuant to Article 6(2) of Council Regulation (EC) No 510/2006 on the protection of geographical indications and designations of origin for agricultural products and foodstuffs 'Redykołka'. Official J Eur Union, Brussels, C 103/21–25, 5.5.2009
European Commission (2019) Quality schemes explained. https://ec.europa.eu/info/food-farming-fisheries/food-safety-and-quality/certification/quality-labels/quality-schemes-explained_en. Accessed 19 July 2019
Fonte M (2008) Knowledge, food and place. A way of producing, a way of knowing. Sociol Ruralis 48(3):200–222
Frączek D (2019) http://szlakrzemiosla.pl/Jolanta-Kramarczyk.html. Accessed 19 July 2019
Fundacja Partnerstwo dla Środowiska (Environmental Partnership Foundation) (2019) Produkt Lokalny z Małopolski (Local Product from Malopolska). http://produktlokalny.pl. Accessed 31 July 2019
IJHARS Inspekcja Jakości Handlowej Artykułów Rolno — Spożywczych (Inspectorate for the Commercial Quality of Agricultural and Food Products) (2019) https://ijhars.gov.pl. Accessed 29 July 2019
Kawęcka A, Krupiński J (2014) Sheep in the Polish Carpathians: Genetic resources conservation of the Podhale Zackel and Coloured Mountain sheep. Geomatics, Landmanagement and Landscape 1:35–45
Kawęcka A, Sosin-Bzducha E (2014) Seasonal changes of the chemical composition of cheese obtained from the milk of indigenous Polish breeds of sheep. J Anim Feed Sci 23(2):131–138
Kędzierska-Matysek M, Florek M, Skałecki P et al (2014) A comparison of the physicochemical characteristics of the regional cheese Oscypek and the traditional cheese Gazdowski from the Polish Podhale. Int J Dairy Technol 67(2):283–289

Kokotkiewicz J, Radzik-Rant A, Rant W (2018) Produkty pochodzenia owczego w systemach jakości żywności (Products of sheep origin in food quality systems). Wiad Zoot LVI(2):85–92
Król J, Kedzierska-Matysek M, Brodziak A et al (2018) The effect of selected factors on yield and protein and mineral retention in traditionally produced tvarog. J Elementol 23(3):959–969
Kruczek Z, Krauzowicz M (2016) Turystyka kulinarna na Podhalu (Culinary tourism in Podhale). Zeszyty Naukowe. Turystyka i Rekreacja 2(18):17–33
Kudełka W (2014) Próba oceny autentyczności produktów tradycyjnych z mleka owczego (An attempt to assess the authenticity of traditional ewe's milk food products). Zeszyty Naukowe Uniwersytetu Ekonomicznego w Krakowie 927(3):21–32
Majcher M, Ławrowski P, Jeleń H (2010) Comparison of original and adulterated Oscypek cheese based on volatile and sensory profiles. Acta Sci Pol Technol Aliment 9(3):265–275
Majcher MA, Goderska K, Pikul J et al (2011) Changes in volatile, sensory and microbial profiles during preparation of smoked ewe cheese. J Sci Food Agric 91(8):1416–1423
Majcher MA, Jelen HH (2011) Key odorants of Oscypek, a traditional Polish ewe's milk cheese. J Agric Food Chem 59(9):4932–4937
Majcher MA, Kaczmarek A, Klensporf-Pawlik D et al (2015) SPME-MS-based electronic nose as a tool for determination of authenticity of PDO cheese, Oscypek. Food Anal Methods 8:2211–2217
MRiRW Ministerstwo Rolnictwa i Rozwoju Wsi (Ministry of Agriculture and Rural Development) (2019) Lista Produktów tradycyjnych (List of Traditional Products). https://www.gov.pl/web/rolnictwo/lista-produktow-tradycyjnych12. Accessed 25 July 2019
Mlekovita Zakopane (2019) Produkty regionalne (Regional products). http://www.mlekovita.com.pl/pl/produkty/zakopianskie-specjaly. Accessed 25 July 2019
Ołdak A, Zielińska D, Rzepkowska A et al (2017) Comparison of antibacterial activity of *Lactobacillus plantarum* strains isolated from two different kinds of regional cheeses from Poland: oscypek and korycinski cheese. Biomed Res Int 2017 Article ID 6820369:1–10
OSM Nowy Sącz (2019) Produkty (Products). http://www.osmnowysacz.pl/s3-produkty.html. Accessed 25 July 2019
Pangallo D, Šaková N, Koreňová J et al (2014) Microbial diversity and dynamics during the production of May bryndza cheese. Int J Food Microbiol 170:38–43
Polska Izba Produktu Regionalnego i Lokalnego (Polish Chamber of Regional and Local Products) (2019) www.produktyregionalne.pl. Accessed 31 July 2019
Regionalny Związek Hodowców Owiec i Kóz (Regional Union of Sheep and Goat Breeders) (2006) Wniosek o rejestrację nazwy pochodzenia produktu rolnego lub środka spożywczego. Bryndza Podhalańska (Application on the registration of geographical indications and designations of origin for the agricultural product or foodstuff. Bryndza Podhalańska). https://ec.europa.eu/agriculture/quality/door/registeredName.html?denominationId=163. Accessed 12 July 2019
Sip A, Olejnik-Schmidt A, Więckowicz M et al (2010) Analiza mikroflory regionalnych serów gołka (Analysis of microflora associated with regional golka cheeses). Acta Sci Polon., Biotechnologia 9(4):25–38
Sip A, Więckowicz M, Olejnik-Schmidt A et al (2012) Anti-*Listeria* activity of lactic acid bacteria isolated from golka, a regional cheese produced in Poland. Food Control 26(1):117–124
Slow Food Foundation for Biodiversity (2019) Gomółka Cheese. https://www.fondazioneslowfood.com/en/ark-of-taste-slow-food/gomolka-cheese/. Accessed 31 July 2019
Surówka K, Rzepka M, Maciejaszek I et al (2016) Jakość i bezpieczeństwo serków wędzonych wytwarzanych w regionie Podhala (Quality and safety of smoked cheeses manufactured in the Podhale region). Zywn Nauk Technol Ja 4(107):102–114
Wszołek M, Bonczar G (2003) Jakość mikrobiologiczna oscypków z mleka owczego, owczo-krowiego i krowiego (The microbiological quality of 'oscypek' cheeses made of ewe's, cow's, and mixed ewe's and cow's milk). Żywność 3(36)Supl: 103–117
Zakład Przetwórstwa Mleczarskiego "Dominik" (2019) Produkty (Products) http://www.zpm-dominik.pl/produkty/. Accessed 25 July 2019
Ziarno M, Lenart A (2016) Traditional polish curd cheeses. In: McElhatton A, El Idrissi MM (eds.) Modernization of traditional food processes and products. Springer, Boston, MA p 3–12

# Chapter 12
# Traditional Unfermented and Fermented Liquid Milk Products from the Malopolska Region

**Jacek Domagała, Dorota Najgebauer-Lejko, and Maria Walczycka**

**Abstract** Polish Red Cattle constitute an important resource in agricultural production in the mountain area of southern Poland and are of high importance for the protection of cultural heritage in the Malopolska Region. This breed is included in the Program of Genetic Resources Conservation. The characteristic features of Polish Red Cattle are: high content of protein and dry matter in milk, advantageous amino acid composition of milk proteins, high resistance to rough environmental conditions, good health and resistance to diseases, very good fertility and high calves' vitality, and extraordinary longevity. Milk from the cows of Polish Red was, in 2012, included in the Ministry of Agriculture and Rural Development's List of Traditional Products. The unique quality of this milk results from both the genetic endowment of the animals and from traditional feeding based on forages obtained from environmentally unpolluted mountain areas with diversified flora. This milk is also an excellent raw material for cheese production. Milk from Polish Red cows may be added to ewe's milk for the production of "Bryndza Podhalanska," "Oscypek," and "Redykołka"—regional products with an EU's Protected Designation of Origin (PDO) status. The Regional Dairy Cooperative in Limanowa-Tymbark, located in the Beskid Wyspowy range, is one of the milk-processing establishments which specializes in the manufacturing of dairy products, included in the list of traditional products, from milk collected from the mountain areas. Among these products, pasteurized milk from Polish Red cows can be distinguished as an example of liquid unfermented products and sour cream as well as curdled milk for fermented products. Besides cow's milk from the Polish Red breed, milk from sheep and goats grazed on the mountain pastures are used. Another traditional product from ewe's milk constituting an important part of the cultural heritage of the Malopolska Region is żentyca—whey

J. Domagała (✉) · D. Najgebauer-Lejko · M. Walczycka
Department of Animal Product Processing, Faculty of Food Technology, University of Agriculture in Krakow, Balicka 122, 30-149 Krakow, Poland
e mail: jacck.domagala@urk.edu.pl

D. Najgebauer-Lejko
e-mail: dorota.najgebauer-lejko@urk.edu.pl

M. Walczycka
e-mail: maria.walczycka@urk.edu.pl

© Springer Nature Switzerland AG 2022 
J. Hernik et al. (eds.), *Cultural Heritage—Possibilities for Land-Centered Societal Development*, Environmental History 13,
https://doi.org/10.1007/978-3-030-58092-6_12

from ewe's milk recorded in the list of traditional products. Żentyca is a by-product from bundz (rennet soft cheese) manufacturing which is carried on in shepherd huts located in the mountain pastures and sold as a fermented or unfermented product.

**Keywords** Polish red cow milk · Sheep milk · Goat milk · Żentyca

## 12.1 Introduction

One of the main elements of cultural heritage protection of Malopolska Region is preservation and development of local animal breeds and their usage for production of traditional food, e.g., unfermented and fermented liquid dairy products. In the era of globalization and systematically growing world demand for food, where animal breeding is affected by the limitation of internationally recognized animal breeds, local breeds are gaining more and more importance. As a result of the last few years of intensive selection of high yield breeds to obtain enlarged milk yields, higher muscle mass gains, or bigger amounts of laid eggs, there are lost, probably irrecoverably, genes determining the high quality of products obtained and important characteristics connected to fertility and health of animals. The local breeds despite their levels of productivity still find a niche in food production processes. These breeds since they are present in a certain area, through decades or hundreds of years, get used to the environment so they are more resistant to diseases, live longer, and the products obtained from them are of better quality. And what is most important in the traditional animal breeds are the genetic reservoir.

The local breeds are frequently the point of reference for former, local traditions such as food production and handicraft products and play an important role in protection of local cultural heritage. They contribute also to a great extent to preservation of typical village landscapes historically connected to breeding and rearing certain animals.

The local animal breeds are connected to culture and local community tradition. The perfect examples of such local breeds, in the Malopolska Region, are the Polish Red Cattle breed (PC) and Polish Mountain Sheep kept in the Beskidy Mountain Region. Also, there are Carpathian goats. About these, there has been an attempt of renovation taken up by National Research Institute of Animal Production at the State Research Institute in Krakow. Shepherding in the Beskidy Region arose with Wallach shepherds. They were shepherds of cattle and sheep westward, along the Carpathian arc and bringing the elements of culture of all populations met on the way: Romanian, Ruthenian, Poles and Slovaks (Barłowska 2011; Sikora and Kawęcka 2015).

## 12.2 Milk of Polish Red Cow

Polish Red Cattle is the native breed kept in relatively large areas of the country, mainly of southern Poland. These cattle, of a milk-meat type, stand out for high vitality, outstanding immunity especially against tuberculosis and perfect adjustment to difficult conditions of existence (Gądek 1998).

Among native breeds, the Polish Red is recognized as the only autochthonic breed in Poland with importance for foreign markets—exported to Europe and South America.

The Polish Red Cattle originated from brachiceric small aurochs (bison)—*bos Taurus brachyceros* (Fig. 12.1) (Holm and Wójcik 2005).

Leopold Adametz is perceived as scientific creator of Polish Red Cattle, and Jan Brandys was the first to introduce regular breeding of Polish Red Cattle. That breed included mainly red animals, but also brownish, pale, and black were included (Szarek and Adamczyk 2005).

The name "das Polnische Rotvieh" translated into Polish as "Polish Red Cattle" was proposed in 1897 by Franz Holdefleiss—the Professor for cattle breeding who worked for the Agricultural Institute in Wrocław (Breslau—at that time belonging to Germany). In 1901, Leopold Adametz—the Professor for cattle breeding working for the Hochschule für Bodenkultur in Vienna after being hired as the head of the Cattle Breeding and Dairy Department in Jagiellonian University Agricultural College in Krakow disseminated the breed name (Staliński 2005). Polish breeders, who founded, in 1894, the Society of Polish Red Cattle Breeders of Western Galicia, Konopiński, and Bormann, applied interchangeably the names "Red Polish Cattle" and "Polish Red Cattle" (Szarek et al. 2002, 2005).

From 1994, the Malopolska Society of Cattle Breeders was advocating the Ministry of Agriculture to accept and start program of preservation of genetic resources of Polish Red Cattle. And in 1999 the Ministry of Agriculture and Rural Development created the legal and financial conditions for such a program. At the early stage of that project, during the years 1999 to 2001, the work was governed by the State Centre of Animal Breeding in Warsaw, whereas from 2002 the program is

**Fig. 12.1** *Bos Taurus brachyceros* (Holm and Wójcik 2005)

headed by the National Research Institute of Animal Production, Polish Academy of Sciences in Krakow. There are two programs in the relevant breeding: the program for genetic improvement of the race (basic) and program for protection of genetic resources. The stock of cows is now estimated at about 30,000 head, whereas, in 2005, only 500 were subject to the program of genetic resources protection. The main aim of this program is the renovation and preservation of the historical PC breed, mainly sustaining existing genetic diversity. The breeding work is intended to sustain the typical characteristic of that breed such as perfect adjustment to difficult environmental conditions; high resistance to diseases, health, and high fertility; easy calving, high calves' vitality, and breeding ease; and high biological value of milk (Stopyra et al. 2005).

In 2009, the program of genetic resources preservation was applied to 1760 cows. In the period between the two world wars, the cattle were bred with breed purity kept whereas animals subjected to milking control nowadays have, in their genotype, noticeable shares of such breeds as Angler, Jersey, Red Belgium, Red Dutch and also Simmental, White-Black and Red-White, and even some traces of Aberdeen Angus and Charolaise. In 2008, the average milk yield of cows subjected to milking yield control was 3927 kg with 3.35% of protein and 4.25% of fat content. The usage for crossbreeding of Angler Breed bulls improved the milking yield for crossbreds, whereas the Red-White breed with meat-milk utility had played the basic role when excluding Polish Red Cattle. That is a pity in that the population of Polish Red Cattle is diminishing in a very rapid manner. The cause is the small milk yield, lowered because of insemination of cows with the semen of bulls in the old type of the Polish Red Cattle breed. The lowering in cows' milk yield causes that breeders to opt out of joining the program of genetic resources protection because the grant money from the project does not refund the losses caused by low milk yield (Adamczyk and Szarek 2009).

The high milk yield cow breeds such as the Holstein-Fresian breed have caused very high demands for breeding and feeding. Such conditions cannot be fulfilled in the mountain and sub-mountain areas. So the local, native breeds such as Polish Red Cattle, preferable, on the one hand, on sustaining so-called biodiversity are subjected to program of genetic resources protection, and, on the other hand, also allow farmers to obtain an outstanding quality of milk and beef meat, because these animals are mainly fed with farm-specific large volume diets (Adamczyk and Szarek 2009).

The results of investigations point to the fact that milk obtained from green-grazed cows is characterized by better parameters for cheese production and is richer in components positively influencing the human health, especially in whey proteins, polyunsaturated fatty acids with CLA (conjugated linoleic acid), and fat-soluble vitamins (Barłowska et al. 2012).

The breed pedigree for Polish Red Cattle as applied in contemporary practices in Poland does not differ from that accepted in the time period between the two world wars and includes the following criteria:

- Usage type: meat-milk.
- Height at back of adult animals (bulls) about 140 cm; cows about 130 cm.

**Fig. 12.2** Cows of Polish Red Cattle breed in the mountains (photo Rychlak M.) (Krupiński 2012)

- Uniform pigment from red to dark red, cherry, dark hooves, and nostrils (the bright muzzle allowed), pale horns with dark ends.
- Characteristics of physical build—properly formed horns and hooves, as the aim is to strive for better shape and build of udder.
- The average milk yield of cows included in the program for genetic resources protection—about 3500 kg per lactation with fat content 4.2–4.5% and protein 3.3–3.6% (Adamczyk and Szarek 2009; Krupiński 2012).

The appearance of the Polish Red Cattle breed is presented in Fig. 12.2.

The PC cattle delivers milk with outstanding quality with high content of protein and fat containing polyunsaturated fatty acids with omega 3 and omega 6 bonds, especially suitable for cheese production (Adamczyk and Szarek 2009).

The contemporary efforts of breeders for breeding and genetics are directed to obtain milk with high protein content and with a rather lower amount of fat. Milk originating from Polish Red Cattle examined by Felenczak et al. (2005) was characterized by high fat content, high total protein content, casein and total solids, and good curdling ability and thermal stability which points to its suitability for technological usage. The short time of milk curdling of these cows is determined to a great extent by the higher total protein and casein contents. In Table 12.1, the average composition and technological properties of milk of the Polish Red breed with BB genotype of κ-casein are presented (Felenczak et al. 2005).

**Table 12.1** The average chemical composition and technological properties of milk of Polish Red cows BBv κ-casein genotype (mean values ± standard deviation) (Felenczak et al. 2005)

| Traits | Mean ± SD |
|---|---|
| Total solids [%] | 13.30 ± 0.83 |
| Total protein [%] | 3.38 ± 0.19 |
| Casein [%] | 2.67 ± 0.16 |
| Fat [%] | 4.26 ± 0.28 |
| Lactose [%] | 4.88 ± 0.30 |
| Ca [mg/dl] | 117.81 ± 12.05 |
| P [mg/dl] | 96.40 ± 7.35 |
| K [mg/dl] | 147.38 ± 11.75 |
| Na [mg/dl] | 59.07 ± 6.02 |
| Density [g/cm$^3$] | 1.0293 ± 0.015 |
| Clotting time [s] | 289.8 ± 39.61 |
| Thermal stability [min] | 3.46 ± 0.36 |
| Cheese yield [%] | 28.07 ± 5.27 |

The examination of basic chemical composition of Polish Red cows in preserved herds and in crossbred herds performed in the 2005–2006 years by Gardzina-Mytar et al. (2007) exhibited the small degree of effectiveness of crossbreeding with European Red breed bulls. The basic composition of cows' milk in preserved herds in dependence of such factors as the way of feeding, age, and lactation phase is presented in Table 12.2.

The outstanding quality of milk of Polish Red cows is a result of both genetic predispositions of these animals and of the traditional way of feeding them based on feeds obtained from ecologically clean mountain areas with differentiated plant cover.

**Table 12.2** The basic composition of cows' milk in preserved (protected) herds with dependence noted on method of feeding, age, and lactation phase (Gardzina-Mytar et al. 2007)

| Parameter | Total solids [%] | Protein [%] | Fat [%] | Lactose [%] |
|---|---|---|---|---|
| Total | 13.33 | 3.33 | 4.50 | 4.72 |
| Summer feeding | 13.26 | 3.28 | 4.33 | 4.67 |
| Winter feeding | 13.40 | 3.35 | 4.59 | 4.78 |
| I lactation | 13.20 | 3.15 | 4.58 | 4.76 |
| II lactation | 13.24 | 3.21 | 4.47 | 4.87 |
| III lactation | 13.52 | 3.28 | 4.80 | 4.67 |
| IV lactation | 13.35 | 3.37 | 4.58 | 4.68 |
| First phase of lactation | 13.12 | 3.13 | 4.58 | 4.77 |
| Second phase of lactation | 13.24 | 3.32 | 4.47 | 4.74 |
| Third phase of lactation | 13.69 | 3.58 | 4.81 | 4.62 |

**Table 12.3** Comparison of the milk yield and average composition of milk derived from Polish Red and Polish Holstein-Friesian cows (Litwińczuk et al. 2012, per 100 mL of milk)

| Milk yield/milk component | Cow breed | |
|---|---|---|
| | Polish red | Polish Holstein-Friesian |
| Daily milk yield (kg) | 12.59 | 25.86 |
| Total solids (%) | 13.30 | 13.27 |
| Lactose (%) | 4.66 | 4.75 |
| Protein (%) | 3.61 | 3.49 |
| Casein (%) | 2.68 must | 2.59 |
| α-lactalbumin (%) | 0.112 | 0.095 |
| β-lactoglobulin (%) | 0.318 | 0.289 |
| BSA (%) | 0.047 | 0.043 |
| Lactoferrin (mg/100 mL) | 11.53 | 9.08 |
| Lysozyme (μg/100 mL) | 1.11 | 0.59 |
| Fat (%) | 4.35 | 4.37 |
| SFA (% fat) | 60.73 | 66.87 |
| MUFA (% fat) | 33.48 | 30.12 |
| PUFA(% fat) | 5.30 | 2.98 |
| CLA (% fat) | 2.24 | 0.34 |
| Ca (mg/ 100 g) | 100.80 | 76.75 |

BSA—bovine serum albumin, SFA—saturated fatty acids, MUFA—monounsaturated fatty acids, PUFA—polyunsaturated fatty acids, CLA—conjugated linoleic acid

Milk obtained from Polish Red given the extensive, traditional way of feeding is characterized by a high content of total solids, especially of protein and fat. Favorable protein-to-fat ratio, high casein and calcium contents, and relatively short coagulation time make the milk from Polish Reds an excellent raw material for cheese production. The confirmation of the above claim is the possibility of that milk's use, when cows are grazed on mountain meadows with some herds of mountain sheep, for production of "Bryndza Podhalańska", "Oscypek", and "Redykołka"—the regional products granted European Commission identification signifying PDO (Protected Designation of Origin). The milk of Polish Red breed according to requirements for oscypek and bryndza podhalańska, PDO, can be added during production of these cheeses in the amount of 40% (to ewe's milk). Milk derived from Polish Red cows when compared to that obtained from Polish Holstein-Friesian (highly productive and kept in the intensive system) is also more abundant in bioactive proteins, such as α-lactalbumin, β-lactoglobulin, lactoferrin, and lysozyme (Table 12.3). It is also characterized by a more favorable, fatty acid profile, from a nutritional point of view, i.e., higher concentrations of mono- and polyunsaturated fatty acids and sevenfold higher content

**Fig. 12.3** Pasteurized consumption milk of Polish Red cows produced by Limanowa-Tymbark Milk Factory (www.limanowa.in/aktualnosci/mleko-trafilo-na-polki-almy-w-calym-kraju/15789)

of CLA. The outstanding characteristics of milk of Polish Red cows were confirmed by its assignment on December 14, 2012 to the countrywide Traditional Products List (Litwińczuk et al. 2012, www.ec.europa.eu/agriculture/quality; www.gov.pl/rolnictwo/lista-produktow-tradycyjnych).

For Polish Red breed promotion, there are exhibitions of Polish Red Cattle organized in Szczyrzyc near Limanowa. A distinct matter is the promotion of dairy food products as so-called "ecological," regional products. A good example of such promotion is pasteurized milk from Polish Red cows produced by Limanowa-Tymbark Milk Factory (Fig. 12.3). The promotion of the breed is also the promotion of the region, because the history of the Polish Red breed's origin is inter-connected with longstanding traditions and the culture of Malopolska (Adamczyk and Szarek 2009).

## 12.3 Traditional Fermented Products of Cow's Milk Obtained in Malopolska Region

The Limanowa-Tymbark Milk Factory also produces some other dairy products assigned to the countrywide Traditional Products List. There are fermented products such as "Śmietana z Limanowej" (sour cream from Limanowa) and "Zsiadłe Mleko z Limanowej" (lactic acid fermented milk from Limanowa) added, on February 17, 2017, to the List in the category of "dairy products from Malopolska Voivodeship" (www.gov.pl/web/rolnictwo/lista-produktow-tradycyjnych).

On the Traditional Products List, authorized by the Ministry for Agriculture and Rural Areas Development, are assigned products with characteristics resulting from application of traditional methods of production, and which are elements of the region's cultural heritage, where their production continues and which are part of the identity of local society. As traditional methods of production are defined, these must have been used for at least 25 years in production processes (Kuźnicka and Zajączkowska 2009).

According to tradition, the raw material for sour cream from Limanowa is milk obtained from farms where the cows are grazed in mountain and sub-mountain meadows (rich in herbs) of Limanowski, Wielicki, and Bocheński Districts. There during summer, the base sources for feeding cows are green pastures whereas in winter the best feeds are used, prepared individually by farmers, characterized by high levels of nutritional components. The beginnings of obtaining sour cream are dated very long ago. Formerly, the cream was obtained at farms where it was used only for farm residents' own purposes. The raw milk, directly after milking of cows, was poured into clay pots and left in the kitchen, at room temperature. After a few hours, sweet cream on the surface of the milk was gathered. Subsequently, after a longer period, as a result of fermentation, which was induced by milk microflora naturally present, the cream had soured. Farmers' housewives gathered the soured cream and used it as an additive for different meals. The cream was applied, among others, to whiten soups, for baking of cakes, and poured into wooden churns for preparing butter. Nowadays, the character of production has changed a little because during milk centrifugation (at a dairy factory) the sweet cream is obtained which is then passed on to other stages of production. Thanks to appropriate processes of production, it is possible to maintain longer storage of cream with consequent preservation of its quality and flavor characteristics. The cream is a creamy, dense suspension uniform in the whole mass, with clean, slightly acidic taste (www.gov.pl/web/rolnictwo/smietana-z-limanowej).

The lactic acid fermented milk, said to be "curdled," is a traditional Polish drink produced from raw, not pasteurized, cow's milk. The beginning of that drink production was ages ago when this type of product was produced from raw cow's milk at any individual farm. The raw milk directly after milking was poured into clay pots and left at room temperature for spontaneous fermentation. After several hours, thanks to the presence of native bacteria, the milk was curdled. Such "curdled" milk was tasty when served cooled and also was used as a drink served with warm potatoes smeared with fried bacon with onion. The "curdled" milk was consumed to refresh oneself during peasant work in the fields and also as an everyday meal. That milk was also mixed with fruit. The perfect, specific taste is not the single benefit of "curdled" milk. It possesses also some pro-health and nutritive values. This is a rich source of calcium, contains large amounts of vitamins, e.g., A, B, K, and the most important point is that it is very advantageous for human digestive tract functions. Because of many benefits "curdled" milk has returned again on the tables of the Limanowa and Tymbark Regions and is very welcome to be consumed by inhabitants. This is like a suspension product with tight consistency which can be cut out by a knife with a shallow layer of cream at the top, a light yellow-creamy color, and clear, slightly acidic taste typical for lactic acid fermented products (www.gov.pl/rolnictwo/zsiadle-mleko-z-limanowej).

## 12.4 Ewe's Milk

In Poland, sheep are bred mainly in mountain areas and also in Great Poland and Podlasie Voivodeships. The stock of sheep, in 2011, was at about 223,000. The ewe's milk production yield is estimated as 1000 tons per year. The sheep is milked in a traditional way, during the grazing at Podhale, Beskidy, and Bieszczady mountains and sub-mountain areas (Danków and Pikul 2011b).

As bred sheep, in Poland, there are present: Great Poland sheep (Wielkopolska), Olkusz sheep, Leszczynska sheep, Pomeranian sheep (Pomorska), Polish Merino sheep, Świniarka sheep, Heather sheep (Wrzosówka), and in the mountains and sub-mountain areas Polish Mountain sheep and Color Mountain sheep (Bonczar et al. 1998).

Polish Mountain sheep—Zakiel is an old primitive breed of white and colored sheep present formerly in the Southern Carpathian and partially in the Balkans. Zakiels were introduced into Poland when Wallachian-Ruthenian shepherd tribes were wandering there, i.e., along the Carpathian mountain ranges, between the fourteenth and sixteenth centuries. Zakiels are closely related to life and culture among the Carpathian Gorals (Highlanders). That sheep meat and milk were used for cheese production: oscypek, bunc, bryndza, and as a part of meals, and also the leather and wool were used for clothing and handmade art pieces. Zakiels are the sheep with a build shaped by generations of living in the mountain areas. They are small with harmonic-shaped trunk, and rather thin but strong legs, with flock pelage that allows for protection of the animal against severe mountain climate. The rams possess long screw horns and ewes are with or without horns. That sheep yields thick wool, mixed in type, white and black in color or pale red or grayish. Zakiels originating from Beskid Śląski and Podhale were the basis for crossbreeding and creation of the noble Polish Mountain Sheep pedigree of breed.

The crossbreeding efforts improved the body weight of adult ewes—at about 50 kg, the yield and properties of pelage and milk yield which, with the 150 days of grazing, is at about 60–80 L of milk. It was managed in mass stock, with the improvement of production parameters of the breed, to keep the precious characteristics of the animals which display marvelous adjustment to the environmental conditions, i.e., good health, long life, resistance (immunity), and acceptable at the given breeding environment—milking yield and fertility. There is also the taste of young lambs' meat which is an export good meeting the demands of the Italian market. That phenotype adjustment to the Podhale sub-mountain environment tends

**Fig. 12.4** Polish Mountain sheep (http://www.pzow.pl/rasy-owiec.html)

to confirm the native character of the genotype of this population—very precious, excellently domesticated in our culture and regional landscape.

The sheep breed grazed in Podhale Region is fed with very specific, diversified flora (specific botanic composition of meadow plants), which influences the taste and composition of milk and as a consequence gives to the product obtained of that milk, an outstanding taste and aroma (www.ec.europa.eu/agriculture/quality). In Fig. 12.4, the appearance of animals of the Polish Mountain Sheep breed is shown.

The Polish Mountain Sheep color variety is also the lineal descendant of an old, primitive, and numerous breed group of Zakiel, living, for ages, in the Southern Carpathian region and in part of the Balkans. Moving into the Polish part of the Carpathians the Zakiel Color Breed wandered together with the white variety (above described). The color variety of Zakiel was kept by Gorals (Highlanders) because of the dark color of wool and skin and could be used for production of regional traditional clothes and/or decorative pieces. The colored mountain sheep with the sister white variety have become inseparable elements of the economy and culture built by the mountaineers. Undeniably there is an advantageous influence of sheep on shaping of the landscape and its preservation especially in such poor biotopes as the mountain areas are. The colored mountain sheep is covered with mixed, two fractions of wool of dark russet color which goes grayish and reddish with time (age). Typical is the presence of a white blotch (star) or bald spot on the head and of the white end of a tail. The rams possess horns and their weight is about 50 kg, whereas ewes

**Table 12.4** The average basic composition and density of sheep milk (Bonczar 2001) and milk of Polish Mountain Sheep (Konieczny 2009)

| Component | Sheep milk | | Milk of Polish Mountain Sheep | |
|---|---|---|---|---|
| | Average | Range | Average | Range |
| Total solids [%] | 18.45 | 15.40–21.50 | 20.82 | 19.12–22.52 |
| Total solids without fat [%] | 11.40 | 10.30–12.50 | 11.32 | 10.93–11.71 |
| Protein [%] | 5.85 | 4.90–6.80 | 4.23 | 4.01–4.44 |
| Fat [%] | 7.05 | 5.10–9.00 | 9.50 | 8.19–10.81 |
| Lactose [%] | 4.80 | 4.20–5.40 | 6.24 | 6.07–6.41 |
| Mineral compounds [%] | 0.86 | 0.82–0.95 | n.d | n.d |
| Density [g/cm$^3$] | 1.0335 | 1.0290–1.0380 | 1.0345 | 1.0330–1.0360 |

n.d.—data not available

can have horns or be without horns and their weight is about 40 kg. Nowadays, the number of stock of this variety, in the Polish part of the Carpathians, is estimated to 500–800 and is declining steadily. In the program of genetic resources preservation of sheep, five stocks are included—154 ewes (counted for the year 2005). That breed is kept on farms mainly in the Malopolska (Lesser Poland) Region (Krupiński 2012).

The ewe's milk in comparison to cow's contains much more of total solids, total protein, casein, and also mineral components. The milk is also characterized by higher calorific value than cow's and is richer in vitamins especially these soluble in water (Bonczar 2001).

The basic composition and density of average ewe's milk and of Polish Mountain Sheep ewe's milk, with oscillation during lactation phases, are presented in Table 12.4.

In the fat of ewe's milk, there are present more dienes of conjugated linoleic acid (CLA) than in goat's and cow's milk. The fat of ewe's milk contains more free fatty acids than that of cow's milk. The average value of cholesterol in ewe's milk is 20 mg/100 g of milk (Bonczar 2001). The light yellow-creamy color of ewe's milk is caused by larger amounts of carotene dissolved in fat and of riboflavin dissolved in water phase than in cow's and goat's milk. The ewe's milk shows also better bacteriostatic characteristics than cow's and goat's milk—it needs for spontaneous fermentation, a few days at room temperature. Thanks to the above raw material, dairy production can be obtained in field conditions—in pastures and meadows without fear of rapid acidification (spoilage) (Danków and Pikul 2011b).

Because of its rich chemical composition ewe's milk is a perfect raw material for cheeses and fermented milk products production. About 70% of ewe's milk of mountain sheep is processed into oscypek type cheeses. That milk has almost two times more yield than cow's milk when used for cheese production. The processing of ewe's unpasteurized milk is often done in farm conditions. The obtained products serve for the farmer's own consumption or are sold at free stall markets halls or at small neighbor shops. Ewe's milk is also very useful for production of different kinds of fermented milk. The large share of total solids (at about 18%) does not need supplemental thickeners in the fermented milk production which is needed when yoghurt and kefir are produced from cow's and goat's milk. Thanks to the above fermented drinks made of ewe's milk, are totally natural and free of additives such as milk powder or stabilizers. The time of lactic acid fermentation during production of yoghurt and kefir from ewe's milk is much shorter in comparison to the same products when obtained from cow's and goat's milk (Danków and Pikul 2011b).

The sheep is a perfect tool which can serve to stimulate rural area development. Their usage can be multidimensional and products yielded by them can be market and tourist assets. The sheep shape and maintain the landscape widening the area for tourism and recreation development. They create new job opportunities not only for shepherds, but they also allow society to keep and sustain a cultural heritage which is based on a living tradition of ewe's milk and other of sheep-related raw material processing. They are a determinant of region identity and sometimes lifestyle, especially for people who want to manage a farm in agreement with nature. Unfortunately, in Poland it is much rarer than formerly to see grazing herds of sheep. Even in the mountains which were always a sheep-breeding bastion their herds become a rarity—the sheep is vanishing from the Polish landscape. The lowering of sheep population is confirmed by statistics which prove that during the last two decades heads of sheep stock dramatically diminished. Each year, in the most recent decade in Poland, they vanished on an average of 10,000 sheep (Mroczkowski 2011).

## 12.5 Goat's Milk

The rearing and breeding of goats in Poland, after long years of stagnation, have been developed remarkably from the 80 s of the twentieth century. An interest in the dietetic value of goat's milk has caused the growth of demand for goat's milk products. The main consumers' group which emerged were people with allergies, and mainly these who suffered from protein defect—children, and also convalescents and elder people. Also the taste value of goat's products was wider appreciated, especially of cheeses. Within the space of last 15 years the goats' breeding undergo different phases of development.

In 2007, there was observed a breakdown in goat breeding. That was a result of decision of the Ministry of Agriculture and Rural Development, in the framework of the Biological Progress Fund, which stopped the subvention for the supervision by breeders of milk productivity and fertility of goats, with the result that the breeders gave up that supervision and at the same time stopped their breeding work. As a consequence of above decisions, there was a drastic drop in the number of herds subjected to milk yield control. In subsequent years, there was observed a further drop of estimated (preserved) heads of stock. Parallel to a diminishing number of bred animals also the total number of goat heads at stock dropped down. From 2002 when there were 193,000 of goats, the slow drop down in their population was ongoing to reach at about 117,000 in 2010—a loss of almost 39% (Sikora and Kawęcka 2015).

The largest numbers of goats are kept in southeast part of Poland. There exist small herds—from 1 to 3 animals. In 2002, in Malopolska were bred 24,914 heads of goats. Above amount of goats stand for 14.4% of country goat stock herds, and this is placed mainly in Malopolska, as the leading breeder of these animals. In 2010, the number of stock further diminished to 17,949, while the largest amounts were bred in Limanowski, Tarnowski, Myślenicki, Nowotarski, and Gorlicki Districts.

In breeding herds, among Polish breeds, White Noble Goats (41%) and Color Noble Goats (19%) dominate. The breeds represented in smaller amounts are White No-breed (9%) and Color No-breed (8%). In the imported breeds, breeds such as Saanen Goat (18%) and Alpine Goat (5%) dominate. The native breeds are present in rather small numbers, home breeds of goats, such as Carpathian Sandomierz (Sandomiria) and Kazimierz (Casimiria) (Danków and Pikul 2011a; Sikora and Kawęcka 2015).

The encouragement for taking up of larger stock herds of goats breeding can be the initiative of National Research Institute of Animal Production, Polish Academy of Science in Krakow. There has been an attempt of revitalization of the Carpathian Goat breed which was a typical mountain breed of goat. Nowadays, in the territory of Malopolska, there exist only two herds of the Carpathian Goat pure breed under genetic survillance (Sikora and Kawęcka 2015).

The goat's milk because of its content of basic components is similar to cow's milk but differs in the quality composition of fat and protein. The differences are a result of the different build, composition, and dimensions of fat globules and casein micelles; from other proportions of individual protein fractions and of a larger, in goat's milk, amount of nitrogen non-protein compounds, and minerals (Wszołek 2001). The comparison of goat's and cow's milk composition is presented in Table 12.5.

Goat's milk is intended for direct consumption and/or to process into different dairy products such as liquid consumption milk (pasteurized and UHT), condensed milk, milk powder, rennet and tvarog cheeses, and fermented milks such as yoghurt, kefir, butter milk, or cream. The last one direction of goat's milk processing seems to be especially desired because it allows to combine the great nutritional value; easy digestibility; assimilability of components; and antioxidant, therapeutic, and

**Table 12.5** Chemical composition of goat and cow milk (Pandya and Ghodke 2007)

| Components | Goat milk | Cow milk |
|---|---|---|
| Total solids (%) | 11.50–13.20 | 12.30–13.50 |
| Total protein (%) | 2.90–3.76 | 3.20–3.50 |
| Casein (%) | 2.60–2.90 | 2.50–2.70 |
| Whey proteins (%) | 0.30–0.86 | 0.70–0.80 |
| Fat (%) | 3.07–5.10 | 3.40–4.20 |
| Lactose (%) | 4.10–4.50 | 4.60–4.70 |
| Mineral compounds (%) | 0.71–0.87 | 0.65–0.81 |
| Cholesterol (mg/100 g) | 11 | 14 |
| Energy in 100 g (kcal/kJ) | 69/290 | 61/257 |

antiallergenic properties of that milk with the specific role of fermented milk in the human nutrition and body. Because the curd made of goat's milk is characterized with lower thickness and density when compared to cow's or ewe's milk, in order to obtain the desired consistency of fermented products of goat's milk there is a recommended application of additional technologies during the production process (Domagała 2005; Danków and Pikul 2011a).

The production of goat's milk and its by-products is in Poland a niche of a sub-market of dairy products. It is estimated that at present in Malopolska Voivodeship, about 9.8 million kg of milk per year with the average goat's milking yield of 542 kg per year is obtained. The small amount of producers distributes their goat's dairy product to the ecological or specialty shops (only cheeses) in larger towns. More frequently the production and sale of milk and goat's dairy products takes place at agro-tourism farms. That very split production is not noticeable at the market and does not have connections with the consumers' market. There is a lack of well-distributed publicity of key products with repeatable sensory quality and this does not develop this part of agricultural market. There is not present any goat's milk originating product at the Traditional Products List of Malopolska Region.

To recapitulate the breeding and rearing of goats in Malopolska Voivodeship, similarly to the rest of the country exists, is stagnating. The factors which can lead to economic success in this niche involves persuading a larger group of consumers of the taste and dietetic values of meat, milk, and dairy products obtained from goats (Sikora and Kawęcka 2015).

## 12.6 Żentyca—Non-fermented or Fermented Product Obtained from Ewe's Milk

Żentyca, the subsequent ewe's milk dairy product, is also an important part of the cultural heritage of the Malopolska Region. Żentyca (or żętyca) is the by-product obtained during production of bundz and oscypki cheeses from ewe's milk. This product is called "ewe's whey" although the name is not quite proper and can lead a consumer to a mistaken judgment. The reality is that it is a kind of cheese of a ricotta type similar to the original Italian Ricotta cheese with admixture of larger or smaller amounts of de-proteinated ewe's whey so-called "zwarnica." The way of production of żentyca is as follows: the whey obtained is poured into the kettle and heated slowly over the watra (the Gorals' name for bonfire). Żentyca originates through pre-cooking (heating to the first symptoms of the start of cooking process) of whey obtained after rennet curdling of the casein fraction of proteins from ewe's milk. At the temperature over 90°C, the coagulation of whey proteins takes place which forms a delicate milk skin floating on the surface of the heated whey. The żentyca is obtained by churning and picking by ladle, floating at the upper part of kettle coagula. The amount of coagulated whey proteins in żentyca depends on the person who is obtaining the drink and of the amount of the liquid part of heated ewe's whey "zwarnica." So the density and composition of żentyca differs from one shepherd hut to another. Żentyca is the safest to be consumed at the shepherd hut because there it corresponds with the quality of pasteurized products. It can be consumed as warm (just freshly made) or it can be left and mixed with żentyca obtained on previous days. When mixed the products goes on the spontaneous fermentation which is caused by natural acidifying microflora present on wooden equipment and pots present in the shepherd hut. The characteristic acid-tart taste of żentyca is obtained after a few or a several days of fermentation depending on the shepherd hut temperature. The thick żentyca is one of the main meals of shepherds. It is consumed as an additive to potatoes. dumplings, or bread. Guests are offered żentyca in wooden scoop pots. The shepherds of Szczawnica District brought żentyca to spa facilities where it was served to bathers with therapeutic mineral waters (Drożdż 2007; Kuźnicka and Zajączkowska 2009).

Żentyca was listed in the traditional products list of the Ministry of Agriculture and Rural Development on September 28, 2005. It should contain water content ranging from 60 to 70%, salt content up to 0.5%, and fat content ranging 3–4% (ewe's żentyca) and 2–3% (żentyca from mixed cow's and ewe's milk). The color of that liquid should be white or slightly creamy and the taste is sweet (fresh żentyca) or acid (żentyca fermented after a few days). As the producers of żentyca shepherds claim, it is a good remedy for diseases of the stomach, guts, upper respiratory tract, and even for Alzheimer (Bonczar 2006).

## 12.7 Conclusions

Traditional liquid milk products from Malopolska include raw or pasteurized cow's, ewe's, and goat's milk, curdled milk, sour cream, and a beverage made from ewe's whey called żentyca. Their outstanding quality and unique characteristics, as described above, result from the use of milk from indigenous animal breeds fed with forages derived from meadows and mountain pastures that are unpolluted and rich in herbs and produced by traditional processing methods. Rearing of the local animal breeds, such as Polish Red Cattle, Polish Mountain Sheep, and Carpathian goats, and processing of the milk obtained from them have a long history deeply embedded in local traditions. Therefore, as an important part of the cultural heritage of the Malopolska Region, they as an important part should if possible be protected for future generations. To attain this goal, many efforts should be made such as promotion of these products and dissemination of the knowledge regarding their high nutritional and unique sensory qualities, so as to improve the likelihood which, in turn, will make the production profitable to the producers. All this would probably require agricultural policy coordinating cooperative efforts by government and non-governmental organizations (of course, including privately owned farms operating under a market framework).

## References


Adamczyk K, Szarek J (2009) Bydło polskie czerwone – nauka na przyszłość (Polish Red Cattle-the science for future). Przegląd Hodowlany 8:9–12

Barłowska J (2011) Znaczenie lokalnych ras zwierząt w produkcji żywności tradycyjnej oraz przekazie tradycji i kultury regionu (The meaning of local animal breeds in traditional food production and tradition and culture message of the region). Przegląd Hodowlany 9:1–5

Barłowska J, Chabuz W, Król J, Szwajkowska M, Litwińczuk Z (2012) Wartość odżywcza i przydatność technologiczna mleka produkowanego w systemie intensywnym i tradycyjnym w trzech rejonach wschodniej Polski. (Nutritional value and technological suitability of milk produced in intensive and traditional systems in 3 regions of Eastern Poland). Żywność. Nauka. Technologia. Jakość 4(83):122–135

Bonczar G (2001) Znaczenie mleka owczego w żywieniu człowieka. (The meaning of ewe's milk in human nutrition). Przegląd Mleczarski 3:125–128

Bonczar G (2006) Jakość oszczypków z uwzględnieniem oceny mleka owczego i żętycy. Materiały szkoleniowe programu „Owca Plus" (The quality of oscypki with consideration of ewe's milk and żentyca assessment. The education materials of the project "Sheep Plus"); Wydawnictwo AR Kraków

Bonczar G, Ciuryk S, Frajdenberg I, Pastuszka E (1998) Ocena przydatności różnych ras owiec do produkcji bundzu (The assessment of suitability of different sheep breeds for bundz production). Zeszyty Naukowe AR Kraków; seria Technologia Żywności. 347(10):5–14

Danków R, Pikul J (2011a) Przydatność technologiczna mleka koziego do przetwórstwa. (Technological suitability of goat milk for processing). Nauka Przyroda Technologia 5(2):6

Danków R, Pikul J (2011b) Przydatność technologiczna mleka owczego do przetwórstwa. (Technological suitability of sheep milk for processing). Nauka Przyroda Technologia 5(2):7

Domagała J (2005) Zmiany tekstury i mikrostruktury jogurtu z mleka koziego pod wpływem wybranych czynników. (Changes in the texture and microstructure of goat's milk yoghurt as induced by selected factors). Zeszyty Naukowe AR Kraków: Rozprawy 425. 309

Drożdż A (2007) Żentyca – karpacka odmiana włoskiej ricotty (Żentyca- the Carpathian variety of Italian ricotta). Przegląd Hodowlany 9:30–32

Felenczak A, Ormian M, Adamczyk K (2005) Skład i właściwości mleka krów rasy polskiej czerwonej i czerwono-białej z uwzględnieniem polimorfizmu białek. (Composition and properties of milk of polish red and red –white cows with regard to protein polymorphism). Wiadomości Zootechniczne R.XLIII. 2:69–72

Gardzina-Mytar E, Węglarz A, Felenczak A, Ormian M, Makulska J (2007) Wydajność i skład mleka krów rasy polskiej czerwonej utrzymywanych w stadzie zachowawczym i doskonalonym. (Yield and composition of milk from Polish Red cows maintained in conservation and improved herds). Roczniki Naukowe Zootechniki 34(2):3–10

Gądek M (1998) Miejsce rasy polskiej czerwonej w hodowli bydła w Polsce południowej (The place of Polish Red Cattle breed in cattle breeding of southern Poland). Biuletyn Informacyjny Instytutu Zootechniki 36(1):15–22

Holm L, Wójcik P (2005) Charakterystyka innych ras czerwonych w Europie zrzeszonych w ERDB. (Characteristics of the European red breeds affiliated with the ERDB). Wiadomości Zootechniczne R.XLIII. 2:144–148

Konieczny M (2009) Wpływ fazy laktacji na skład chemiczny i parametry fizykochemiczne mleka polskiej owcy górskiej utrzymywanej w warunkach chowu ekologicznego (The influence of lactation phase on chemical composition and physico-chemical parameters of ewe's milk obtained from Polish Mountain Sheep reared in the ecological breeding way). Roczniki Naukowe Zootechniki 36(1):25–30

Krupiński J (Ed) (2012) Polskie rasy zachowawcze - Atlas zwierząt gospodarskich objętych programem ochrony w Polsce (Polish preserved breeds – Atlas of farm animals included into state program of preservation and protection in Poland). Instytut Zootechniki PIB. Kraków

Kuźnicka E, Zajączkowska K (2009) Tradycyjne wyroby regionalne z mleka owczego i koziego jako element dziedzictwa kulturowego wsi. Ochrona ich nazw, promocja produktów oraz wsparcie producentów (Traditional regional dairy products made of ewe's and goat's milk as a part of cultural heritage of villages. Protection of their names. promotion of products and support for producers). Przegląd Hodowlany 11:18–22

Litwińczuk Z, Barłowska J, Chabuz W, Brodziak A (2012) Nutritional value and technological suitability of milk from cows of three Polish breeds included in the genetic resources conservation program. Ann Animal Sci 12(3):423–432

Mroczkowski S (2011) Ginące owce (Dying - endangered sheep). Przegląd Hodowlany 1:1–3

Pandya AJ, Ghodke KM (2007) Goat and sheep milk products other than cheeses and yoghurt. Small Ruminant Res 68:193–206

Sikora J, Kawęcka A (2015) Aktualny stan krajowej hodowli i chowu kóz ze szczególnym uwzględnieniem województwa małopolskiego. Wiadomości Zootechniczne R.LIII. 4:76–82

Staliński Z (2005) Przyczynek do historii bydła rasy polskiej czerwonej (A contribution to the history of polish red cattle). Wiadomości Zootechniczne R.XLIII. 2:16–21

Stopyra R. Kowol P. Majewska A (2005) Perspektywy rozwoju hodowli bydła rasy polskiej czerwonej z uwzględnieniem krów objętych programem ochrony zasobów genetycznych. (Prospects for development of polish red cattle breeding with regard to cows included in the genetic resources conservation program). Wiadomości Zootechniczne R.XLIII. 2:137–143

Szarek J, Adamczyk K Zarys historyczny hodowli bydła polskiego czerwonego. (A historic outline of Polish Red Cattle breeding). Wiadomości Zootechniczne R. XLIII. 2:3–12

Szarek J, Adamczyk K, Felenczak A (2002) Polish Red Cattle breeding: past and present. Animal Genetic Resour Inform (AGRI) 35: 21–35

Wszołek M (2001) Przydatność technologiczna mleka koziego. Przegląd Mleczarski 3:12–14

## *Internet Sources*

http://ec.europa.eu/agriculture/quality/door/documentDisplay.html?chkDocument=579_1_pl. Accessed 17 July 2019

https://www.gov.pl/web/rolnictwo/lista-produktow-tradycyjnych12. Accessed 17 July 2019

http://www.pzow.pl/rasy-owiec.html. Accessed 17 July 2019

https://limanowa.in/aktualnosci/mleko-trafilo-na-polki-almy-w-calym-kraju/15789. Accessed 1 August 2019

https://www.gov.pl/web/rolnictwo/smietana-z-limanowej. Accessed 2 August 2019

https://www.gov.pl/web/rolnictwo/zsiadle-mleko-z-limanowej. Accessed 2 August 2019

# Chapter 13
# Farm Animals and Traditional Products of the Carpathian Mountains

**Wladysław Migdał, Maria Walczycka, Łukasz Migdał, and Sylwester Tabor**

**Abstract** The traditional native breeds of animals and seem to be rich sources of valuable, basic nutritional components. The ecological way of breeding supports land and landscape preservation. Understanding the interactions of humans and domesticated animals is vital for understanding cultural heritage and specific region developmental potential. The breeding of animals, grazing, and the traditional products of the Carpathian Mountains are the results of the so-called Wallachs colonization. The Wallach shepherds pastured sheep, goats, less often cattle customized for hard mountain conditions. The typical mountain sheep stayed almost unchanged in their genotype till today and the meat—"Jagnięcina Podhalańska"—obtained from Polish Mountain Sheep and Zakiel from October 11, 2012 has been registered as PGO (Protected Geographical Origin) by EU Regulations. The meat itself is rather fatty and contains less protein than similar "old genotype" sheep, but the animals still stay adapted to difficult climate conditions. Traditional rennet cheese—"Oszczypek" obtained from that sheep milk also is registered as a PGO (February 2, 2007) product and still is a popular staple food in the Polish part of the Carpathian Mountains. The Carpathian sheep are accompanied by Carpathian goats and Polish Red Cattle, whereas goat herds are rather small and disappearing in the Polish mountains; Red Cattle herds are renowned and appreciated for very tasty, fatty milk. Also, there is

W. Migdał (✉) · M. Walczycka
Department of Animal Product Processing, Faculty of Food Technology, University of Agriculture in Krakow, Balicka 122, 31-149 Kraków, Poland
e-mail: wladyslaw.migdal@urk.edu.pl

M. Walczycka
e-mail: maria.walczycka@urk.edu.pl

Ł. Migdał
Department of Animal Genetics, Breeding and Ethology, University of Agriculture in Krakow, Mickiewicza 24/28, 30-059 Kraków, Poland
e-mail: lukasz.migdal@urk.edu.pl

S. Tabor
Department of Production Engineering, Logistics and Applied Computer Science, University of Agriculture in Krakow, Balicka 116b, 30-149 Kraków, Poland
e-mail: sylwester.tabor@urk.edu.pl

© Springer Nature Switzerland AG 2022

J. Hernik et al. (eds.), *Cultural Heritage—Possibilities for Land-Centered Societal Development*, Environmental History 13,
https://doi.org/10.1007/978-3-030-58092-6_13

an attempt to crossbreed such cattle to better their muscle—meat potential. Some of the traditional products obtained from above-described animal breeds are finally processed with traditional smoking (especially cheeses), and there exist doubts about PAH (Polycyclic Aromatic Hydrocarbons) formation. In all examined product types, the smoked meat products did not meet the standards for PAH levels.

**Keywords** Carpathian sheep · Carpathian goat · Polish Red cattle · Heritage · Tradition · PAHs

## Abbreviations

| | |
|---|---|
| IZ PIB Balice | National Research Institute of Animal Production (Balice, Poland) |
| PAHs | Polycyclic aromatic hydrocarbons |
| PWB | Polish White Black (cattle) |
| PR | Polish Red (cattle) |
| PWR | Polish White Red (cattle) |
| WBw | White Back (cattle) |
| MLL | *Musculus longissimus lumborum* |
| MST | *Musculus semitendinosis* |
| PFHBiPM | Polska Federacja Hodowców Bydła i Producentów Mleka (Polish Federation of Cattle Breeders and Dairy Farmers) |

## 13.1 Introduction

Hunting, fishing, and farming were developed as early activities in the area of the Carpathian Mountains, for which the land was obtained by exploitation of forests with slash-and-burn agriculture. The breeding of animals, grazing, and the traditional products of the Carpathian Mountains are the results of the so-called Wallachs colonization. Starting from the twelfth century, Wallach shepherds (Wallachs were the ethnic group originating from the Balkans Peninsula, living to the south of the Danube River) started to move with their animal herds (mainly of sheep and goats; which needed pastures) to the north, along the Carpathian range reaching the Oravia and Moravia and into the "Polish" area, the contemporary Beskids. The Mongol invasion and then the expansion of Turks to the Balkans forced part of the Wallach clans to wander and settle at available mountain areas of the Carpathian Mountains. The Polish land was reached by Wallach at the beginning of the thirteenth century, when the large clans settled at Red Ruthenia, Podole, and Lubelszczyzna (land near contemporary Lublin), and accepted the Polish Nobles culture but continued themselves to be occupied with shepherding and cattle breeding (Czajkowski 2008). In 1373, Walachs reached the Pogórze Karpackie region (near the village of Dynów at the banks of the

San River—the settlement was called Wolosze), and in 1395 a Wallach clan reached Nowy Sącz. At the end of the fifteenth century, the wandering Wallach shepherds stopped in the Żywiecki region near Wadowice. In Silesian Beskid where, at that time, the agriculture was well developed and land densely populated, there was no place for pastures and the Wallach western movement stopped. The farthest at the west and the latest occupied by Wallach settlement was a region of the Carpathian Mountains called Valašsko (the region of Czech Republic called now Moravia). The Wallachs were a wandering shepherd tribe which, because of continuous movement, was dispersed and favored the formation of a cultural community with Russian, Polish, and Slovak highlanders groups leaving the mountain pasturing economic system called the "chalet/hut economy" based on summer mountain pasturage, with associated vocabulary, rites, and culture. The Wallachs worked at mainly transhumance pasturage based on the year round pasturing of a few hundred to a few thousand sized cattle herds of so called "Wallach's cattle"—sheep and goats (less frequently of real cattle) and on cyclical wandering between high mountain located pastures in the summer (alps and polonina) and valleys, where, thanks to the milder climate, they were able to feed their herds during winter. Thanks to these conditions, the higher pastures and poloninas were used with the support by supply networks at lowlands and mountain basins where the feed for animals was gathered for winter season. Also the Wallachs were engaged in trade of animals from Ukraine to western Poland, where cattle were sold at fairs. The King of Poland Casimir the Great when he attached Halicka Ruthenia to Poland in 1340, gave land and settlement privileges, in those territories, to Wallachs who were distinguished fighters in the Ruthenian wars. The first settlement privilege, with a related so-called "Wallachs law" was issued in 1377 at Hodle Pole. That was the beginning of villages located in accordance to the Wallachs law, and new settlements were created in the area of the Muszyna village in Sądecki Beskid (Szczawnik, Zubrzyk, Krynica, Izby) and villages at Rock Podhale and in Silesian Beskid where the first mountain shepherds villages, at Cieszyński Silesia, were Mosty Jabłonkowskie, established in 1590 (Popiołek 1939). The majority of shepherds' villages, in that area, were created later, during the years from 1621 to 1654, when 10,000 sheep and goats were pastured there and 858 sticks of cattle and 30 shepherds' huts were there (Popiołek 1939). The historical records of transhumance pasturage were written for Cieszynski Silesia at the beginning of the twentieth century, where the Gorols drove out their herds to so-called "pits". Transhumance pasturage include not only the movement of herds "vertically" in the scope of a certain mountain range but also seasonal movements for long distances between different geographic land areas. The Wallachs' settlements were based on grubbing of forest and on building of villages on so-called "raw roots" or involved relocation of existing older cottages onto Walachs' law rules. The settlement took place, most frequently, in sub-mountain areas, where the low-quality soils and hard climate conditions caused old cottages to succumb to economic disaster. A parallel development was the growth of sheep and goat pasturage in the area of the Carpathian primeval forest, "cerhlenie" (cutting out) of trees and forests and burning influenced on formation of mountains slopes' meadows and pastures.

The Wallach settlers mixed with the local population. Their contemporary ancestors are Hutsuls, Boykos, Babia Góra Lemkos, Czadec Gorals, Pieniny Mountains Gorals, Nowy Sącz Gorals, Silesian Gorals, Żywiec Gorals, Kliszacy, Oravians, Podhalanie (Highlanders), Spisz (Zips), and Zagórzanie (Szlak kultury wołoskiej 2018).

## 13.2 Carpathian Sheep

The Walach shepherds pastured sheep, goats, less often cattle customized for hard mountain conditions. In difficult mountain conditions, at high located hala (meadows), the specific breed of sheep adjusted to the mountain climate named "Zakiel" had flourished, which was connected closely to the life-style and culture of Carpathian Gorals. The primitive sheep "berka" with villus structure of wool cover, protecting the animal against unfavorable influence of precipitation, provided milk, wool and leather skin, was an ancestor for development of the noble breed of Polish Mountain Sheep. The Polish Mountain Sheep originated as a crossbreed of native ewes with rams of Transylvania cakiel (Zakiel) (Kawęcka 2007). Zakiels are the numerous group of primitive sheep of versatile utility dispersed in all southern Europe: e.g., in Greece, and in the Carpathian and Balkan Mountains regions. Anyone can list a few of local varieties of Zakiel: Walachs' sheep, Transylvania Sheep, Tatra (mountain)-Podhale sheep, Racka (crayfish) sheep, Curkana sheep. These sheep are accustomed to long walks and kosar living (living in specially built wooden fences in pasture), pasturing on steep and rocky slopes and hardly available meadows. The long neck and narrow muzzle allow them to eat scarce plants from difficult, inaccessible places. The primitive sheep, perfectly adjusted to hard conditions, resistant to illnesses with balanced demand for food and with flocked pelage, straight hair that protects it against unfavorable effects of precipitation, provided milk, meat, wool, and leather. There were two varieties of Podhalanski Zakiel—the western (white) popular at Podhale and in the western Carpathians and the eastern variety (dark brown or black) living from eastern Malopolska to Transylvania and Bessarabia (Kawęcka 2007). Before World War II, the dark Zakiels were kept mainly at Hutsuls land areas. The Zakiel sheep were also kept at Lower Silesia in Bystrzyca, Kłodzko, Dzierżoniów, Jelenia Góra, and Kamienna Góra counties. Professor Czaja distinguished two types of Polish Zakiel: Beskid Zakiel (larger with meat utility predominat) and Podhale-Tatra Zakiel (smaller with milk utility predominant) (Kawęcka 2007). At the turn of the nineteenth and twentieth centuries, there were a few attempts to better the utility of Zakiel sheep by bringing, to the Podhale region, different breeds of sheep. In 1835, the Merino Negretta was brought to Podhale. The attempt had not been successful because of extinction of the crossbreeds, which were not able to adapt to the hard climate conditions. At the end of the nineteenth century, there were brought Gypsy Sheep (Cajga) from Hungary and Black-Headed Sheep from England, which were crossbreeded. The crossbreeds suffered from liver fluke and sheepball. Good results were obtained when the native Zakiel was ennobled with Transylvania Zakiel

(1911–1913). There was failure, however, of the attempt (undertaken during 1927 to 1934) of ennobling of primitive Zakiel with Fresian Sheep and Pomeranian Fagas Sheep (Fagas Sheep—Żuławy Sheep belonging to Dutch colonists who settled at the fork delta of the Vistula and Nogat rivers), because of lowered immunity of Zakiel-Fresian crossbreed sheep (Kawęcka 2007). The crossbreeding with meat Wensleydale (large shape, long wool sheep) worsened wool utility, fertility, and immunity. In 1935, due to the initiative of Professor Różycki Transylvania, Zakiel (150 ewes and 70 rams) from the Region of Sibiu in Romania were imported. The sheep acclimatized well to the new conditions, the bettering of wool pelage occurred and slight worsening of milk yield (Kawęcka 2007). After World War II, there was introduced a new type of Polish Mountain Sheep for breeding (Polish Mountain Sheep) bred at the Experimental Plant in Grodziec Śląski, led by Professor Czaja, which was the crossbreed of Podhalanski Zakiel ewes with Transylvania Zakiel rams (imported from Romania) and Fresian Sheep rams (Kawęcka 2007). The Slovak counterparts of Polish Mountain Nowosadecki Sheep are the following breeds: Valaška, Zošl'achtená Valaška, and Cigája, whereas the Hungarian counterparts are: Racka sheep (Hortobágy Racka Sheep) with long spiral horns; and primitive Transylvania Zakiel.

The process of improvement of Podhalanski Zakiel and then of Polish Mountain Sheep caused the change of their genotype and led to the extinction of primitive Zakiel traits. As a result, Podhalanski Zakiel, for years, was not distinguished as a separate breed but was included with some ennobled sheep under the common name of Polish Mountain Sheep (Kawęcka 2007). The program of genetic resources protection as regards Podhalanski Zakiel started in 2007. The number of sheep in the program is 7874 in 119 bred herds. The breed is raised at Malopolska Voivodeship (7323 sheep in 116 herds) and at Podkarpackie voivodeship (551 sheep in 3 herds). The Zakiels are sheep with wool and milk direction of utility. They provide milk for the production of traditional products (bundz, bryndza, oscypek, redykolka, zentyca) and also wool, leather or skins, and tasty meat. The average weight of rams is 65 kg and of ewes is 45 kg. The fertility of the breed ranges between 125 and 130%, whereas the milk yield of one lactation is about 60 L (Rasy rodzime 2019).

The colorful mountain sheep is a native variety of an old, primitive breed group of Zakiel. In the area of Polish part of Carpathian Mountains, there was introduced the white variety of Zakiel, which wandered with the Wallach ethnic group moving along the mountains. After World War II, the name Polish Mountain Sheep was adopted for both varieties. The colored variety of Polish Mountain Sheep is covered with a mixed, two fractions pelage (wool), of dark brown color. Typical is the presence of a white dot (star) or bald spot at the head and a white end of tail. The rams have horns, and their weight can reach 50 kg, whereas ewes 40 kg. The fertility of the colored variety of Polish Mountain Sheep can reach 120%. The meat is characterized by a specific taste close to the taste of game meat. In the program of genetic resources protection of this variety, there are 1970 ewes kept in 34 herds in the areas of Silesia Voivodeship, Malopolska Voivodeship, Podkarpackie Voivodeship, ŚwiętokrzyskieVoivodeship and Podlaskie Voivodeship (Rasy rodzime 2019).

The meat—"Jagnięcina Podhalańska"—obtained from Polish Mountain Sheep and Zakiel from October 11, 2012 has been registered as PGO (Protected Geographical Origin) by EU Regulations. The EU Certificate can be granted only to lamb meat born and bred at Nowotarski and Tatrzański counties (the whole area) and at some municipalities in those counties: Cieszyński (Istebna municipality), Żywiecki (Milówka, Węgierska Górka, Rajcza, Ujsoły, Jeleśnia, Koszarawa municipalities), Suski (Zawoja, Bystra, Sidzina municipalities), Limanowski (Niedźwiedź, part of Kamienica—municipalities and the municipality of Mszana Dolna; and in these villages: Olszówka, Raba Niżna, Łostówka, Łętowe, and Lubomierz), and in the county of Nowosądecki (Piwniczna-Zdrój, Muszyna, Krynica-Zdrój municipalities). The "Jagnięcina Podhalańska" is a lamb meat of Polish Mountain Sheep and Zakiel breeds, of which the age does not exceed 60 days and the carcass weight ranges from 4 to 8 kg (Regulation of EU EC 929/2012).

Lamb meat from native breeds can be treated as a high-quality and nutritious food because of the high protein and mineral content with relatively low-fat content. The characteristic property of Zakiel Podhalanski lamb meat, as the most primitive of native breeds, is low content of total solids and proteins (Table 13.1) (Zając et al. 2019).

**Table 13.1** The chemical composition of lamb meat depending on the breed and type of feeding system (Zając et al. 2019)

| Breeds of lambs | Feeding systems | Dry mass [%] | Protein [%] | Fat [%] | Ash [%] |
|---|---|---|---|---|---|
| Uhruska sheep UHR | KW | 23.47 ± 0,15 | 20.66 ± 0.48 | 1.96 ± 0.2 | 1.10 ± 0.01 |
| | GM | 23.91 ± 0,07 | 20.66 ± 0.11 | 2.23 ± 0.18 | 1.11 ± 0.02 |
| Cakiel podhalański CKP | KW | 20.65 ± 0,33 | 16.36 ± 3.46 | 2.22 ± 0.25 | 1.10 ± 0.07 |
| | GM | 19.61 ± 015 | 15.98 ± 0.71 | 1.76 ± 0.24 | 1.16 ± 0.06 |
| Pomeranian sheep POM | KW | 24.37 ± 0.55 | 20.83 ± 2.66 | 1.82 ± 0.49 | 1.14 ± 0.08 |
| | GM | 23.39 ± 0.15 | 19,06 ± 0,12 | 1.27 ± 0.1 | 1.34 ± 0.09 |
| Świniarka SWIN | KW | 26.56 ± 0.12 | 20.48 ± 0.23 | 5.46 ± 2.38 | 1.07 ± 0.08 |
| | GM | 23.95 ± 0.05 | 19.27 ± 3.86 | 4.60 ± 0.13 | 1.05 ± 0.05 |
| Wrzosówka WRZOS | KW | 24.86 ± 1.18 | 21.78 ± 1.06 | 2.10 ± 0.41 | 1.20 ± 0.18 |
| | GM | 2239 ± 0.34 | 19.12 ± 1.75 | 1.74 ± 0.08 | 1.10 ± 0.03 |
| Black-headed sheep CZGL | KW | 24.69 ± 0.67 | 21.17 ± 1.98 | 2.22 ± 0.25 | 0.98 ± 0.09 |
| | GM | 24.08 ± 0.25 | 22.16 ± 1.75 | 1.76 ± 0.24 | 1.15 ± 004 |
| Old Polish Merino sheep MST | KW | 27.05 ± 2.69 | 20.84 ± 1.01 | 5.01 ± 0.23 | 1.23 ± 0.08 |
| | GM | 23,.90 ± 2,98 | 19.56 ± 4.50 | 4.62 ± 0.63 | 1.11 ± 0.01 |
| Wielkopolska sheep WLKP | KW | 25.23 ± 0,.9 | 20.25 ± 2.75 | 7.47 ± 0.60 | 1.09 ± 0.01 |
| | GM | 25.58 ± 0.19 | 18.98 ± 1.89 | 9.14 ± 0.38 | 1.05 ± 0.07 |

Mean values ± standard deviation values
GM—fattening methods used in maternity farms
KW—fattening methods used in IZ PIB ZD Kołuda Wielka

During pasturage of sheep on mountain meadows, contemporary shepherds produce, as the Walachs formerly did, rennet cheese—bundz—the clump of cheese from which, after proper processing and smoking, the whole array of Wallach cheeses is obtained—oscypek, oštiepok, redykołka. The bundz cheese is also a raw material for bryndza, korbacz, parenica, Klenovecký syrec, zázrivské vojky (cheese threads) i zázrivský korbáčiki (cheese braid) and ovčí hrudkový syr—salašnícky (Salasnicky sheep cheese) [www.bioroznorodnosc.izoo.krakow.pl/.../polska_czerwona].

Initially, sheep cheese production was a side activity, nevertheless very important, because cheeses were the staple base of nutrition for baca and juhas (shepherds) during the 5 months of sheep pasturage at the mountain slopes. Today, the production of these cheeses is the most important shepherds' activity because that is the base purpose of sheep pasturage. The rennet cheese bundz (bunc, gruda, udój) in the Slovak part of the Carpathian Mountains is named hruda and in Hungary gomolya. The components for its production beside ewes' milk are the rennet added to sweet and hot milk, directly after milking. The rennet causes curdling and settlement of cheese curd/gel. Formerly, the rennet was obtained from dried lamb's stomachs but nowadays a ready to use, commercial formula is employed. The treatment of ewe milk with rennet is named "klagowanie" (klogowanie) and is prepared in a big wooden pot—puciera. Then the cheese gel is mixed with a special, wooden stick, named a ferule (beteło, beater—trzepaczka, pocisk-missle). The next stage is "pucenie", the process of squeezing of coagulated cheese curd to expel "żentyca" (whey), and formation with soaking in hot water followed by soaking in salt water ("rosół"). After formation and soaking, the cheese is put into oscypiorki (wooden shaping devices) to obtain the proper shape and specific outer decoration, which patterns are typical and characteristic for the individual "baca" (shepherd). The last stage is a gentle smoking of cheeses placed on special wooden hangers, under the roof of the shepherd's hut, with cold smoke. The best cheeses are obtained in the second half of the summer, when the ewe's milk is fattier and contains more nutritive components. Bundz produced of only the ewe's milk can be very rarely encountered, because the ewe's milk typically is rather mixed with cow's or goat's milk. The favorable taste results from a 10% additive of cow's or goat's milk and EU regulations allow for 40% of cow's milk additive to regional "Oscypek" cheese but only cow's milk from Polish Red Cattle and for production of these cheeses in Malopolskie and Silesia Voivodeships. The "Oscypek" cheese beginning on February 2, 2007 is registered on the EU regional product list (PDO). The small, remaining pieces of ewe's cheese curd are processed into bruski or "Redykołki" (PDO). The "Redykołka" is mostly of small sheep, cock, deer, or duck shape and formerly was given by shepherds to their children after coming back from pasturage season in the mountains, as a symbolic toy. The Hutsuls handed redykołka to beggars to pay them for prayers for dead relatives and family members. The prevailing influence as to milk taste in "Oscypki" production and as a consequence as to cheese taste results from very differentiated flora — the specific botanic composition of mountain meadows and pastures where the sheep is grazed. The grazing season at hala (mountain meadow) starts on Saint Adalberts Day (23 of April) and ends on Saint Michaels Day (29 of September). The special ceremony for shepherds and sheep starting their departure for pasturage in mountain meadows is called redyk wiosenny (spring parade), and the return is called redyk jesienny (autumn parade). The redyk (parade) is a relic of Transhumance in Poland. It was perceived as the greatest festivity of a whole village, in the shepherds' culture, but today it is rather a tourist attraction. In 2013 there was organized the Carpathian Redyk when the route led through all Carpathian countries: Romania, Ukraine, Poland, Slovak Republic, Czech Republic. The herd then had about 300 sheep, shepherds' dogs, horses, and donkeys. The shepherds from Poland, Ukraine, and Romania wandered from 11 of May to 14 of September 2013 along a route of about 1200 km. The organizers of that event (in Poland—the Foundation of Transhumance Pasturing) announced that the Redyk was organized to celebrate shepherds' walks — wandering of the Wallach shepherds, which led to settlements in the Carpathian Mountains and to the growth of a common highlander shepherds' culture connecting the communities living in the whole Carpathian Mountains range [www.transhumance.pl].

## 13.3 Carpathian Goat

The Carpathian goat is the old Polish breed of goats. At the turn of the nineteenth and twentieth centuries that breed was frequently present in the mountain areas among others at Podkarpacie. Later it was removed from breeding because of the influx of breeds with higher milk yield. Nowadays, it is practically extincted. In 2005, there were found two small herds in Lower Silesia—in Bogatynia village, that is, herds of Carpathian goats. The animals were bought and placed in Odrzechowa Experimental Facility. A year later in the Wrocław zoo, there was found one goat ram and nine female goats, which were also brought to Odrzechowa. The animals are the core of the protection of genetic resources project for Carpathian goats. The Carpathian goat is characterized by high immunity and longevity. It is perfectly adjusted to hard mountain climate conditions, unrefined in feed choice. The milking yield during 8 months lactation is 500 kg of milk on average with 4% of fat, that type of goat's yield is 50–60% lower than the milking yield of so-called goats of noble breeds. Nowadays, the breed is being reproduced in Odrzechowa. There also exists the colored variety of Carpathian goat, which is commonly present in Romania, Ukraine, and the Slovak Republic.

## 13.4 Native Carpathian Cattle

Also in the Carpathian Mountains, the autochthonous cattle breeds are present, which include: Red Cattle, Carpathian Red Cattle, Grey-russet Cattle, and Steppe Grey Cattle. Polish Red Cattle originated from a small, wild, brachyceric (short-horn) cattle, living in Scandinavia and the middle regions of Europe. The spreading of the Red Cattle breed is connected to demographic movements at the beginning of the sixteenth century. That cattle is characterized by high immunity against illnesses, is long-living, and has a good fertility. The autochthonous cattle was formerly used for multiple purposes—for milk, meat production, and as a work animal. Nowadays, the Polish Red Cattle is used for milk obtaining (for production of mixed cheeses—golka, pucek; up to 40% additive in oscypek curd) and for meat production (Polish Red Cattle meat line). Population of native cattle breeds significantly decreases, therefore in 1999, Lesserpoland (Malopolska) Cattle Breeders' Association suggested to establish rules and financial support for the protection of genetics resources of Polish Red cattle. Since 2003 similar program was launched for Polish Whiteback cattle (Krupiński et al. 2012).

The scientific creator of Polish Red Cattle breed is thought to be Professor Leopold Adametz, whereas the first breeder who introduced that breed into breeding practice was Jan Brandys (Adamczyk et al. 2008). That breed included mainly not only red pigmented animals but also dark brown, pale, and black (Szarek and Adamczyk 2005). The name "das Polnische Rotvieh"—Polish Red Cattle was proposed by Franciszek Holdefleiss in 1897 (Staliński 2005). Polish breeders—Konopiński and Bormann, who founded, in 1894 the Society of Polish Red Cattle Breeders in western Galicia, used both the names "Polish Red Cattle" and "Red Polish Cattle". In 1876, the first barns in Limanowski County started and the place is regarded as the homeland for the Polish Red Cattle sub-mountain variety. In 1895, the Society of Red Cattle Breeders in Krakow's Agricultural Society started and it is assumed that from that moment, the methodical breeding work on that breed had started. The Society took steps toward the promotion of that valuable native breed through exhibitions, shows, auctions of cows, bulls, calves, and heifers. At that time cattle was used as a working agricultural resource and as a source of meat, and also milk from which butter was the main product. In 1910, the Society elaborated norms and rules for estimations and zoometric measurements and established breed templates for individual Red Cattle categories. After World War I, starting from 1921, the renewal of breeding work was continued and continues to be observed in Malopolska. In 1912–1913, the cows of Polish Red breed gave milk with 3.29–4.10% of fat, whereas in 1924, from 2.70 up to 5.00%. Before World War II, almost 8000 barns with 105 054 Polish Red cows were subjected to milking yield control, where the average milk yield was 2639 kg with 3.99% of fat. Polish Red Cattle was at that time the oldest pedigree breed of cattle in Europe, with a share of 25% of all cattle herds in Poland. There were villages where that cattle was the base of existence of an entire farmer family—Krzesławice, Zegartowice, Gruszów, Góra Jana, Komorniki, Jodłownik, Dąbie, Raciechowice, Tymbark, Łososina, Łapanów, Gdów (Szarek et al. 2002).

After World War II and even in the late 60 s, Polish Red Cattle, in the area of Krakow's Voivodeship, accounted for 60–70% of cattle stock. The cattle were kept in the middle and southern part of the voivodeship, mainly in Limanowski, Myślenicki, Nowosądecki, Nowotarski counties. The year 1894 is considered as the starting point for regular breeding work on Polish Red Cattle, when the Society of Breeders of Polish Red Cattle was established but the regular control of milking yield was reintroduced in 1921 by the Malopolska Agricultural Society. In order to perform some research on that breed in a mountain environment, the Experimental Facility in Raba Wyżna was found. The main aim was to better the breed to obtain a final milking yield of 4000 kg of milk with 4.2% of fat and average animal body weight 500 kg. The Polish Red cows should be first of all: healthy, properly built, long living providing properly fed with good feedobtained from mountain pastures and fields. The work started in 1951, on 65 cows with 2150 kg milking yield with 3.65% of fat. During 2 years, the average milking yield rose to 3011 kg with 3.80% of fat. At the beginning of the 60 s, the milking yield obtained was 4000 kg with 4.05% of fat. Dutch Red Cattle were chosen for crossbreeding to obtain the above-mentioned results (Żukowski and Trela 2005).

In 1983, there were established three protected herds of Polish Red Cattle in Hańczowa (near Uście Gorlickie), in the Institute of Genetics and Animal Breeding Polish Academy of Sciences in Jastrzębiec (barn Popielno) and in Ełk. These herds contained altogether 280 cows. Nowadays, the protected genetic reserve is the herd in the Experimental Facility of Ecological Agriculture and Animal Breeding in Popielno and about 500 cows selected from peasant farms and multiple herd barns located in the sub-Carpathian Mountain area—in Jodłownik and Szczyrzyc located in Limanowski County (Żukowski and Trela 2005). During 1977–1966, the genetic reserve of Polish Red Cattle was built with a method ex situ by freezing, in liquid nitrogen, of semen of about 70 bulls, and more than 1500 fetuses obtained from several pedigree cows which all were representative of the old variety of Polish Red Cattle. That stock, at present, is located in the Biological Materials Bank of Reproductive Physiology in Balice.

From 1994, Malopolska Society of Cattle Breeders was applying to Ministry of Agriculture and Rural Development to accept and introduce of the program for protection of genetic resources of Polish Red Cattle. The financial and law regulation were there introduced as late as 1999. At the beginning of program — 1999–2001, the project was led by State Centre of Animal Breeding in Warsaw, starting from 2002, the program is led by Institute of Zootechnics in Krakow. There are two projects existing in the program: the project of genetic improvement of breed (basic) and the project of genetic resources protection. Today, the headstock of Polish Red Cattle is estimated for about 30,000, where, in 2005, only 500 were subjected to the project of genetic resources protection. As the main task of above project is to renovating and keeping up Polish Red Cattle with existing genetic diversity protection. The breeding work is mainly focused on keeping up of typical characteristics of these cattle such as ability of perfect accommodation to hard environmental conditions, high immunity and healthy, very good fertility, easy birth giving, high vitality and easy breeding of calves, and the high biological value of milk (Stopyra et al. 2005).

In Poland, there are existing four programs of genetic resource protection, which concern four Polish, native breeds of cattle: Polish Red (PR, from 1999), White Back (WB, from 2003), Polish Red-White (PRW, from 2007), and Polish Black White (PBW, from 2008). The projects of protection of genetic resources concern altogether (data on March 15, 2018) 9000 cows kept on about 800 farms. The coordination of projects is based on work by the Institute of Zootechnics—State Research Institute with the help of parties that manage breeders' books for individual breeds (Polish

Federation of Cattle Breeders and Milk Producers that protects PR, PRW, and PBW), whereas the University of Sciences in Lublin takes care of the WB breed.

In 2009, the program of genetic resources protection concerned 1760 Polish Red breed cows. Before, between the periods of the two world wars, Polish Red Cattle were bred as purebred cattle, whereas nowadays, animals subjected to milking yield control are crossbreds. They have some shares of the genes of Angler, Jersey, Belgium Red, Dutch Red, and also Simmental (black–white and red–white) and even Aberdeen Angus and Charolaise. To return to the pure breed traits, the cows of contemporary living Polish Red Cattle were crossed with bulls of the old Polish Red Cattle breed type. The decision caused a rapid diminishing in PR stock number because the result of the above-mentioned "back crossbreeding" caused a drastic lowering in milking yield. The low milking yield causes farmers not to want to breed such cows even if they are offered some extra money from the genetic resources protection project, since the losses are not covered (Adamczyk and Szarek 2009).

Nowadays, the project of genetic resources protection accepts cows with 131–136 cm at the back, with well-outlined muscle profiles, delicate, soft head, well-pronounced chest with flat ribs, strong legs—properly built and with strong bones and well-expressed joints. The udder should be capacious, with back high and wide suspension and with front suspension moved forward and connected well to abdominal coatings. The pigmentation should be uniform from light red—cherry to dark red, small light stars on the udder or stomach acceptable, dark hooves and nostrils, light muzzle acceptable, pale horns with dark endings. The average yield of all Polish Red cows (2773 in 404 herds), in 2017, was 3656 kg of milk with 4.27% of fat and 3.37% of protein (PFHBiPM 2018). As of March 15, 2018, the qualification for the program resulted in 2419 milking cows (259 herds) and 577 of meat utility type (40 herds). The Polish Red Cattle as the local, native breed adjusted to the hard mountain environment promotes on one hand preservation of biodiversity—protected by the genetic resources protection project, and on the other hand allows for obtaining of outstanding quality of milk and beef, because these animals are fed with household produced roughage (Adamczyk and Szarek 2009). The results of assessments showed that milk obtained from naturally grazed cows is characterized by better parameters for cheeses production and is rich in components favorable to human health especially in whey proteins, polyunsaturated fatty acids, including CLA, and fat-soluble vitamins (Barłowska et al. 2012). Milk obtained from Polish Red Cattle was included in the List of Traditional Products, on December 14, 2012, as "milk from Polish Red Cow".

Recently also the second goal for Polish Red Cattle usage is being promoted—to obtain valuable beef meat. On May 10, 2016, the Polish Society of Breeders and Producers of meat type of Polish Red Cattle was registered. Breeding work was started with the aim of obtaining beef meat with the possible best characteristics. The comparison of chemical composition of samples from *musculus longissimus lumborum* and *musculus semitendinosus* collected from young bulls from the native Polish Red breed and intensively feed crossbreeds are presented in Table 13.2.

The level of protein in muscle is between 19% and 23%. Crossbreed young bulls from intensively feed were characterized by lean meat—a statistically lower content

**Table 13.2** Chemical composition of *lumborum* part of muscle *m. longissimus dorsi lumborum* (MLL) and semitendinosus muscle *m. semitendinosus* (MST) bulls (Migdal et al. 2019)

| Chemical component [%] | *lumborum* part of muscle *m. longissimus dorsi lumborum* (MLL) | | | *semitendinosus* muscle *m. semitendinosus* (MST) | | |
|---|---|---|---|---|---|---|
| | Breed of cattle | | | Breed of cattle | | |
| | Polish Red (RP) | White-backed (BG) | Cross-breeds PHFxCh | Polish Red (RP) | White-backed (BG) | Cross-breeds PHFxCh |
| Water | 74.51 ± 1.25 | 75.39 ± 1.52 | 76.11 ± 1.99 | 74.33 ± 2.05 | 74.58 ± 1.01 | 76.07 ± 1.50 |
| Total solids | 25.49 ± 1.28 | 24.61 ± 1.56 | 23.89 ± 2.01 | 25.67 ± 2.08 | 25.42 ± 0.98 | 23.93 ± 1.43 |
| Protein | 21.97 ± 2.85 | 21.27 ± 3.07 | 21.05 ± 2.12 | 22.11 ± 1.98 | 21.95 ± 2.12 | 21.33 ± 1.96 |
| Fat | $2.30^{a} \pm 0.31$ | $2.08^{b} \pm 0.24$ | $1.57^{c} \pm 0.19$ | $2.31^{r} \pm 0.18$ | $2.16^{r} \pm 0.25$ | $1.28^{s} \pm 0.11$ |
| Ash | 1.11 ± 0.05 | 1.09 ± 0.08 | 1.15 ± 0.04 | 1.15 ± 0.1 | 1.13 ± 0.07 | 1.19 ± 0.02 |
| Carbohydrates | 0.11 ± 0.01 | 0.17 ± 0.02 | 0.12 ± 0.01 | 0.10 ± 0.018 | 0.18 ± 0.01 | 0.13 ± 0.009 |
| Thermal losses % | 41.93 ± 3.87 | 39.76 ± 3.86 | 39.62 ± 2.68 | 36.66 ± 4.02 | 39.98 ± 3.66 | 40.16 ± 3.12 |
| Color<br>L*<br>a*<br>b* | $37.94^{a} \pm 3.91$<br>$11.15^{a} \pm 1.57$<br>$10.43^{a} \pm 1.33$ | $45.74^{b} \pm 3.03$<br>$8.03^{b} \pm 1.10$<br>$11.07^{a} \pm 1.21$ | $34.58^{a} \pm 2.76$<br>$16.49^{c} \pm 1.96$<br>$7.42^{b} \pm 1.01$ | $44.28^{r} \pm 3.24$<br>$7.41^{r} \pm 1.36$<br>$13.51^{r} \pm 1.13$ | $49.10^{r} \pm 2.79$<br>$8.88^{r} \pm 1.02$<br>$14.95^{r} \pm 1.14$ | $36.22^{s} \pm 2.74$<br>$18.89^{s} \pm 1.01$<br>$8,91^{s} \pm 0.89$ |

[a, b, c, r, s]Mean values in lines marked with different letters differ statistically significantly at $P \leq 0.05$
PHFxCh crossbreed of Polish Holstein-Friesian and Charolaise

of intramuscular fat compared with native polish cattle. Muscles from Polish Red and Polish Whiteback were characterized by higher fat levels and favorable fatty acids profile in comparison to crossbreds (PHFxCh). However, also higher thermal losses were reported. Meat from Polish native breeds seems to be good quality material for the production of traditional meat products. There even previously was Serwolatka sausage (fr. ***cervelas***) a smoked, beef sausage, popular at the Polish market in the 70's of the twentieth century due to low prices for beef meat at that time.

## 13.5 Traditional Products Obtained From Milk and Meat of Native Animals' Breeds

Products obtained in a traditional way from native raw materials such as milk and meat of Polish native animal breeds were assessed and classified in many different ways up to today. Still, one aspect of that assessment seems to be the key one for the safety of consumption of traditional and regional products. According to recent EU Regulations and demands, several traditional meats and dairy products were analyzed for possible content of PAHs (EC Regulation no 835/2011; UE Regulation no 1327/2014)

The outstanding taste of smoked ewe's milk cheeses and all handmade cheeses obtained from native animals' milk as raw material despite their undoubted quality and nutritional value can also cause some health threats because of the possible presence of polycyclic aromatic hydrocarbons (PAHs) in the product. Among several hundred smoke aroma compounds, some PAHs are also present. In Table 13.3, there are presented selected results for PAHs levels in traditionally made cheeses.

There were also analyzed some smoked meat products of lamb meat (Table 13.4) and Polish Red Cattle beef (Table 13.5).

As can be observed, some of the analyzed products were a threat to human health due to high levels of PAHs present. The traditional smoking process of small scale to obtain traditional and regional products should be modified in a way to omit too high temperature of the process to avoid the formation of PAHs.

## 13.6 Conclusion

The traditional native breeds of animals and products obtained seem to be richer sources of valuable, basic nutritional components. The ecological way of breeding supports land and landscape preservation. Understanding the interactions of humans and domesticated animals is vital for understanding cultural heritage and certain, specific region developmental potential.

Also, there is a strong need to sustain and protect the values and substance, which we inherited for further development and growth. Somebody has said that the nation

**Table 13.3** The level of polycyclic aromatic hydrocarbons in traditionally smoked cheeses (μg/kg)

| Polycyclic aromatic hydrocarbons | Ewe's milk cheeses | | | Goat's milk cheeses | | Cow's milk cheeses | |
|---|---|---|---|---|---|---|---|
| | Oscypek 1 | Oscypek 2 | Small cheeses | Walachs' cheese | | Cheese 1 | Cheese 2 |
| Naphtalene | 54.0 ± 16.2 | 70.0 ± 21.0 | 1200 ± 360 | 970 ± 291 | 210 ± 63.0 | 50.0 ± 15.0 | 120 ± 36.0 |
| Acenaphtylene | 130 ± 39.0 | 96.0 ± 28.8 | 1200 ± 360 | 840 ± 252 | 200 ± 60.0 | 64.0 ± 19.2 | 180 ± 54.0 |
| Acenaphthene | 3.70 ± 1.11 | 4.80 ± 1.44 | 38.0 ± 11.4 | 8.0 ± 14.4 | 9.30 ± 2.79 | 3.60 ± 1.08 | 8.00 ± 2.40 |
| Fluorene | 50.0 ± 15.0 | 31.0 ± 9.3 | 400 ± 120 | 140 ± 42.0 | 100 ± 30.0 | 37.0 ± 11.1 | 66.0 ± 19.8 |
| Phenanthrene | 170 ± 51.0 | 60.0 ± 18.0 | 790 ± 237 | 330 ± 99.0 | 120 ± 36.0 | 100 ± 30.0 | 120 ± 36.0 |
| Anthracene | 48.0 ± 14.4 | 14.0 ± 4.20 | 190 ± 57.0 | 87.0 ± 26. | 32.0 ± 9.60 | 25.0 ± 7.50 | 37.0 ± 11.1 |
| Fluoranthene | 19.0 ± 5.7 | 6.50 ± 1.95 | 94.0 ± 28.2 | 20.0 ± 6.0 | 7.30 ± 2.19 | 26.0 ± 7.80 | 19.0 ± 5.70 |
| Pyrene | 14.0 ± 4.20 | 4.40 ± 1.32 | 67.0 ± 20.1 | 6.0 ± 4.80 | 6.00 ± 1.80 | 21.0 ± 6.30 | 15.0 ± 4.50 |
| Benzo(a)anthracene | <0.85 | <0.61 | 8.60 ± 2.58 | 0.70 ± 0.51 | <0.80 | 2.10 ± 0.63 | 2.30 ± 0.69 |
| Chrysene | <0.82 | <0.61 | 7.30 ± 2.19 | 1.60 ± 0.48 | <0.80 | 2.00 ± 0.60 | 1.90 ± 0.57 |
| Benzo(b)fluoranthene | <0.82 | <0.61 | 4.50 ± 1.35 | 0.97 ± 0.291 | <0.80 | 0.91 ± 0.27 | 1.20 ± 0.36 |
| Benzo(k)fluoranthene | <0.82 | <0.54 | 2.10 ± 0.63 | <0.59 | <0.74 | < 0.56 | < 0.80 |
| Benzo(a)pyrene | <0.82 | <0.61 | 4.50 ± 1.35 | 0.85 ± 0.255 | <0.80 | < 0.85 | 1.20 ± 0.36 |
| Indeno(1,2,3-cd)pyrene | <0.82 | <0.61 | 2.30 ± 0.69 | 0.78 ± 0.234 | <0.80 | < 0.56 | < 0.80 |
| Dibenzo(a,h)anthracene | <0.39 | <0.067 | <0.61 | <0.63 | <0.14 | < 0.29 | < 0.41 |
| Benzo(g,h,i)perylene | <0.82 | <0.61 | 1.90 ± 0.57 | 0.81 ± 0.243 | <0.80 | <0.56 | < 0.80 |
| Sum of: benzo(a)pirene, benzo(a)antracene, benzo(b)fluorantene and chryzene | <3.31 | <2.44 | 24.90 ± 2.60 | 5.12 ± 0.72 | <3.20 | <5.86 | 6.60 ± 0.72 |

**Table 13.4** Chemical composition of sheep meat products traditionally smoked

| Characteristics | Produkt | | | |
|---|---|---|---|---|
| | Lamb sausage ("Wrzosówka" sheep meat) | Lamb ham ("Wrzosówka" sheep meat) | Lamb loin ("Wrzosówka" sheep meat) | Lamb sausage |
| Energy value Directive EC 1169/2011 | 1176.1 kJ 282.9 kcal | 626.37 kJ 148.79 kcal | 697.0 kJ 165.40 kcal | 979.77 kJ 234.89 kcal |
| Total fat PN-ISO 1444:2000 | 20.98 g | 4.79 g | 4.76 g | 14.81 g |
| In: Saturated fatty acids (gas chromatography) | 10.75 g | 2.46 g | 2.44 g | 8.46 g |
| Carbohydrates PN-A-82100:1985 | 0.74 g | 0.80 g | 0.10 g | 0.23 g |
| Total protein PN-A-04018:1975/Az:2002 | 22.78 g | 25.62 g | 30.54 g | 25.17 g |
| Salt (as chlorides) PN-EN-15505:2008 | 1.8 g | 1.7 g | 2.5 g | 2.1 g |
| Water content PN-ISO 1442:2000 | 52.98 g | 66.12 g | 5995 g | 56.19 g |
| Ash content PN-ISO 936:2000 | 2.52 g | 2.67 g | 4.65 g | 3.59 g |
| *Polycyclic aromatic hydrocarbons—content µg/kg* | | | | |
| Benzo(a)pyrene | 7.20 ± 2.16 | 18.00 ± 5.40 | 15.60 ± 3.30 | 5.10 ± 1.53 |
| Chryzene | 12.0 ± 3.60 | 23.00 ± 6.90 | 20.00 ± 4.90 | 11.0 ± 3.30 |
| Benzo(a)antracene | 19.0 ± 5.70 | 42.00 ± 12.60 | 36.50 ± 10.10 | 11.0 ± 3.30 |
| Benzo(b)fluorantene | 7.00 ± 2.10 | 16.00 ± 4.80 | 10.00 ± 4.10 | 7.00 ± 2.10 |
| Sum of four basic PAHs | 45.20 ± 5.30 | 99.00 ± 12.30 | 82.10 ± 9.80 | 34.10 ± 3.4 |
| Comments | Product **DOES NOT** fulfill demands of Regulation EC 835/2011 and regulation of EU 1327/2014 | Product **DOES NOT** fulfill demands of Regulation EC 835/2011 and regulation of EU 1327/2014 | Product **DOES NOT** fulfill demands of Regulation EC 835/2011 and regulation of EU 1327/2014 | Product **DOES NOT** fulfill demands of Regulation EC 835/2011 and regulation of EU 1327/2014 |

without history does not exist, and this not only concerns the great movements or conquests but also the details which build a very substantial and basic background of everyday life. Such a base is the fruitful promise of development and enables establishing of each individual human life conditions and prosperity.

**Table 13.5** Chemical composition cattle meat products traditionally smoked

| Characteristics | Product | | | |
|---|---|---|---|---|
| | Beef ham Gawor (Polish Red) | Beef ham MEDES | Beef kabanos Gawor | Serwolatka sausage |
| Energy value<br>Directive EC 1169/2011 | 510.16 kJ<br>120.37 kcal | 694.28 kJ<br>165.16 kcal | 1395.57 kJ<br>333.89 kcal | |
| Total fat<br>PN-ISO 1444:2000 | 1.13 g | 6.12 g | 18.7 g | 9.26 |
| In: Saturated fatty acids (gas chromatography) | 0.69 g | 3.72 g | 11.48 g | |
| Carbohydrates<br>In: Sugars<br>PN-A-82100:1985 | 0.01 g | 0.10 g | 1.51 g | 0.89 |
| Total protein<br>PN-A-04018:1975/Az:2002 | 27.54 g | 27.42 g | 39.73 g | 19.51 |
| Salt (as chlorides)<br>PN-EN-15505:2008 | 1.3 g | 2.1 g | 2.1 g | |
| Water content<br>PN-ISO 1442:2000 | 68.60 g | 62.10 g | 36.18 g | 66.94 g |
| Ash content<br>PN-ISO 936:2000 | 2.70 g | 4.26 g | 3.81 g | 3.40 |
| *Polycyclic aromatic hydrocarbons—content µg/kg* | | | | |
| Benzo(a)pyrene | 1.80 ± 0.54 | 0.89 ± 0.27 | 8.90 ± 2.67 | 1.3 ± 0.15 |
| Chryzene | 1.70 ± 0.51 | 2.10 ± 0.63 | 21.0 ± 6.30 | 5.2 ± 2.3 |
| Benzo(a)antracene | 5.30 ± 1.59 | 2.60 ± 0.78 | 26.0 ± 7.80 | 2.9 ± 0.3 |
| Benzo(b)fluorantene | 2.60 ± 0.78 | 0.910 ± 0.273 | 9.50 ± 2.85 | < 0.6 |
| Sum of four basic PAHs | 11.4 ± 1.64 | 6.50 ± 0.78 | 65.40 ± 8.14 | 10.0 ± 1.9 |
| Comments | Product fulfill demands of Regulation EC 835/2011 | Product fulfill demands of Regulation EC 835/2011 | Product **DOES NOT** fulfill demands of Regulation EC 835/2011 and regulation of EU 1327/2014 | Product fulfill demands of Regulation EC 835/2011 |

**Acknowledgements** Project "Cultural Heritage of Small Homelands" (CHSH) under the International Partnership Academic Program by the National Agency of Academic Exchange NAWA/CHSH 6918 Project *"The uses and the conservation of farm animal genetic resources under sustainable development"* co-financed by the National Centre for Research and Development within the framework of the strategic R&D program "Environment, agriculture and forestry" – BIOSTRATEG, contract number: BIOSTRATEG2/297267/14/NCBR/2016.

## References

Adamczyk K, Szarek J (2009) Bydło polskie czerwone – nauka na przyszłość. (Polish Red Cattle – "lesson" for the future), Przegląd Hodowlany 8:9–12

Adametz L (1901). Studien tiber das Polnische Rotvieh. Oesterreichischen Molkerei Zeitung, 133 Wien

Adametz L (1936). Właściwe kierunki hodowli zarodowej bydła czerwonego-polskiego na tle ogólnych stosunków ekonomicznych w Europie. (Proper directions of breeding Red-Polish cattle against the background of general economic relations in Europe) Rocznik Nauk Rolniczych i Leśnych. Tom XXXVII, Poznań

Adametz L (1925) Hodowla ogólna zwierząt domowych (General breeding of domesticated animals), Kraków

Barłowska J (2011) Znaczenie lokalnych ras zwierząt w produkcji żywności tradycyjnej oraz przekazie tradycji i kultury regionu. (The significance of native animal breeds in the production of traditional food and the transmission of tradition and culture of the region). Przegląd Hodowlany, 9:1–5

Barłowska J, Chabuz W, Król J, Szwajkowska M, Litwińczuk Z (2012) Wartość odżywcza i przydatność technologiczna mleka produkowanego w systemie intensywnym i tradycyjnym w trzech rejonach wschodniej Polski. (Nutritional value and technological suitability of milk produced in intensive and traditional systems in 3 regions of Eastern Poland). Żywność. Nauka. Technologia. Jakość 4(83):122–135

Bonczar G, Ciuryk S, Frajdenberg I, Pastuszka E (1998) Ocena przydatności różnych ras owiec do produkcji bundzu. (The assessment of different ewe breeds for bundz production), Zeszyty Naukowe AR Kraków, Technologia Żywności 347(10):5–14

Choroszy Z, Choroszy B (2005) Możliwości wykorzystania bydła rasy polskiej czerwonej do produkcji mięsa wołowego. (The possibility of using Polish Red Cattle for culinary beef production). Wiadomości Zootechniczne XLIII(2):73–78

Choroszy Z (1987) Przydatność buhajków mieszańców pochodzących z krzyżowania krów rasy pc z buhajami rasy Charolaise, Piemontese, Limousine, Blonde d'Aqitiaine i simentalskiej w opasie intensywnym. (Usefulness of hybrid bulls derived from the crossing of polish red cows with Charolaise, Piemontese, Limousine, Blonde d'Aqitiaine and Simmental bulls in intensive fattening). Pr. doktorska, maszynopis, IZ Balice (manuscript)

Commission Regulation (EU) No 835/2011 of 19 August 2011 amending Regulation (EC) No 1881/2006 as regards maximum levels for polycyclic aromatic hydrocarbons in foodstuffs

Commission Regulation (EU) No 1327/2014 of 12 December 2014 amending Regulation (EC) No 1881/2006 as regards maximum levels of polycyclic aromatic hydrocarbons (PAHs) in traditionally smoked meat and meat products and traditionally smoked fish and fishery products

Czaja H (1991) Polskie bydło czerwone – wielowiekowa historia bez "happy end'u" (Polish red cattle - a long history without a "happy end") IZ PIB Balice

Czaja M (1952) Polska owca górska. (Polish Mountain Sheep), Roczniki Nauk Rolniczych, 63

Czajkowski J (1999) Studia nad Łemkowszczyzną.(Studies on the Lemkowyna region). Sanok, 225

Czajkowski J (2008) Czy Wołosi mogli mieć wpływ na rozwój budownictwa w Karpatach? [w:] Richter L. (red.) Wołoskie dziedzictwo Karpat. (Could Wallach have had an influence on the development of construction in the Carpathian Mountains?) [In:] Wallachian heritage of the Carpathian Mountains, 69–124, Czeski Cieszyn

Dobrowolski K (1961) Studia nad kultura pasterską Karpat Północnych [w:] *Wierchy*, (Studies in the pastoral culture of the Northern Carpathian Mountains part), 29, 33

Dobrowolski K (1962) Migracje wołoskie na ziemiach dawnego państwa polskiego. *[w:]* Pasterstwo Tatr Polskich i Podhala. *(*Wallach migrations on the territory of former Poland. [In:] Shepherding of the Polish Tatra and Podhale), 4, 89–121, Wrocław

Domaradzki P, Florek M, Staszowska A, Litwińczuk Z, (2016b) Evaluation of the mineral concentration in beef from polish native cattle. Biol Trace Elem Res 169:328–332

Drozdowski A (1947). Znaczenie cakla siedmiogrodzkiego dla terenów górskich. (The importance of the Transylvanian Zackiel for mountain areas), Przegląd Hodowlany, 1

Drożdż A (2007). Żentyca – karpacka odmiana włoskiej ricotty. (Żentyca - a Carpathian variety of Italian ricotta), Przegląd Hodowlany 9:30–32

Drożdż A (2006) Węgierskie rasy zachowawcze zwierząt gospodarskich w Hortobagy (Hungarian conservative breeds of farm animals in Hortobagy), Wiadomości Zootechniczne 44(4):61–65

Florek M, Litwińczuk Z, Domaradzki P, Chabuz W, Żółkiewski P, Jankowski P (2017). Produkty regionalne z wołowiny pochodzącej z bydła ras rodzimych. (Regional products from beef of native cattle breeds), Wiadomości Zootechniczne, LV 5:123–133

Kawęcka A (2007) Program konserwacji zasobów genetycznych cakla podhalańskiego (Podhale Zackel genetic resources preservation program), Wiadomości Zootechniczne, R. XLV 4:23–26

Kopczyńska-Jaworska B (1951) Gospodarka pasterska w Beskidzie Śląskim 1950/1951. *[w:]* Prace i Materiały Etnograficzne. (Shepherd economy in the Silesian Beskids), 8–9, 155–322, Łódź-Lublin

Kopczyńska-Jaworska B (1962) Szałaśnictwo w Karpatach Polskich w świetle prac zespołowych w r. 1960.(Szałaśnictwo in the Polish Carpathian Mountains in aspect of team work in 1960). [w:] *Etnografia polska*, 6, 321–329. Instytut Archeologii i Etnologii PAN; Instytut Archeologii i Etnologii PAN, Instytut Historii Kultury Materialnej PAN

Kopczyńska-Jaworska B (1961) Owcze sery zdobione z Karpat. (Sheep decorated cheeses from the Carpathian Mountains), Etnografia Polska 5:197–226

Krupiński J (Ed) (2012) Polskie rasy zachowawcze - Atlas zwierząt gospodarskich objętych programem ochrony w Polsce. (Polish conservation breeds - Atlas of farm animals covered by the protection program in Poland) IZ PIB, Balice

Krupiński J, Martyniuk E, Krawczyk J, Baran L, Bielański P, Bobak L, Calik J, Chełmińska A, Kawęcka A, Kowalska D, Majewska A, Obrzut J, Pasternak M, Piórkowska M, Polak G, Puchała M, Sikora J, Sosin-Bzducha E, Szyndler-Nędza M, Tomczyk-Wrona I (2017) 15th anniversary of coordination of animal genetic resources conservation programmes at the National Research Institute of Animal Production . Przegląd Hodowlany 4:30–36 (In Polish)

Litwińczuk Z (2017) Domestication and breeding of livestock as a crucial element of the development of civilization. Przegląd Hodowlany 2:30–32 (In Polish)

Litwińczuk Z, Chabuz W, Domaradzki P, Jankowski P (2012) Slaughter value of young Polish Blackand-White, White-Backed, Polish Holstein-Friesian-Friesian and Limousin bulls under semi-intensive fattening. Ann Animal Sci 12(2):159–168

Litwińczuk Z, Domaradzki P, Florek M, Żółkiewski P (2016) Chemical composition, fatty acid profile, including health indices of intramuscular fat, and technological suitability of the meat of young bulls of three breeds included in a genetic resources conservation programme, fattened in a low-input system. Annals Animal Sci Papers Reports 34:387–397

Litwińczuk Z, Żólkiewski P, Florek M, Chabuz W, Domaradzki P (2014) Semi-intensive fattening suitability and slaughter value of young bulls of 3 Polish native breeds in comparison with Polish Holstein-Friesian-Friesian and Simmental. Annals Animal Sci 2:453–460

Majewska A (2019) Krótka historia dziejów programów ochrony genetycznej dla bydła i ras objętych tymi programami oraz stan obecny tych chronionych populacji. A short history of conservation programs for cattle and the breeds under these programs, and the current status of these populations. Wiadomości Zootechniczne, R. LVII 1: 84–101

Olszański TA (2000) [W: Płaj] Wołosi zapomniany lud Bałkanów. (Wallach, the forgotten Balkan tribe) 21:73–90
Ocena i hodowla bydła mlecznego. Dane za rok 2018. Polska Federacja Hodowców Bydła i Producentów Mleka, 2018, www.pfhb.pl
PN-75/A-04018:1975/Az3:2002, Agricultural food products. Nitrogen contents determination with Kjeldahl's method and recalculation into protein. (In Polish)
PN-ISO 936:2000. Meat and meat products. Determination of total ash content. (In Polish)
PN-ISO 1442:2000, Meat and meat products. Water contents determination. (In Polish)
PN-ISO 1444:2000, Meat and meat products. Free fat contents determination. (In Polish)
Popiołek F (1939) Historia osadnictwa w Beskidzie Śląskim. Wydawnictwa Instytutu Śląskiego, Katowice
Pruski W (1975) Hodowla zwierząt gospodarskich w Galicji w latach 1772–1918 (Livestock breeding in Galicia in 1772-1918),Warszawa
Rasy rodzime i ich wybrane produkty. Praca zbiorowa (2019) (Native breeds and their selected products), Stowarzyszenie na Rzecz Rozwoju i Promocji Podkarpacia PRO CARPATHIA, ISBN 978–83-65752-14-7
Reinfuss R (1959) Problem Karpat w badaniach kultury ludowej, *[w:]* Polska Sztuka Ludowa, (The problem of Carpathian mountains in research of popular culture. [In]: Polish Popular Art) Warszawa, 13, 1–2, 3–11
Reklewski Z (2005) Hodowla zachowawcza bydła rasy polskiej czerwonej. (Conservation breeding of Polish Red Cattle), Wiadomości Zootechniczne, R. XLIII 2:98–101
Rostafiński J (1921) Owce. Pochodzenie i rasy. (Sheep. Origin and breeds), Warszawa
Regulation UE EC 929/2012, issued at 8th October 2012 registering PDO (protected geographical origin) – Jagniecina Podhalanska, Commission Implementing Regulation (EU) No 929/2012 of 8 October 2012 entering a name in the register of protected designations of origin and protected geographical indications (Jagnięcina podhalańska (PGI))
Sawicki L (1919) [W:] Materiały Antropologiczno-Archeologiczne i Etnograficzne: Szałaśnictwo w Górach Żywieckich (Szałaśnictwo in the Żywiec Mountains), Kraków 14:137–183
Staliński Z (2005) Przyczynek do historii bydła rasy polskiej czerwonej (A contribution to the history of Polish Red cattle). Wiadomości Zootechniczne R .XLIII 2:16–21
Stopyra R, Kowol P, Majewska A (2005) Perspektywy rozwoju hodowli bydła rasy polskiej czerwonej z uwzględnieniem krów objętych programem ochrony zasobów genetycznych. (Prospects for development of Polish Red cattle breeding with regard to cows included in the genetic resources conservation program). Wiadomości Zootechniczne R .XLIII 2:137–143
Szlak kultury wołoskiej, 2018. https://szlakwoloski.eu
Szarek J, Adamczyk K (2005) Zarys historyczny hodowli bydła polskiego czerwonego. (A historic outline of Polish Red Cattle breeding). Wiadomości Zootechniczne R . XLIII 2:3–12
Szarek J, Adamczyk K, Felenczak A (2002) Przeszłość i teraźniejszość hodowli bydła polskiego czerwonego. (Polish Red Cattle breeding: past and present). Animal Genetic Resour Inform (AGRI) 35:21–35
Zając M, Tkaczewska J, Kulawik P, Guzik P, Borys B, Migdał W (2019) Comparing the chemical composition of the lamb meat of various native breeds. Proceedings 12th international symposium modern trends in livestock production 9–11 October 2019, Belgrade, Serbia, 610–617
Zdebska B (2005) Historia oceny użytkowości mlecznej bydła polskiego czerwonego w Małopolsce. (History of evaluating milk performance of Polish Red cattle in the Malopolska region). Wiadomości Zootechniczne, R .XLIII 2:118–125
Żukowski K, Trela J (2005) Zmiany w populacji bydła rasy polskiej czerwonej na przestrzeni lat. (Changes in the population of Polish Red Cattle over the years), Wiadomości Zootechniczne, R.XLIII 2:36–39

## *Internet*

http://ec.europa.eu/agriculture/quality/door/documentDisplay.html?chkDocument=579_1_pl. (opened 17.07.2019)
https://www.gov.pl/web/rolnictwo/lista-produktow-tradycyjnych12 (opened 17.07.2019)
http://www.pzow.pl/rasy-owiec.html (opened 17.07.2019)
https://limanowa.in/aktualnosci/mleko-trafilo-na-polki-almy-w-calym-kraju/15789(opened 01.08.2019)
https://www.gov.pl/web/rolnictwo/smietana-z-limanowej (*opened* 2.08.2019)
https://www.gov.pl/web/rolnictwo/zsiadle-mleko-z-limanowej (opened 2.08.2019)
http://skansen-studzionki.pl/wolosi-w-ochotnicy/ Wołoscy pasterze w Ochotnicy oraz tutejsze nazwy pochodzenia wołoskiego (opened 31.10.2019)
www.bioroznorodnosc.izoo.krakow.pl/.../polska_czerwona(opened 31.10.2019)
www.transhumance.pl /.../REDYK KARPACKI TRANSHUMANCE 2013 (opened 31.10.2019)
http://www.bioroznorodnosc.izoo.krakow.pl/bydlo/dokumenty/Program ochrony zasobów genetycznych bydła polskiego czerwonego (2011)

# Chapter 14
# Traditional Crops Cultivated in Southern Małopolska

**Wiktor Berski, Gabriela Zięć, and Marcin Łukasiewicz**

**Abstract** Southern Malopolska covers a fragment of the Carpathian Mountains. The rising elevation above sea level is negatively correlated with the increase in the economic potential of agriculture and the facilitation of people's lives regarding transport and construction. Living is more expensive, partially due to a longer winter heating period than in the lowlands. Through the centuries, this territory was subjected to different rulers. They were the poorest and most economically backward areas in Europe, which affected the escalation of social conflicts in connection with national disputes. An additional problem was a dramatic overpopulation that resulted in extreme poverty. Agricultural production carried out in the Polish Carpathians underwent many transformations over a period of several hundred years. The Carpathian region is rich in many traditional plants, but these three: potato, oats and rutabaga were the most popular in southern Malopolska. Ease of cultivation, low climatic and soil requirements, the possibility of cultivation in areas located in the mountain range, and many potential ways of utilization made them perfect plants for hard conditions and times.

**Keywords** Southern małopolska · Galicia · Potatoes · Oats · Rutabaga

## 14.1 Introduction

Southern Malopolska covers a fragment of the Carpathian Mountains. There are 19.3 thousand km$^2$ in Poland (9.63% of the total area of the Carpathians, 6.3% of the country). About 56% of the area of the Polish Carpathians is occupied by

W. Berski (✉) · G. Zięć
Department of Carbohydrate Technology and Cereal Processing, Faculty of Food Technology, University of Agriculture in Krakow,
ul. Balicka 122, 30-149 Krakow, Poland
e-mail: rrberski@cyf-kr.edu.pl

M. Łukasiewicz
Department of Engineering and Machinery for Food Industry, Faculty of Food Technology, University of Agriculture in Krakow, ul. Balicka 122, 30-149 Krakow, Poland

© Springer Nature Switzerland AG 2022

J. Hernik et al. (eds.), *Cultural Heritage—Possibilities for Land-Centered Societal Development*, Environmental History 13,
https://doi.org/10.1007/978-3-030-58092-6_14

mountainous areas, and the rest by foothills. The rising elevation above sea level is negatively correlated with the increase in the economic potential of agriculture and the facilitation of people's lives regarding transport and construction. Living is more expensive, including due to a longer winter heating period than in the lowlands (Jóźwiak 2012). The Carpathians are not one uniform mountain range but rather consist of many mountain ranges (Guzik 2001; Zemanek 2009). They take the characteristic shape of a strongly bent arch surrounding the Pannonian Basin and the Transylvanian Upland. The length of the Carpathian range measured along the dorsal line is 1300 km, while the extent of the range varies from 100 to 300 km (Zemanek 2009). These mountains are not particularly high—the highest peak, Gerlach, rises to 2.655 m, and areas above 2.000 m are relatively few, and domination of hills from 1000 to 1500 m is observed. The Carpathians are geologically linked with the lowering belt of Podkarpacie, South Romanian Plains and the Pannonian Valley.

Altitudinal zonation of vegetation is a typical phenomenon occurring in the mountains, although this condition has been disturbed due to human economic activity, which has usually been conducted for centuries. The dominant layer of vegetation in the Carpathians is the forest one, which is caused by their relatively low height. Difficult terrain conditions to some extent limited the conversion of forests into arable fields or pastures, making the degree of forest cover in the Carpathians, one of the highest in Europe (Guzik 2001; Zemanek 2009; Musiał 2018).

In EU agricultural policy, issues related to agriculture in economically problematic areas, including mountain areas, constitute an important issue of interest for Member States (Musiał 2010). The main features of mountain and foothill areas next to the harsh climate are the diverse terrain, which is characterized by large leveling and steepness of slopes, and the occurrence of many terrain obstacles and natural boundaries associated with the network of ravines, watercourses and mosaic of arable land. All unfavorable features characterizing the economy conditions occur in this area: difficult terrain in terms of agriculture production, poor soils, harsh climate and small farms with land in a large chessboard, which affects, among other issues, the creation of large areas with non-agricultural land. As the height above sea level increases, the snow cover stays longer, the growing season is shortened, rainfall and wind speeds are higher. In the mountains, relative humidity indicators and higher intensity of solar radiation are also observed. Temperature inversions that are present there and local strong winds are the obstacles to agricultural activity. Soil covers also exhibit altitudinal variation. As the inclination of slopes increases, erosion processes intensify.

## 14.2 Historical Background and Introduction to Agricultural Production of Malopolska

At the end of the eighteenth century, as a result of the Poland Partitions, Malopolska and some lands belonging at present to Ukraine were incorporated into the Austrian

Empire creating the Kingdom of Galicia and Lodomeria and the Grand Duchy of Kraków with the Duchies of Auschwitz and Zator. The set was the poorest and most economically backward areas in Europe, which affected the escalation of social conflicts in connection with national disputes (Guzik 2001; Davies 2005; Potocki et al. 2012; Mizgalska 2014). These areas were inhabited by three major ethnic groups: Poles, Ruthenians and Jews. Galicia's main problem was its dramatic overpopulation, reaching 78 people per $km^2$ (Hołub 2013). An excess labor force was not absorbed by an underdeveloped industry, and the labor market in cities was not much different from the rural one (Davies 2005; Błąd 2009; Mizgalska 2014). In fact these areas of the Austrian Empire operated in a similar way to colonies. The liquidation of serfdom (1848) was carried out in a such way that the peasants were granted with relatively small pieces of land that prevented independent subsistence. Agricultural overcrowding in the agrarian system led to a decrease in the size of farms (the collapse of larger farms and the creation of medium ones) (Błąd 2009; Davies 2005). As a result, it weakened the peasant economy. The effects of this fragmentation are still visible, because in south-eastern Poland, there is the largest number of small farms (creating so-called "checkboard") (Guzik 2001; Popławski 2009; Ostrowska 2012).

The situation began to improve gradually, at the end of the nineteenth century, as drainage works were started, fertilization of fields began, and industrial plants were built up and a quite dense railroad network was created. This led to a sharp improvement in the quality of the inhabitants' lives (Davies 2005; Taylor 2007). However, it was only in the 1870s that crop rotation on peasant farms began to be widely applied (Mizgalska 2014).

Agricultural production carried out in the Polish Carpathians underwent many transformations over a period of several hundred years. It relied on settling, by nomadic shepherd people mainly from the Balkans or the Southern Carpathians, where the open mountain spaces were suitable for animal grazing. The importance of new pasturing areas and those useful for agricultural production increased, so they were obtained through clearing of forests. Such places were also colonized by the newcomers from populations living north of the Carpathians. The gradual saturation of mountain areas by people caused a relative surplus of manpower at the turn of the nineteenth and twentieth centuries. This led to regional food shortages. This was caused by the chronic technological backwardness of farms, that did not offer economic opportunities nor did the people feel motivated to switch to more intensive, newer and more efficient production technologies. Generationally inherited land was becoming increasingly fragmented (chessboard), becoming the main brake on the rationalization and economic improvement of production. The food shortage resulting from the overpopulation of villages and the lack of production and economic possibilities to increase production also caused several waves of long-term and seasonal migrations. Emigration resulted in negative selection of village residents, as usually more industrious, more entrepreneurial and braver people decided to leave. Returning to their previous homes contributed both to mental changes (including the way the world was perceived) as well as to the development of farms, and the creation of tourist and service facilities (Musiał 2017).

In the past, agriculture was the main area of economic activity and the main source of income for most inhabitants of mountain and foothill areas. To this day, field crops reach the height of 1000–1050 m (Ząb near Zakopane). The average size of private farms in the Polish Carpathians is about 5 ha. In mountain areas, as much as 89.2% were farms with arable land not exceeding 5 ha, including 34.8% of farms whose arable land did not exceed 1 ha. Agricultural land is split up, even up to 42 plots, of which only half are arable. The average area of an arable plot of land is just over 1/10 ha (farms with plots of land of 6 or more plots accounted for nearly 24.7%). But there are regional differences in this respect (Anonim 2003; Ostrowska 2012).

It should be mentioned that arable farming on mountain slopes is in conflict with the rational use of water reservoirs built in the mountains (Ostrowska 2012; Czudec 2013). Fortunately, recently, there is a growing tendency of abandoning the cultivation of agricultural plots, especially those with low soil quality, and of these located at large distances from significant levels of human habitation (Guzik 2001; Ostrowska 2012). Still, a high percentage of the population in Polish mountain areas works in agriculture −20.3% of the total number of those who are actively working, while in most countries, this rate does not exceed 10%. Nearly 3/4 of individual farms in mountain areas showed that the percentage of family income from agricultural activity did not exceed 30%. Agri-tourism was an active interest of 0.7% of the total number of those living in mountain areas. The greater number of farms shows signs of decline (Jóźwiak 2012).

Agriculture and forestry cover about 1.76 million ha, i.e. about 91% of the area of the Polish Carpathians. Compared with the whole country, the forest cover of this area is higher by about 8%, but the share of arable land is smaller by about 8%. The latter in the middle of the last decade was distinguished by a greater share of permanent grassland and fallowing and set aside of arable land, and a smaller share of arable land used for agriculture and orchards (Jóźwiak 2012). Agricultural farms located in mountain areas were characterized by a definitely smaller share in the total area of: farmland (76.8%) and arable land (39.5%). In these areas, there were significant proportions of meadows and pastures (26.0%), fallowed land (15.6%) and forests (18.2%). In mountainous areas, abandonment of crops is a very serious problem, hence the large share of fallows.

The share of cereals in the totality of sown crops in mountain areas was 67.6%, while the national average was 77.0%. The share of potato and fodder plant growing in mountain areas accounted for 17.2% and 10.4% of the total sown area (the national average was 7.5% and 4.9%, respectively). Considering the fact that in these areas, the sown area did not exceed 30% of the total land used, it can be stated that the production potential of organizational entities managing mountain areas is very small (Anonim 2003). In 1931, oats prevailed in crops (47% of crops). The next crops were: potatoes (18% sown) and clover (14%) as well as barley and rye. Together, these five plants accounted for almost 94% of the entire sown area. At the beginning of the 1990s, clover became the most important crop (48% of the planted area). The next significant crops were oats and potatoes—(17% of the total crop). The other three cereals—barley, rye and wheat are also more important in sowing (Górz 1994). Currently in the Carpathians in the totality of crops, cereals (63.5%) and potatoes

(17.9%) predominate, maize (green fodder) occupies 1.7%, and root fodder less than 0.5% (Matyka 2013).

Between 1925 and 1935, there was a visible reduction in the share of crops such as cereals and potatoes, while clover increased its share. This direction of changes in crops became permanent after World War II. The decreasing tendency was also the decrease in the total crop area related to the previously described trend for agriculture to decline in difficult-to-cultivate lands combined with afforestation.

The contribution of individual cereals to crops depends primarily on the terrain and elevation of the area above sea level. Oats reach the highest—up to 1000 m above sea level. Similarly high–rye. Whereas the other two cereals, especially wheat, are usually grown below 650 m above sea level. A little higher, even over 700 m above sea level, barley is produced. Therefore, the majority of arable land was also used for feed production. In the interwar period, the relevant area was 77 thousand ha, i.e. 61% of total arable land, in 1960–90 thousand ha (79% of arable land), while in 1986–76 thousand ha, which, however, already constituted 81% of the total arable land (Górz 1994).

It should also be mentioned that during famine periods or food shortage, the population also consumed wild plants (Łuczaj 2007a, b, 2008, 2011). These plants were used to prepare a wide spectrum of dishes: soups and bows, bread and flatbreads, fillings for dumplings or as spices. As to plants such as blueberries or blackberries, their consumption was probably so obvious that people even did not mention them. Some wild plants can be consumed not only during periods of hunger but also as attractive additives enriching the diet of rural inhabitants (Łuczaj 2008; Köhler 2018).

## 14.3 Potatoes

Potatoes (*Solanum tuberosum* L.), along with wheat and rice and corn, are the most popular food source for the world's population. They are grown on all continents, most in Asia and Europe. These two continents account for more than 80% of the world's potato growing area, with their production becoming more limited in Europe and increasing in Asia (Birch et al. 2012; Dzwonkowski 2017).

The origins of the modern cultivated potato can be dated back almost 8,000 years to the South America Andes, on the border located between Bolivia and Peru. It is believed that domestication of wild potato plants took place 3.800 m above sea level, around the shores of Lake Titicaca. In the 1570s, cultivated potatoes were introduced into Europe and from there, distributed throughout the world starting from the late seventeenth century (Birch et al. 2012).

Over time, it has become something more important than just a source of food, because this plant contributed to the demographic explosion. It can be said that the introduction of potatoes into widespread cultivation laid the foundation for the industrial revolution in the nineteenth century. Potatoes began to provide much more food per hectare than previously grown plants, including cereals, and could compete with them as the primary source of protein (Mazoyer and Roudart 2006; Birch et al. 2012), especially among the poorest people. Potato has become the basic food product that has allowed overcoming of the problem of hunger and scurvy in Europe, especially among its lowest social strata. Vitamin C (preventing scurvy development) content in potatoes tubers is not particularly high, but through their significant consumption potatoes can be recognized as a good source of vitamin C (Magiorkinis et al. 2011). In addition, it became the basic feed for pigs, of which the numbers increased significantly. Potato is also a good fore crop for cereals, as well as legumes and industrial plants, which enabled progress in agrotechnology (Mazoyer and Roudart 2006).

The potato has high nutritional and low energy values. Its nutritional value is primarily due to the chemical composition of tubers, their main ingredients include: starch (up to 16%), protein rich in exogenous amino acids (about 2%), dietary fiber, numerous macroelements and microelements (1%), vitamins C, $B_1$, $B_2$, $B_6$, polyphenols, carotenoids. Consumption of about 200 g of potato covers the daily requirement of an adult human body for some vitamins: C in about 50%, $B_6$ in 25%, and other vitamins about 10–15% and minerals about 12–30%. As compared with other plant products, potato, due to its low-fat content, accumulates small amounts of heavy metals, nitrates and residues of plant protection agents, and culinary treatment significantly reduces their content. Nutritionists believe that the nutritional value of potatoes is so great, that it can, for some time, be the only ingredient in human diet without compromising anyone's health (Leszczynski 2012; Zgórska 2013; Ezekiel et al. 2013; Akyol et al. 2016).

Potato production in Poland has dropped dramatically in recent years, from about 48 million tonnes in the 70s, when we were among the world's leading producers, to less than 9 million tonnes in 2017. The main reason for that decrease was the abandonment of using potatoes for feeding animals (for this purpose, 50% of potatoes were used), as well as changes in dietary habits (decrease in human consumption) as well as less use of potatoes for the production of starch or alcohol (Nowacki 2017). After 1990, maintenance and extension of cultivation of this tuber took place in the area with the largest fragmentation of farms in the south-eastern voivodeships, especially in Podkarpackie, where potatoes largely serve self-supplying purposes (Kulikowski 2013).

The consumption of potatoes in Poland is also decreasing. Currently, about 84.4 kg of potatoes and 16.6 kg of potato products are consumed annually (2014/15) per capita, while at the beginning of the twenty-first century, it was 122.3 and 11.5 kg, respectively (Nowacki 2017). The peak high potato consumption in the twentieth century occurred after the end of World War II, in 1946, in a situation of food shortage, when potato consumption per capita was 320 kg (Leszczynski 2012). The rule is that the decrease in potato consumption is positively correlated with the increase in the society's wealth and greater access to competitive food substitutes in the group of vegetables and cereal products (bread, groats, pasta). The level of consumption of various food products, including potatoes, depends on the traditions cultivated in communities, changing fashion in the nutrition of societies, the promotion and marketing of specific products, and the changing work situations of consumers. Mostly, potatoes are consumed in agricultural families and among pensioners (Nowacki 2017).

In Polish lands, which were annexed by Austria as a consequence of the Partitions, no increase in interest in potato cultivation was observed until the end of the eighteenth century. They could only be found as one of many garden plants. Steady expansion of potato cultivation in these areas arose in the nineteenth century and actually the 60s of twentieth century, which was a consequence of among other things introduction of a new agricultural system (moving away from the triple field, the development of animal husbandry and the consequent necessary animal feed needs to cause an increase in potato crops) (Badyna 2015), although information about a much earlier increase in prevalence of potato cultivation in these areas can also be found (Franaszek 2016). The deteriorating situation of the peasants who lost part of their income as a result of increasing numbers of land plots played an important role. It is worth noticing that Galicja was typically agricultural country, as around ¾ of the population was employed in agriculture and forestry, while in the same period in industry and construction −9.5%, and in trade and transport −9.6% (Kuklo et al. 2014).

Potato worked out well as part of the diet to facilitate survival in the pre-harvest period. The growing importance of this plant in Galicia is reflected by the size of the acreage designated for its cultivation over just one decade. In 1876, it was about 430 thousand ha, and in 1886 already over 550 thousand ha (Badyna 2015). In subsequent years, a further increase in the cultivation area was observed to about 619 thousand ha (1897), 714 thousand ha (1913) and 714 thousand ha (1913) (Kuklo et al. 2014). The increase in the area intended for growing potatoes resulted in an increase in their production. In Galicia, in the years 1871–73, about 2125.1 thousand tons were collected, 1880/82 2433.3 thousand tons, 1890/92 3461.8 thousand tons, 1895/97 4326.7 thousand tons and 1906/1910 5402.0 thousand tons. Just before the beginning of the Great War (1911/13), 5,657,000 tons of potatoes were harvested in Galicia (Kuklo et al. 2014). The largest increase was recorded after the end of the potato blight, which overtook over a large part of Europe in the middle of the nineteenth century. This increase can also be associated with the recognition of the importance of this plant for food safety (Badyna 2015). On the scale of potato production, apart from peasant crops, it was due to the fact that granges, which were lost in the crisis related to the reduction of grain export, switched to animal production and increased their share in the propination industry (Badyna 2015). From the beginning of the nineteenth century, potato consumption began to spread in Galicia. In a relatively short time, potatoes became a basic dish on the peasants' menu, regardless of their wealth. They constituted the basis of food for the peasant population in both western and eastern parts of Galicia (Franaszek 2016). Their dissemination significantly contributed to the reduction in the peasant's diet of the presence of other vegetables, which as a result worsened the quality of food. The dominance of the potato brought with it a particularly dangerous effect for the year's crop with the possible occurrence of potato blight, which could cause waves of famines. This situation took place in Galicia in the second half of the 1840s.

Potatoes were eaten in very different forms. Most often they were cooked, although sometimes they were also toasted. Raw, grated potatoes were often used to make pies. Potato soup was a frequent food. In addition, pickling of chopped potatoes was widespread, by which the potato was further cooked after being pickled. Still another way of using potatoes was to add them to flour products, especially bread intended for one's own consumption. Thanks to this, flour was saved. Usually, partly spoiled and frozen potatoes were used for this purpose. The spoiled potatoes were also used to make pies, known as "moskale". Potatoes were eaten two or even three times a day.

According to various estimates, depending on the level of wealth and the region of Galicia, the daily amount of potatoes consumed per member of the peasant family ranged from about 1 kg to less than 4 kilograms. On average, for the entire province, this value was about 475 kg per year per capita (Franaszek 2016). According to other authors (Potocki et al. 2012; Kuklo et al. 2014), the average annual consumption of potatoes in Galicia around 1885 was 310 kg per capita, whereas at the same time throughout the Austro-Hungarian Monarchy, their consumption was 100 kg per year.

## 14.4 Rutabaga

Rutabaga (*Brassica napus* L. *var. Napobrassica* L. Rchb.) belongs to the *Brassicaceae* (cruciferous, brassica) family. It comes from the Mediterranean. This plant has been known since antiquity. In Poland, it is widely cultivated in foothill and mountain ranges for fodder and as a vegetable plant. It is an autumn-winter vegetable. Rutabaga is a biennial plant. It produces a rosette and fleshy root in the first year and branched flower shoots in the second. The optimum temperature for growth is 15–20 °C. It is a plant resistant to cold and drought. Soil requirements are not high, but the best yield of rutabaga is obtained from moist soil rich in minerals, as well as in a cool and humid climate. Rutabaga became popular in Europe at the end of the seventeenth century. Initially, rutabaga was cultivated by the inhabitants of Scandinavia. It was a vegetable consumed by humans and used as animal feed in the northern part of Europe. Two hundred years later, rutabaga got to Canada and the USA, where it is still very popular nowadays (Lim 2015).

The short vegetation period and resistance to frost made it a rescue plant in the event of natural disasters. Rutabaga was treated as food for poor people for years. This plant was mainly associated with starvation, poverty and World War II, during which rutabaga was the main ingredient of soup in the concentration camps. It was planted in greater amounts up to the 1970s. Recently, rutabaga has become increasingly popular for its taste and health-promoting properties, and in many countries, it is even a delicacy (Tomf et al. 2015).

It is a vegetable that is increasingly appreciated because of its dietary and health-promoting values especially in diseases of the gastrointestinal tract and bile ducts. There are three species of rutabaga: Kaszubska and Saba, referred to as fodder varieties, and Nadmorska, which is described as edible, in the national register of varieties of crop species (COBORU) (Anonim 2019). This vegetable is a valuable source of pharmacologically active substances, i.e. phytochemicals: phenols, vitamins, minerals and glucosinolates (Tomf et al. 2015).

Rutabaga root is the edible part, which contains about 14% dry matter, including 1.1% protein and 8.9% carbohydrates (Świetlikowska 2008). Rutabagas root is a low-calorie vegetable (36 kcal/100 g edible product). Moreover, these roots contain in 100 g–0.77 g of minerals, including: sodium 5–15 mg, potassium 200–227 mg, magnesium 11 mg, calcium 40–55 mg, phosphorus 31 mg, iodine 4 μg and iron 400–500 pg. Rutabagas root is also a valuable source of vitamins: vit. C 33-44 mg, vit. B1 50 μg, vit. B2 58 μg, vit. B6 200 μg, vit. K 21 μg and beta-carotene 99 μg (Puupponen-Pimiä et al. 2003; Tomf et al. 2015). In addition, a high content of 41–47% essential amino acids, i.e. lysine, leucine, isoleucine, threonine and glutamic acid, is found in such roots (Prośba-Białczyk 1995).

Rutabaga is characterized by the presence of antioxidant compounds, which have beneficial effects on human health. The basic antioxidants present in the roots of rutabaga are: vitamin C, vitamin E, carotenoids and polyphenols. The content of polyphenols is on average about 42 mg/100 g of fresh mass, while antioxidant activity (DPPH) −1.87 μmol/trolox/g of fresh mass (Sikora et al. 2008; Tomf et al. 2015).

In southern Malopolska, due to the already harsh climatic and soil conditions, rutabaga, along with turnips and potatoes, was the most commonly grown and eaten root crop. Rutabaga and other vegetables (beets, turnips, onions and garlic) played an important role in Galician peasants' menu. The introduction of potatoes reduced its role. It was far more popular in Western Galicia than in Eastern. Rutabaga, otherwise known as carpel or corpel, was once common in Polish cuisine, but eventually it has been completely replaced by potatoes and carrots, parsley etc. (Tomf et al. 2015; Franaszek 2016). In Poland, rutabaga has been traditionally grown and consumed since the seventeenth century, especially in the southern regions. Less wealthy people ate rutabaga instead of bread. It was served after slicing and boiling in water with salt addition, sometimes seasoned with flour and fat. Sometimes meat was stewed in rutabaga. Also, mashed rutabaga was consumed with potatoes or, more rarely, in soup with potatoes. In starvation years, rutabaga was eaten cooked, only slightly scraped (Franaszek 2016).

## 14.5 Oats

Oat (*Avena sativa* L.) belongs to the group of secondary plants developed from weeds grown with barley, wheat and rice. It probably originates from the Middle East and neighboring areas of the Mediterranean (Sadiq Butt et al. 2008). The oldest European oat traces are dated from the Bronze Age. Available archaeological evidence shows that oats were cultivated in Europe as early as the Bronze and Iron Age (Gąsiorowski 1995; Paczos-Grzęda 2003). It was also well known in ancient Greece and Rome. In Poland, the name oat is already mentioned in the fourteenth century (Gąsiorowski 1995). For centuries, up to the eighteenth century, our ancestors highly valued this plant, as food, feed and medicinal plant. Oatmeal as well as bread with the addition of oat flour was daily food especially for the poorer people in nineteenth century society. However, this cereal was replaced by more fertile potatoes (Gąsiorowski 1995; Bartnikowska 2003).

Due to the fact that oats are cereals with low soil requirements and have high thermal tolerance, they are cultivated on all continents, even near the Arctic Circle (Gąsiorowski 1995; Bartnikowska et al. 2000). The largest producers of oats include Russia, the USA, Canada and Northern European countries (Gąsiorowski 1995; Lange 2010). Currently, around 525–577 thousand hectares in Poland are devoted to different varieties of oat cultivation, and most of the cultivated area is located in the northern and southern regions.

Except potatoes, the second main food for Galician peasants was cereals, and above all bread. At the turn of the nineteenth and twentieth centuries, in the structure of crops of five crop groups (cereals, potatoes, legumes, industrial and fodder plants) in Eastern Galicia, cereals covered almost 60% of the area, of which nearly 29% was used for oat cultivation, rye—over 26%, barley – 16%, and wheat only less than 13.7%. The basic flour used to make bread was rye flour, which was usually mixed with flours of other, inferior cereals, e.g. oats or barley. In Eastern Galicia and in the foothill areas, oat flour was the basic one (in the foothill areas its daily average consumption exceeded even 1 kg for person). Also, it is worth notice that unleavened pies (from water and flour, usually oat or barley) were prepared as bread substitutes. As previously mentioned, to save flour, potatoes, legumes (especially broad beans), vetches, ground beech seeds and even wood ash were added to the bread. Adding boiled potatoes allowed the bread to be kept moist longer (Franaszek 2016). Oats were also used to prepare groat, borsch. and it could also be used to produce the so-called "bryja" (i.e. small amounts of flour dispersed in boiling water) (Franaszek 2016).

As in other cereal grains, the main component of oats is carbohydrates—primarily starch and non-starch polysaccharides (NSP), such as: cellulose, hemicelluloses (pentosanes and hexoses) and pectin substances, which together with lignin create dietary fiber(DF) (Sadiq Butt et al. 2008; Haskå et al. 2008; Lange 2010; Shewry et al. 2010). However, oat grain is distinguished, by the lowest saccharide content, mainly starch, which content in hulled grain is about 10% lower, as compared with other cereals. Oat grain is also an important source of DF in the human diet. On average, the whole, unhulled grain contains 32% of DF, and after separating of the husk, this amount decreases to about 11–14%, including an insoluble fraction (IDF)—over 6%, and soluble fiber fraction (SDF)—close to 8%. Similar content and fractional composition of the fiber are present in oat flakes. Such a high level of SDF is higher than in other cereals. The distribution of DF in the oat grain is not uniform. Pentosans, as in the grain of most cereals (especially rye), dominate over hexosans, while in the oat grain, it is the opposite—hexosans predominate, among them the water-soluble fraction, the main component of which is β—D—glucan (Sadiq Butt et al. 2008; Haskå et al. 2008; Shewry et al. 2010).

Moreover, oat grains are characterized by a large quantity of protein (11–15%) with high biological value (Hüttner et al. 2010). They are rich in essential amino acids (41%). However, the most important amino acid limiting the biological value of these proteins is lysine, as in other cereal proteins. Due to the biological value, cereal proteins can be arranged in the following order (Kawka 2010; Lange 2010):

oats > rye > barley > corn > wheat

Oat grains are characterized by an extremely high content of lipids, 3–5 times higher as compared with other cereals. The content of lipids in unhulled varieties of oat grains, cultivated in Poland, is on average 5.2%, in dehulled, and on average 7.9%, and in nude varieties −9% db. Lipids in oat grains are distributed in a rather uniform manner, which is in opposition to other cereals, where they are mainly located in the embryo and the aleurone layer, while starchy endosperm is virtually depleted of them (Kawka 2010).

The content of minerals (ash) in the hulled oat grain is higher (2–3.4% d.m.) than in the unhulled grain (about 3.1% dm), so the husk contains small amounts of these compounds. Oatmeal contains much more minerals than wheat, rye or barley flakes. As in other cereals, in the pool of minerals, dehusked oat grains contain the following elements in the greatest amounts: phosphorus, potassium, magnesium and calcium. Moreover, oatmeal preparations are also a good source of manganese, copper, zinc and iron and have very low sodium content, which can be important information for people suffering from cardiovascular disease (Gibinski et al. 2005; Kawka 2010).

Oat grain is also a very good source of thiamine (vitamin $B_1$), pantothenic acid (vitamin $B_5$) and vitamin E. Consumption of 100 g of oat flakes satisfies about 40% of human daily demand for thiamine (vitamin $B_1$) and about 20% for vitamin E (Gibinski et al. 2005).

In addition to vitamins, oat grains also contain other bioactive substances, which include polyphenolic compounds, such as phenolic acids, flavonoids and phytoestrogens. Other substances are also present, such as tocochromanols (tocopherols, tocotrienols), melatonin, sterols and aventramides (Zielinski et al. 2001; McKevith 2004; Sadiq Butt et al. 2008). Therefore, oat grain and its products are products that are a good source of compounds with antioxidant properties, i.e. they act bacteriostatically and pharmacologically, improving the work of the heart and bloodstream and preventing chronic inflammation, as well as cancer (Peterson 2001; Finley 2004).

Oats and their preparations are rich in various biologically active compounds, i.e. water-soluble β-glucans, tocols, avenantramides, polyphenolic acids, phytosterols and polyunsaturated fatty acids (Zielinski et al. 2001; McKevith 2004; Sadiq Butt et al. 2008). Hypocholesterolemic and antihypertensive effects of oat preparations, as well as their effect on reducing postprandial glycemia, favoring weight reduction, have been found in numerous studies, including clinical studies. Therefore, oat grain can be classified as a raw material with beneficial functional properties and is even considered an outstanding twenty-first century cereal (Sadiq Butt et al. 2008; Kawka 2010; Lange 2010).

## 14.6 Conclusion

The Carpathian region is rich in many traditional plants, but these three: potato, oats and rutabaga were the most popular in southern Malopolska. Ease of cultivation, low climatic and soil requirements, the possibility of cultivation in areas located in the mountain range have meant that these plants have accompanied man for a long time. Their described chemical composition and health properties mean that dishes from them are becoming more frequent on tables.

## References


Akyol H, Riciputi Y, Capanoglu E et al (2016) phenolic compounds in the potato and its byproducts: an overview. Int J Mol Sci 17. https://doi.org/10.3390/ijms17060835

Anonim (2003) Rolnictwo na terenach górskich i terenach o słabszych warunkach glebowych (Agriculture in mountainous areas and areas with poorer soil conditions), GUS, Warszawa

Anonim (2019) Odmiany w rejestrze (Registered varieties). http://www.coboru.pl/Polska/Rejestr/odm_w_rej.aspx?kodgatunku=BRK. Accessed 27 Aug 2019

Badyna P (2015) Wybrane historie kartofla—Fryderyk II i ziemie polskie od XVII do połowy XIX stulecia (The Histories of the Potato: Frederick II and Polish grounds between the seventeenth and nineteenth century). Kultura - Historia Globalizacja 18

Bartnikowska E (2003) Przetwory z ziarna owsa jako źródło ważnych substancji prozdrowotnych w żywieniu człowieka (Oat products as a source of important bioactive substances in human nutrition). Biuletyn IHAR 229:235–245

Bartnikowska E, Lange E, Rakowska M (2000) Ziarno owsa - niedocenione zrodlo skladnikow odzywczych i biologicznie czynnych. Czesc I. Ogolna charakterystyka owsa. Bialka, tluszcze (Oat grain - an underestimated source of nutrients and biologically active ingredients. Part I. General characteristics of oats. Proteins, lipids). Biuletyn Instytutu Hodowli i Aklimatyzacji Roślin 215:209–222

Birch PRJ, Bryan G, Fenton B, et al (2012) Crops that feed the world 8: Potato: are the trends of increased global production sustainable? Food Sec 4:477–508. https://doi.org/10.1007/s12571-012-0220-1

Błąd M (2009) Rolnictwo jako "przechowalnia" nadwyżek siły roboczej w okresie transformacji systemowej w Polsce (Agriculture as a "Repository" for Excessive Workforce during the Period of Systemic Transformation in Poland). Wieś i Rolnictwo 144–156

Czudec A (2013) Wielofunkcyjność rolnictwa górskiego i podgórskiego (na przykładzie Bieszczadów i Beskidu Niskiego) (Multifunctional agriculture in mountain and semimountain areas (case study of bieszczady and beskid niski)). Polish J Agron 3–9

Davies N (2005) God's Playground A History of Poland: Volume II: 1795 to the Present. OUP Oxford

Dzwonkowski W (2017) Ewolucja produkcji ziemniaków w Polsce i UE (Evolution of Potato Production in Poland and the EU). Zeszyty Naukowe Szkoły Głównej Gospodarstwa Wiejskiego w Warszawie 17:71–80. https://doi.org/10.22630/prs.2017.17.3.54

Ezekiel R, Singh N, Sharma S, Kaur A (2013) Beneficial phytochemicals in potato—a review. Food Res Int 50:487–496. https://doi.org/10.1016/j.foodres.2011.04.025

Finley JW (2004) Phenolic antioxidants and prevention of chronic inflammation. Food Technol

Franaszek P (2016) Dieta chłopów galicyjskich w drugiej połowie XIX w. i na początku XX w. (Diet of Galician peasants in the second half of the 19th century and at the beginning of the 20th century). Roczniki Dziejów Społecznych i Gospodarczych 76:289–313. https://doi.org/10.12775/rdsg.2016.10

Gąsiorowski H (1995) Owies. Chemia i technologia (Oat. Chemistry and Technology). Państwowe Wydawnictwo Rolnicze i Leśne w Poznaniu, Poznań

Gibinski M, Gumul D, Korus J (2005) Prozdrowotne właściwości owsa i produktów owsianych (Health promoting properties of oat and oat products). Żywność Nauka Technologia Jakość 45:49–60

Górz B (1994) Rolnictwo Podhala (Podhale agriculture). In: Studia nad przemianami Podhala (Studies on the transformation of Podhale). Wydawnictwo Naukowe Wyższej Szkoły Pedagogicznej, Kraków

Guzik CZ (2001) Rola rolnictwa górskiego w gospodarce kraju (na przykładzie Polskich Karpat) (Mountain Agriculture in National Economy (Polish Carpathians Example)). In: Człowiek i przestrzeń (Man and space). IGiGP UJ, Kraków, pp 173–184

Haskå L, Nyman M, Andersson R (2008) Distribution and characterisation of fructan in wheat milling fractions. J Cereal Sci 48:768–774. https://doi.org/10.1016/j.jcs.2008.05.002

Hołub B (2013) Studium historyczno-geograficzne narodowości w Galicji Wschodniej w świetle spisów ludności w latach 1890–1910 (Historical-geographical study of the nationalities in Eastern Galicia in the light of the population censuses in the years 1890-1910). ANNALESUNIVERSI TATIS MARIAE CU RIE - SKŁODOWSKA LUBLIN – POLONIA LXVIII:15–40

Hüttner EK, Bello FD, Arendt EK (2010) Rheological properties and bread making performance of commercial wholegrain oat flours. Journal of Cereal Science 52:65–71. https://doi.org/10.1016/j.jcs.2010.03.004

Jóźwiak W (2012) Polskie rolnictwo i gospodarstwa rolne w pierwszej i drugiej dekadzie XXI wieku (Polish agriculture and farms in the first and second decade of the 21st century). Instytut Ekonomiki Rolnictwa i Gospodarki Żywnościowej, Warszawa

Kawka A (2010) Współczesne trendy w produkcji piekarskiej – wykorzystanie owsa i jęczmienia jako zbóż niechlebowych (Present trends in bakery production - use of oats and barley as non-bread cereals). Żywność Nauka Technologia Jakość, 3:25–43

Köhler P (2018) Etnobotanika Podhala na podstawie ankiety Józefa Rostafińskiego (1850–1928) z 1883 r. (Ethnobotany of the Podhale region based on Józef Rostafiński's (1850-1928) questionnaire, distributed in 1883). ETNOBIOLOGIA POLSKA 8:39–98

Kuklo C, Łukasiewicz J, Leszczyńska C (2014) Historia Polski w liczbach. Polska w Europie (Polish history in numbers. Poland in Europe). GUS, Warszawa

Kulikowski R (2013) Produkcja i towarowość rolnictwa w Polsce. Przemiany i zróżnicowanie przestrzenne po II Wojnie Światowej (Production and commercialization of polish agriculture changes after world war ii and present spatial differentiation). IGiPZ PNA, Warszawa

Lange E (2010) Produkty owsiane jako żywność funkcjonalna (Oats products as functional food). Żywność Nauka Technologia Jakość 17

Leszczynski W (2012) Żywieniowa wartość ziemniaka i przetworów ziemniaczanych (Przegląd literatury) (Nutrition value of potato and potato products (Review of literature)). Biuletyn Instytutu Hodowli i Aklimatyzacji Roślin 5–20

Lim TK (2015) Edible Medicinal and Non-Medicinal Plants. Volume 9, Modif ed Stems, Roots, Bulbs. Springer, New York

Łuczaj Ł (2007a) Dzikie rośliny jadalne używane w okresach niedoboru żywności we wschodniej części Karpat (powiaty Krosno, Sanok, Lesko, Nadwórna, Kosów i Kołomyja) według ankiety szkolnej z 1934 roku (Wild edible plants eaten during food shortages in the eastern carpathians (Krosno, Sanok, Lesko, Nadwórna, Kosów and Kołomyja regions): an ethnobotanical analysis of a 1934 school questionnaire). Przemyśl-Bolestraszyce, pp 161–182

Łuczaj Ł (2007b) Zapomniane dzikie rośliny pokarmowe południa polski – czyściec błotny, paprotka zwyczajna, bluszczyk kurdybanek i ostrożeń łąkowy (Forgotten wild food plants of southern poland: stachys palustris, polypodium vulgare, glechoma hederacea and cirsium rivulare). Przemyśl-Bolestraszyce, pp 183–200

Łuczaj Ł (2008) Dziko rosnące rośliny jadalne w ankiecie Józefa Rostafińskiego z roku 1883 (Wild edible plants in Józef Rostafiński's questionnaire of 1883). Wiadomości Botaniczne 52

Łuczaj Ł (2011) Dziko rosnące rośliny jadalne użytkowane w Polsce od połowy XIX w. do czasów współczesnych (Wild food plants used in Poland from the mid-19th century to the present). ETNOBIOLOGIA POLSKA 1:57–125
Magiorkinis E, Beloukas A, Diamantis A (2011) Scurvy: Past, present and future. Eur J Int Med 22:147–152. https://doi.org/10.1016/j.ejim.2010.10.006
Matyka M (2013) Rolnictwo na obszarach specyficznych (Agriculture in specific areas). GUS, Warszawa
Mazoyer M, Roudart L (2006) A History of world agriculture: from the neolithic age to the current crisis. Earthscan, London
McKevith B (2004) Nutritional aspects of cereals. Nutrition Bulletin 29:111–142. https://doi.org/10.1111/j.1467-3010.2004.00418.x
Mizgalska M (2014) Uwarunkowania ekonomiczne, społeczne i narodowościowe na terenie Ukrainy Zachodniej w II połowie XIX wieku (The economic, social and national conditions in Western Ukraine in the second half of the nineteenth century). 13:78–85
Musiał W (2018) Przyrodnicze, ekonomiczne i społeczne uwarunkowania przemian w rolnictwie obszarów górzystych na przykładzie polskich Karpat (Natural, economic and social conditions of changes in agriculture in mountainous areas on the example of the Polish Carpathians). Towarzystwo Ekonomistow Polskich
Musiał W (2017) Problemy rolnictwa i drobnych gospodarstw w polskich Karpatach wczoraj i dziś – reminiscencja broszury dla włościan podhalańskich z 1913 r (Problems of the agriculture in the Polish Carpathians, yesterday and today – revisiting a 1913 brochure for the peasants of Podhale). Problemy Drobnych Gospodarstw Rolnych)Problems of Small Agricultural Holdings) 3:97–108. doi:http://dx.doi.org/10.15576/PDGR/2017.3.97
Musiał W (2010) Rozwój rolnictwa w górach z perspektywy prac Parlamentu Europejskiego (Development of agriculture in mountains from the perspective of the european parliament works). Acta Scientiarum Polonorum Oeconomia 205–214
Nowacki W (2017) Historia ziemniaka w Polsce po II WŚ (History of potato in Poland after World War II). Ziemniak Polski 48–60
Ostrowska R (2012) Wpływ scalenia gruntów na rozwój rolnictwa w terenach górskich (The impact of land integration for agricultural development in mountain areas). Infrastruktura i Ekologia Terenów Wiejskich - Infrastructure and Ecology of Rural Areas 1:49–58
Paczos-Grzęda E (2003) Systematyka, Ewolucja i cytogentyka gatunków z rodzaju Avena L. (Systematics, evolution and cytogenetics of genus Avena L. species). Wiadomości Botaniczne 47:7–17
Peterson DM (2001) Oat Antioxidants. Journal of Cereal Science 33:115–129. https://doi.org/10.1006/jcrs.2000.0349
Popławski ZF (2009) Zmiany użytkowania ziemi w Polsce w ostatnich dwóch stuleciach (Land use changes in Poland during last two centuries). Teledetekcja Środowiska T. 42:
Potocki A, Lang-Młynarska D, Wójtowicz D, Zając J (2012) Zmiany sposobu żywienia ludności Polski Południowej (Galicji) na tle przemian polityczno-gospodarczych w XIX i XX wieku (Changes in nutrition among the inhabitants of Southern Poland (Galicia) against a background of economic and political changes between XIX and XX century). Hygeia Public Health 47:518–524
Prośba-Białczyk U (1995) Porównanie składu aminokwasowego białek gatunków okopowych roślin korzeniowych (Comparison of amino acid composition of proteins of root crops species). Rocznik Nauk Rolniczych III:151–159
Puupponen-Pimiä R, Häkkinen ST, Aarni M, et al (2003) Blanching and long-term freezing affect various bioactive compounds of vegetables in different ways. Journal of the Science of Food and Agriculture 83:1389–1402. https://doi.org/10.1002/jsfa.1589
Sadiq Butt M, Tahir-Nadeem M, Khan MK, et al (2008) Oat: unique among the cereals. Eur J Nutr 47:68–79. https://doi.org/10.1007/s00394-008-0698-7
Shewry PR, Piironen V, Lampi A-M, et al (2010) The HEALTHGRAIN Wheat Diversity Screen: Effects of Genotype and Environment on Phytochemicals and Dietary Fiber Components. J Agric Food Chem 58:9291–9298. https://doi.org/10.1021/jf100039b

Sikora E, Cieślik E, Leszczyńska T, et al (2008) The antioxidant activity of selected cruciferous vegetables subjected to aquathermal processing. Food Chemistry 107:55–59. https://doi.org/10.1016/j.foodchem.2007.07.023

Świetlikowska K (2008) Surowce spożywcze pochodzenia roślinnego (Vegetable raw materials). Wydawnictwo SGGW, Warszawa

Taylor Z (2007) Rozwój i regres sieci kolejowej w Polsce (The growth and contractionof the railway network in Poland). IGiPZ PAN, Warszawa

Tomf A, Turek K, Słupski J (2015) Brukiew jadalna - warzywo o właściwościach prozdrowotnych (Rutabaga - vegetables with health promoting properties). In: Wybrane zagadnienia nauki o żywności i żywieniu. Polskie Towarzystwo Technologów Żywności Oddział Małopolski, Kraków, pp 149–159

Zemanek B (2009) Fitogeograficzne problemy Karpat (Phytogeographical problems of the Carpathians). Roczniki Bieszczadzkie 17:43–58

Zgórska K (2013) Wykorzystanie ziemniaka do celów spożywczych i przemysłowych (Use of Potatoes for Food and Industrial Purposes). Inżynieria Przetwórstwa Spożywczego 5–9

Zielinski H, Kozlowska H, Lewczuk B (2001) Bioactive compounds in the cereal grains before and after hydrothermal processing. Innov Food Sci Emerg Technol 2:159–169. https://doi.org/10.1016/s1466-8564(01)00040-6

# Chapter 15
# Fruits of Traditional Varieties

**Jacek Słupski, Piotr Gębczyński, and Emilia Bernaś**

**Abstract** Blooming, tall fruity trees near country houses, in the fields, along roads, have recently been a decoration of the Polish landscape. This includes antonówka, malinówka, reneta kronselka, glogierówka, kosztela, etc., and every type of apple with a different taste, smell, and shape. Pears of various flavors, plums for tasty plum jam, and cherries for juices, tinctures, and preserves, such were traditional Polish home orchards. Today, these old orchards are the treasury of our biodiversity. Many of us remember from childhood these flavors from grandma's orchards. At that time, every farm had many fruit trees whose purpose was to meet the needs of the household members in terms of fruit availability. Many amateur gardeners with fondness remember fruit-rich orchards, which were used to diversify homemade pastries, make tasty compotes, juices, and other preserves for which rural pantries were famous. Old orchards also played a role in a rich ecosystem. We encounter many more species of birds, plants, and insects in one hectare backyard home than in a commercial industrial orchard. Unfortunately, as time goes by, the old trees slowly go away and are replaced by new varieties. Nowadays, when fruits are widely available, and their price is reasonable, it is much easier to buy fruits in a supermarket or bazaar rather than grow them yourself. However, on the store shelves, it is difficult to find the flavors of past centuries. Maybe it's time to pay attention to the old varieties of fruit trees, so common in old orchards, which nobody dealt with too much, and which provided very tasty and healthy fruits. The best traditional products are produced from good quality raw materials, often based on traditional varieties of plants from local sources. From apples formerly cultivated in Poland, marinades, silages, fried, dried, soups, cabbage stew, cheese, ice cream, and donuts were prepared. Many producers

J. Słupski (✉) · P. Gębczyński · E. Bernaś
Department of Plant Products and Nutrition Hygiene, Faculty of Food Technology, University of Agriculture in Krakow, ul. Balicka 122, 30-149 Krakow, Poland
e-mail: jacek.slupski@urk.edu.pl

P. Gębczyński
e-mail: piotr.gebczynski@urk.edu.pl

E. Bernaś
e-mail: emilia.bernas@urk.edu.pl

© Springer Nature Switzerland AG 2022

J. Hernik et al. (eds.), *Cultural Heritage—Possibilities for Land-Centered Societal Development*, Environmental History 13,
https://doi.org/10.1007/978-3-030-58092-6_15

of traditional products also use the richness of wild-growing local plants—forest fruit, mushrooms, and herbs. Our research thus concerns traditional local fruits.

**Keywords** Old apple cultivars · Chemical composition

## 15.1 Introduction

The high quality of traditional products is associated with the use of plants based on traditional varieties from local resources, often organic raw materials—often old varieties of fruit (especially apples and plums), cereals, and native breeds of farm animals (Lodeta et al. 2017; Byszewska 2011; Collective work 2014). Hence, culinary heritage is closely related to the preservation of agricultural biodiversity. High-quality fruit and vegetables create a reputation for products often derived from selected varieties grown in a specific microclimate, such as apples from Łącko, beans from the Lower Dunajec Valley, Suska sechlońska, or Galician garlic (Min Rol 2018). Fruit from home orchards and preparations and products obtained therefrom may be an additional attraction on agri-tourism farms.

The variety of arable crops that have arisen over the past thousands of years has meant that we have had plants that were of different taste, smell, color, and shape (Iaccarino et al. 2019; Lodeta et al. 2017; Mikołuszko 2018). When promoting the preservation of old varieties of arable crops, attention is paid to the protection of genetic resources, to the rescue of disappearing varieties. The preservation of old species and their varieties increases crop diversity, which prevents the simplification of crop rotation and ensures diversity of habitats. Most of the outgoing species have less demanding cultivation requirements, which allows reducing fertilization and the number of treatments with plant protection products. Typically, these species are particularly useful in extensive and ecologically valuable production systems, as well as for maintaining agricultural production in marginal areas (Sazońska 2010). The increase in demand for unconventional and traditional food causes these plants to return to cultivation, although often old varieties do not meet some of the usual contemporary appearance standards (Lončarić et al. 2019; Mikołuszko 2018).

From the beginning of the development of Polish fruit-growing, apple trees and apples have always played an important role in agricultural and horticultural production. Apples are planted around farms and manors, in balks, and along roads. Today, these old orchards are a treasury of our biodiversity, and they are a refuge for birds (Byszewska 2011; Łuczaj 2011, Dziubak 2006). Apples are one of the most important raw materials in domestic fruit processing. In Poland, fruit production, export, processing, or consumption dominate many farms and the national economy. Apples are the basic species of fruit grown in the temperate climate zone, belonging to the most consumed fresh fruit, and are a valued raw material in the processing industry. The production of apples in Poland has been exceeding 3 million tons for several years, which means that in the production of these fruits Poland ranks first in Europe and third in the world after China and the USA (Makosz 2015; Strojewska 2015;

Nosecka and Mierwiński 2016). The harvest of these fruits in Poland accounts for about 70% of the total harvest of all fruit and over 80% in the case of fruit harvest from trees. Over the past 15 years, the volume of apple production in Poland has doubled on average from 1.6 to 3.2 million tons (GAIN Report 2018; GUS 2019).

Apples, as a fruit, are an important component of a fruit and vegetable diet, and their consumption per one inhabitant is several kilograms a year. They can be consumed in fresh or processed form. Desirable economic features of the fruit are, among others: fruit alignment, their purpose, storage capacity, and resistance in transport. The most-grown apple varieties in Poland include Idared, Szampion, Jonagold, Ligol, Gloster, Golden Delicious (Borowska 2013). Antonówka, the most commonly grown old apple variety, ranks tenth, both in terms of crop area and harvest size (GUS 2019; Borowska 2013). The Idared variety has the largest share dedicated to apple growing in Poland so far; however, the importance of commercial varieties such as Jonagold and Szampion as well as varieties relatively recently introduced—Golden Delicious, Gala, and Ligol—have grown significantly from year to year. Older varieties lose their importance, e.g., Jonathan, McIntosh, as well as Antonówka, Bankroft, Spartan, Boskoop, and Wealthy (Borowska 2013). This is due to changing preferences of consumers who are willing to buy large, attractively colored apples, more often red than green, not necessarily with a sweet taste (Borowska 2013).

One of the ways to make apples stand out from the rich variety and assortment offered on the market is to provide products identified by the unique conditions of the region in which they are grown or the traditional nature of the variety, as well as the traceability of the producer. The origin of the product from a known source confirms its credibility on the market; its reputation, history of fruit production at each of its stages, and the knowledge of the fruit grower about cultivation. In addition, the consumer is convinced that the product is not anonymous, and thus indirectly considers it safer (Borowska 2013).

## 15.2 Trends in Apple Processing

In Poland, apples are the most commonly used raw material in the fruit industry. They have a wide range of applications, they are used primarily for the production of drinking juices, concentrated juices, as well as processed into pomace, purees, compotes, marmalades, apple cider vinegar, pasteurized apples, pickled or dried, including freeze-dried (Nosecka and Trojanowicz 2016; Żygała and Piórecki 2016; Makosz 2015; Turek et al. 2015, 2016; Chabłowska et al. 2013; Jarczyk and Płocharski 2010). Significant amounts are also used in the alcohol industry for the production of fruit wine and cider (Kostrz and Satora 2016). In the household, apples were a component of baked goods and traditional dishes, such as 'galas', groats with apples, donuts, pancakes, cabbage with apples, 'suszczanka', or cabbage stew (Szewczyk 2010). In recent years, the increasing use of fruit and vegetables, including apples, is common in so-called Smoothies, i.e., products based on fruit with a semi-liquid, smooth consistency, prepared by mixing juice and fruit in the right proportions, often also with the addition of other products such as milk or yogurt (Markowski et al. 2017; Teleszko and Wojdyło 2014).

The popularity of apples among consumers results from their nutritional, health, and dietary values. These fruits are a valuable source of bioactive ingredients, dietary fiber, pectins, and different classes of phenolics that are essential for the human body, vitamins A, B, PP, C, K, carbohydrates (glucose, fructose), organic acids (apple, citric, tartaric), and minerals (Table 15.1). The advantage of apples is their low calorific value (a medium-sized apple provides about 60 kcal) (Naszkowska 2017; Łata and Tomala 2007; Kunachowicz et al. 2017; Wojdyło et al. 2008).

Apples are good and durable material, so they can be processed immediately after harvesting or after storage. The requirements for the treatment of the raw material depend on the intentions of its processing. For example, in the case of fruit preserves, varietal homogeneity is necessary, in contrast to juices, where it is better to use mixtures of varieties that differ in their chemical composition, appearance, or degree of ripeness, which results in a product with a more harmonized taste. In addition, in the case of juice production, the appearance of the raw material is not so important as its chemical composition. Varieties with a high extract content and acidity and low susceptibility to browning are desirable. The intensity of browning is related to the activity of polyphenol oxidase and the content of substrates of this enzyme. Apples stored in cold storage are less susceptible to browning than apples processed immediately after harvest. The maturity of the raw material should be close to consumption, and the raw material should not contain starch, thanks to which it is possible to obtain adequate juice (e.g., stable turbidity) (Achremowicz et al. 2016; Gasik and Mitek 2012; Jarczyk and Płocharski 2010).

Currently, old or local apple varieties, known for many years, are returning to favor, which had previously been replaced by new ones with improved features, more economically profitable, giving high yields every year. Trees of old varieties have been ousted by new varieties, despite the fact that they grow and yield well, without the use, or with only minimal use of plant protection products, and often older varieties in sensory and nutritional qualities (Lončarić et al. 2019; Donno et al. 2012). Local varieties or those known in cultivation for many years are often particularly valuable because of their proven adaptation to specific environmental conditions. In addition, their characteristic features often include late flowering, a high degree of self-fertilization, and high resistance to major apple diseases (Żygała and Skrzyński 2008). The old varieties include several dozen apple varieties, among others Antonówka and Kosztela, Szara Reneta, Malinówka, Glogierówka, and Kronselka (Arboretum 2019; Poznański 2014; Żygała and Skrzyński 2008).

As Rembiałkowska et al. (2004, 2007) argue, products from the fruit of old varieties from small, home-grown orchards, where no mineral fertilizers or pesticides are used, may have similar nutritional and antioxidant properties to products obtained from organic apples.

**Table 15.1** Nutritional value of apples

| Component | Content [per 100 g edible portion] | Component | Content [per 100 g edible portion] |
|---|---|---|---|
| Water [g] | 80.7–90.8 | Vitamin A [μg] | 4.0–6.8 |
| Energy value [kcal] | 46–54 | Niacin [μg] | 100–500 |
| Soluble solids [g] | 10.5–17.7 | Vitamin C [mg] | 1–32 |
| Protein [g] | 0.06–0.45 | Malic acid [mg] | 420–938 |
| Total sugars [g] | 7.41–11.99 | Citric acid [mg] | 30–175 |
| Fat [g] | 0.06–0.58 | Oxalic acid [μg] | 300–700 |
| Available carbohydrates [g] | 11.4–12.1 | Total polyphenols [mg] | 48–740 |
| Total dietary fibre [g] | 0.64–3.00 | Total flavonols [mg] | 5.6–152.5 |
| Dietary fibre water soluble | 0.36–0.71 | Total anthocyjanins [mg] | 1.0–56.0 |
| Dietary fibre water insoluble | 0.43–0.77 | Chlorogenic acid [mg] | 1.4–119.5 |
| Pectines [g] | 0.64–2.24 | Ferulic acid [μg] | 600 |
| Total acidity [g] | 0.17–1.80 | Caffeic acid [mg] | 2.2–8.0 |
| pH | 2.97–4.34 | Para-coumaric acid [mg] | 2.0 |
| Minerals [g] | 0.26–0.36 | Salicylic acid [μg] | 310 |
| Sodium [mg] | 1–4 | Glucose [g] | 0.94–3.86 |
| Potassium [mg] | 79–134 | Fructose [g] | 3.96–8.52 |
| Calcium [mg] | 4–11 | Sucrose [g] | 0.06–2.90 |
| Phosphorus [mg] | 6–19 | Starch [mg] | 600 |
| Magnesium [mg] | 3.3–5.4 | Sorbitol [mg] | 25–58 |
| Iron [mg] | 0.14–0.85 | Total sterols [mg] | 12 |
| Silicon [mg] | 0.1–1.0 | Total purines [mg] | 2–19 |
| β-carotene [μg] | 24–35 | Kaempferol [μg] | 330 |
| Vitamin E [μg] | 490 | Quercetin [mg] | 0.4–14.0 |
| Thiamine [μg] | 15–60 | Isorhamnetin [mg] | 1.2 |
| Riboflavin [μg] | 20–50 | Glutatione [μg] | 610–1770 |

(Based on: Akagić et al. 2019; Neri et al. 2019; Kiczorowski 2019; Kiczorowski et al. 2018; Kumar et al. 2018; Oszmiański et al. 2018; Bartolini et al. 2017; Jakobek and Barron 2016; Souci et al. 2015; Begić-Akagić et al. 2014; Ceymann et al. 2012; Markowski et al. 2012, Markuszewski and Kopytowski 2008; Żygała ant Skrzyński 2008; Łata and Tomala 2007; Oszmiański and Wojdyło 2006; Łata 2005a, b, 2007, Kunachowicz et al. 2017; Sanoner et al. 1999; Gheyas et al. 1997)

## 15.3 Biologically Active Ingredients of Apples

Apples, like most fruits, consist mainly of water, which accounts for 85% of the weight of the fresh fruit. In the fresh weight of these fruits, the largest group is total sugars—10%, of which about 4–8% is fructose (Table 15.1). However, later ripening varieties are more abundant in extract components and sugars than early varieties (Iaccarino et al. 2019). Apples collected earlier may contain a significant amount of starch, which during the ripening and storage of fruit is converted into simple sugars. The content of sugars and essential oils, and above all the ratio of sugars to organic acids largely determine the taste of apples (Borowska 2013; Solomakhin and Blanke 2010).

The content of organic acids is on average 0.6 g/100 g of fresh matter (mainly malic acid). An important ingredient is a dietary fiber, the content of which is 0.8–2.9 g/100 g of fresh fruit weight (including almost a third or even half which is soluble fiber) (Gorinstein et al. 2001; Kunachowicz et al. 2017; Gheyas et al. 1997). Eating one apple with the skin provides the body with over 10% of the daily requirement for dietary fiber (Oszmiański 2009). Apples can be a good source of minerals, primarily potassium (120 mg/100 g on average), as well as calcium, phosphorus, magnesium, iron, and sodium (Table 15.1). They also contain trace amounts of group B vitamins and provitamin A, and generally a small amount of vitamin C (Souci et al. 2015).

An important group of substances found in apples is compound with antioxidant properties—polyphenols (Chabłowska et al. 2013; Oszmiański 2009). Their content in apples can be 48–740 mg/100 g of fresh fruit weight, with peel even several times more than in the flesh (Lončarić et al. 2019; Łata 2007; Łata et al. 2005b). The most important groups of phenolic compounds include flavonoids and phenolic acids. The most common flavonoids are quercetin in the form of glycosides as well as catechin and epicatechin. In addition to flavonoids and phenolic acids, apples also contain florentine glycosides (dihydrochalcones) (Alberti et al. 2014; Oszmiański and Wojdyło 2006; Markowski and Płocharski 2006). Total polyphenol content has an impact on the antioxidant activity of apples. The higher the amount of these compounds, the more generally apples show greater antioxidant activity; however, this depends on the amount and composition of polyphenols. The content of polyphenols in apples, as well as other biologically active compounds, is influenced by, among others variables, rootstocks, times of ripening, degree of maturity, size and manner of fertilization, soil, tree age and fruit abundance, and weather in a given year (Iaccarino et al. 2019; Kiczorowski 2019; Lončarić et al. 2019; Kiczorowski et al. 2018; Oszmiański et al. 2018; Teleszko et al. 2010; Oszmiański 2007; Łata 2007; Gheyas et al. 1997).

The level of antioxidant capacity in conventional apple cultivars grown in Poland, analyzed by the ABTS and FRAP assay, was 1.8 times lower and 1.5 times higher, respectively, than old cultivars (Oszmiański et al. 2018). According to Ceymann et al. (2012), the concentration of antioxidant capacity in analyzed apple cultivars grown in Germany according to the FRAP assay was two times lower than in the old apple cultivars. Furthermore, similar relationships were demonstrated by Panzella et al. (2013) and Feliciano et al. (2010). All the old apple cultivars from Southern Italy and Portugal showed higher antioxidant capacity than the commercial fruits. As Fołta et al. (2007) showed, apples of old varieties were characterized by a higher content of antioxidant components, and winter varieties also contained more of these compounds. As reported by Oszmiański and Wojdyło (2006), the content of phenolic compounds in apples can range from 100 mg to 500 mg in 100 g of fresh fruit, with the skin containing about seven times more of these valuable ingredients than the flesh. And Jakobek and Barron (2016) found even 15 times more of these ingredients in the skin than in the flesh. Apples are considered a fruit that provides significant amounts of phenolic compounds due to their high content in the fruit, and this motivates the significant consumption of fruit and preparations by consumers. Mitek and Gasik (2007) claim polyphenols delivered to the body from apples accounted for one-fifth of the total amount of polyphenolic compounds consumed.

According to Eberhardt et al. (2000), the total antioxidant activity of apples usually consumed with the peel was approximately 83 μmol vitamin C equivalents, which means that the antioxidant activity of 100 g apples (about one serving of apple) is equivalent to about 1500 mg of vitamin C. However, the amount of vitamin C in 100 g of apples is usually not larger than the 32 mg/100 g edible portion (Table 15.1).

Antioxidant compounds present in fruits (polyphenols, carotenoids, tocopherols, vitamin A, ascorbic acid, organic acids, phytates, thiocyanates, calcium and selenium, and others) are an important component of the human diet due to their role in the prevention of such diseases as atherosclerosis, diabetes, Alzheimer's disease, and many others classified as lifestyle diseases. Fruit, including apples, owes its antioxidant properties mainly to compounds belonging to polyphenols (Szajdek and Borowska 2004). Half of these compounds are procyanidin polymers that give fruit astringency, while the other half are chalcones, anthocyanins, quercetin glycosides, and phenolic acids (Teleszko et al. 2010). Among phenolic acids, apples contain caffeic and p-coumaric acid, mainly in the esterified form with quinic acid, chlorogenic acid (Malik et al. 2009). In apples there are also flavonoids, e.g., quercetin 3-glycoside flavan-3-ol derivatives, catechin and epicatechin, as well as procyanidins B2 and C1, as well as flavanols, among them quercetin 3-galactoside, phloridzin (located mainly in seeds), and phlorentine xyloglucosides (Akagić et al. 2019; Oszmiański et al. 2007, 2018; Sieliwanowicz et al. 2005). Most polyphenols are found in the skin of apples. Flavonol cyanidin 3-galactoside is responsible for its red color. Flavonoids from the flavonol group are found only in the skin of these fruits (Akagić et al. 2019).

Phenolic acids, especially chlorogenic acid, are precursors of flavor in fruits and vegetables and they exhibit anticarcinogenic, antimutagenic, and antioxidant properties in vitro, and scavenge reactive oxygen species. Chlorogenic acid is also a substrate of polyphenol oxidases, and under the influence of oxygen and polyphenol oxidase it is oxidized to o-quinone, reacting later with other phenolic compounds; it is responsible for the color change in fruit preparations during their production. Unfortunately, in the human body chlorogenic acid is metabolized mainly by colonic microflora, because it is very poorly absorbed in the human body (Sato et al. 2011).

The content of vitamin C in whole fruits according to Planchon et al. (2004) was between 2.9 and 25.6 mg/100 g fresh weight for whole fruits of 30-old Belgian apple cultivars. Similarly, Varming et al. (2013) stated that the ascorbic acid content of the 71 analyzed apple cultivars ranged from less than 1 to 27 mg/100 g. Kumar et al. (2018) gave 19–32 mg/100 g fresh weight slightly higher values. According to Planchon et al. (2004), the examined six old apple varieties had three to seven times more ascorbic acid than two commercial varieties. According to Drogoudi et al. (2008) and Pissard et al. (2018) apple peel contains 1.5–9.2 times greater total antioxidant activity and 1.2–8.0 times greater total phenolic content compared with flesh. On the other hand, the content of individual groups of polyphenols in apples can vary significantly; e.g., fruit peel contained much less chlorogenic acid than flesh (Akagić et al. 2019). Statistical analysis showed that a more nutritious peel may be darker, redder, and bluer, while a more nutritious flesh may have a lighter color and lower soluble solid content (Drogoudi et al. 2008).

According to Hallmann and Rembiałkowska (2007) and Rembiałkowska et al. (2007) apple purée made from fruits of six old cultivars showed significantly higher levels of dry matter, phenolic acids, flavonols, and vitamin C than three new conventional cultivars (Lobo, Idared, and Jonagold). However, pasteurization caused the decrease of bioactive compounds in apple purée. The biggest drop in flavonols and phenolic acids was noted in conventional apple purée, vitamin C in organic foods, and the dry matter in the puree from old apple cultivars.

According to Bassi et al. (2017), among 64 apple varieties from South Tyrol, all were characterized by a higher content of ascorbic acid in the peel 2.7–56.0 mg/100 g fresh weight than in the tissue of these fruits 0.1–13.9 mg/100 g fresh weight. Similarly, Łata et al. (2005a) gave significantly higher values for the peel of freshly harvested four commercial apple varieties than for tissue 21–74 and 7–10 mg/100 g fresh weight, respectively. Therefore, in the processing of apples, fruit should be used with the skin. However, this is often impossible due to the bright color of the products (Jarczyk and Płocharski 2010).

From the seventeenth century the saying ‘An apple a day keeps the doctor away’ has been a saying (Oszmiański 2009). These fruits are a rich source of active substances. Based on epidemiological studies, it appears that apples may play a large role in reducing the risk of a wide variety of chronic diseases and maintaining a healthy lifestyle in general. Apple consumption has been most consistently associated with reduced risk of cancer, cardiovascular disease, asthma, and type II diabetes, and may have beneficial effects on vascular function, blood pressure, lipids, inflammation and hyperglycemia, increased lung function, and increased weight loss when compared to other fruits and vegetables and other sources of flavonoids (Chabłowska et al. 2013; Boyer and Liu 2004, Bondonno et al. 2017).

## 15.4 Summary

Old apple varieties can be a promising source of bioactive compounds with potential health benefits, and quality features meet the preferences of demanding consumers for tasty and healthier food who appreciate local fruits with a high content of bioactive ingredients like polyphenols. They can also be a good source of genetic variability, provide resources for appropriate varieties for selective growth to produce new varieties with a higher content of bioactive compounds, that improve human disease resistance and preserve the biodiversity of the area in which they are grown.

## References


Achremowicz B, Oszmianski J, Puchalski C (2016) Antyoksydanty owoców i warzyw. Cz. 2. Antyoksydanty jabłek. Przem Ferm Owoc-Warzyw 4 (60):22–26

Akagić A, Vranac A, F Gaši, Drkenda P, Spaho N, Oručević Žuljević O, Kurtović M, Musić O, Murtić S, Hudina M (2019) Sugars, acids and polyphenols profile of commercial and traditional apple cultivars for processing. Acta Agric Slov 113(2):239–250

Alberti A, Zieliński AAF, Zarco D, Demiate IM, Nogueira A, Mafta LI (2014) Optimisation of the extraction of phenolic compounds from apples using response surface methodology. Food Chem 149:151–158

Arboretm (2019) Stare odmiany jabłoni. Arboretum i Zakład Fizjografii w Bolestraszycach. https://bolestraszyce.com.pl/kolekcje/stare-odmiany-jabloni-2/

Bartolini S, Ducci E (2017) Quality evaluation of local apple varieties: physicochemical and antioxidant proprieties at harvest and after cold storage. Agron Res 15(5):1866–1877

Bassi M, Lubes G, Bianchi F, Agnolet S, Ciesa F, Brunner K, Guerra W, Robatscher P, Oberhuber M (2017) Ascorbic acid content in apple pulp, peel, and monovarietal cloudy juices of 64 different cultivars. Int J Food Prop 20(sup3):S2626–S2634

Begić-Akagić A, Spaho N, Gaši F, Drkenda P, Vranac A, Meland M, Salkić B (2014) Sugar and organic acid profiles of the traditional and international apple cultivars for processing. J Hyg Eng Des 7:190–196

Bondonno N, Bondonno CP, Ward NC, Hodgson JM, Croft KD (2017) The cardiovascular health benefits of apples: Whole fruit vs. isolated compounds. Trends Food Sci Technol 69:243–256

Borowska A (2013) Zmiany na rynku jabłek w Polsce z uwzględnieniem jabłek regionalnych. Roczniki Naukowe Ekonomii Rolnictwa i Rozwoju Obszarów Wiejskich, 100(1):152–167

Boyer J, Liu RH (2004) Apple phytochemicals and their health benefits. Nutr J 3.1:1–15

Byszewska I (2011) Produkty lokalne i regionalne: wyroby o szczególnej jakości. FAOW Forum Aktywizacji Obszarów Wiejskich. In: Kamiński R, Mierzejewski M, Woźniak A (Ed) Produkty lokalne – szanse i wyzwania. http://mazowieckie.ksow.pl/fileadmin/user_upload/mazowieckie/pliki/FAOW_Produkty_lokalne.pdf

Ceymann M, Arrigoni E, Schärer H, Nising AB, Hurrell RF (2012) Identification of apples rich in health-promoting flavan-3-ols and phenolic acids by measuring the polyphenol profile. J Food Comp Anal 26(1):128–135

Chabłowska B, Piasecka-Jóźwiak K, Rozmierska J, Szkudzińska-Rzeszowiak E. Kliszcz M (2013) Fermentacja mlekowa jabłek z upraw ekologicznych sposobem na otrzymanie nowego asortymentu produktów – biosoków. J Res Applic Agric Eng 3:71–77

Collective work (2014) Polsko-słowackie smaki, Pol'sko-slovensk´e chute, Polish and Slovak tastes. Sucha Beskidzka: Stowarzyszenie Lokalna Grupa Działania "Podbabiogórze"; Tuchów: Pogórzańskie Stowarzyszenie Rozwoju; Oravsk´a Polhora: Miestna Akˇcn´a Skupina Babia Hora

Donno D, Beccaro GL, Mellano MG, Torello Marinoni D, Cerutti AK, Canterino S, Bounous, G (2012) Application of sensory, nutraceutical and genetic techniques to create a quality profile of ancient apple cultivars. J Food Qual 35(3):169–181

Drogoudi PD, Michailidis Z, Pantelidis, G (2008) Peel and flesh antioxidant content and harvest quality characteristics of seven apple cultivars. Sci Hortic 115(2):149–153

Dziubiak M (2006) O dawnych odmianach uprawnych jabłoni i ich pochodzeniu. Rocznik Dendrologiczny 54:51–66

Eberhardt M, Lee C, Liu RH (2000) Antioxidant activity of fresh apples. Nature 405:903–904

Feliciano RP, Antunes C, Ramos A, Serra AT, Figueira ME, Duarte CM, de Carvalho A, Bronze MR (2010) Characterization of traditional and exotic apple varieties from Portugal. Part 1—Nutritional, phytochemical and sensory evaluation. J Funct Food 2(1):35–45

Fołta M, Krośniak M, Piotrowska A, Zachwieja Z, Błaszczyk J (2007) Właściwości antyoksydacyjne oraz zawartość polifenoli w sokach z jabłek starych i nowych odmian. Żyw Człow Metab 34:344:549

GAIN Report 2018. Fresh Deciduous Fruit Annual. EU-28. Global Agricultural Information Network, Report Number: AU1809. USDA Foreign Agricultural Service, https://gain.fas.usda.gov/Recent%20GAIN%20Publications/Fresh%20Deciduous%20Fruit%20Annual_Vienna_EU-28_10-30-2018.pdf

Gasik A, Mitek M (2012) Przydatność technologiczna owoców do produkcji soków. In: Collective work. Przetwórstwo owoców na poziomie gospodarstwa. Centrum Doradztwa Rolniczego w Brwinowie, Radom. http://ksow.pl/fileadmin/user_upload/ksow.pl/Projekty_z_konkursu_I_KSOW/Przetw%C3%B3rstwo/przetworstwo_owocow_poprawa_1.pdf

Gheyas F, Blankenship SM, Young E, McFeeters R (1997) Dietary fibre content of thirteen apple cultivars. J Sci Food Agric 75(3):333–340

Gorinstein S, Zachwieja Z, Folta M, Barton H, Piotrowicz J, Zemser M, Weisz M, Trakhtenberg S, Màrtín-Belloso O (2001) Comparative contents of dietary fiber, total phenolics, and minerals in persimmons and apples. J Agric Food Chem 49(2):952–957

GUS (2019) Production of agricultural and horticultural crops in 2018. Statistical information. Główny Urząd Statystyczny, Statistics Poland, Warsaw. https://stat.gov.pl/download/gfx/portal informacyjny/pl/defaultaktualnosci/5509/9/17/1/produkcja_upraw_rolnych_i_ogrodniczych_w_2018_.pdf

Hallmann E, Rembiałkowska E (2007) Influence of thermal processing on bioactive compounds content in apple puree prepared from organic fruits of old and new apple cultivars. Po J Nat Sci Suppl 4:37–42

Iaccarino N, Varming C, Petersen, MA, Viereck N, Schütz B, Toldam-Andersen TB, Randazzo A, Engelsen SB (2019) Ancient Danish apple cultivars—a comprehensive metabolite and sensory profiling of apple juices. Metabolites 9(7), 139:1–17

Jakobek L, Barron AR (2016) Ancient apple varieties from Croatia as a source of bioactive polyphenolic compounds. J Food Comp Anal 45:9–15

Jarczyk A, Płocharski W (2010) Technologia produktów owocowych i warzywnych. Tom 1. Wyższa Szkoła Ekonomiczno-Humanistyczna. Skierniewice

Kiczorowski P (2019) influence of NPK minerals and biostimulants on the growth, yield, and fruit nutritional value in cv.'Šampion' apple trees growing on different rootstocks. Acta Sci Pol, Hort Cult 18(1):197–205

Kiczorowski P, Kiczorowska B, Krawiec M, Kapłan M (2018) Influence of different rootstocks on basic nutrients, selected minerals, and phenolic compounds of apple CV.'Šampion'. Acta Sci Pol, Hort Cult 17(4):167–180

Kostrz M, Satora P (2016) Jabłko - surowiec do produkcji napojów alkoholowych. In: Duda-Chodak A, Najgebauer-Lejko D, Drożdż I, Tarko T. Rola procesów technologicznych w kształtowaniu jakości żywności. Oddział Małopolski Polskiego Towarzystwa Technologów Żywności, Kraków, pp 118–6

Kumar P, Sethi S, Sharma RR, Singh S, Saha S, Sharma VK, Verma MK, Sharma SK (2018) Nutritional characterization of apple as a function of genotype. J Sci Technol 55(7):2729–2738

Kunachowicz H, Przygoda B, Nadolna I, Iwanow K (2017) Tabele składu i wartości odżywczej żywności. PZWL Wydawnictwo Lekarskie, Warszawa
Lodeta KB, Vujević B, Čmelik Z, Kereša S (2017) Morphological traits of five traditionally grown domesticated apple varieties in Bjelovar Bilogora County. Proceedings of 52nd Croatian and 12th International Symposium on Agriculture 12th-17th February 2017, Dubrovnik, Valamar Lacroma, Croatia p 597–601
Lončarić A, Skendrović Babojelić M, Kovač T, Šarkanj B (2019) Pomological properties and polyphenol content of conventional and traditional apple cultivars from Croatia. Food in Health and Disease, Scientific-Professional Journal of Nutrition and Dietetics 8(1):19–24
Łata B (2007) Relationship between apple peel and the whole fruit antioxidant content: year and cultivar variation. J Agric Food Chem 55(3):663–671
Łata B, Przeradzka M, Bińkowska M (2005a) Great differences in antioxidant properties exist between 56 apple cultivars and vegetation seasons. J Agric Food Chem 53(23):8970–8978
Łata B, Tomala K (2007) Apple peel as a contributor to whole fruit quantity of potentially healthful bioactive compounds. Cultivar and year implication. J Agric Food Chem 55(26):10795–10802
Łata B, Trąmpczyńska A, Oleś M (2005b) Antioxidant content in the fruit peel, flesh and seeds of selected apple cultivars during cold storage. Folia Hort 17(1):47–60
Łuczaj Ł (2011) Dziko rosnące rośliny jadalne użytkowane w Polsce od połowy XIX w. do czasów współczesnych. Etnobiologia Polska 1:57–125
Makosz E (2015) Przyszłość polskich jabłek. Biul Inf ARR 3:10–15
Malik A, Kiczorowska B, Zdyb J (2009) Zawartość kwasów fenolowych w częściach jadalnych wybranych odmian jabłek. Roczn PZH, 60(4):333–336
Markowski J, Celejewska K, Rosłonek A, Kosmala M (2017) Impact of different thermal preservation technologies on the quality of apple-based smoothies. LWT-Food Sci Technol 85:470–473
Markowski J, Płocharski W (2006) Zmiany składu związków fenolowych przy przerobie jabłek na soki i przeciery. Przem Ferm Owoc-Warz 4:33–36
Markowski J, Płocharski W, Pytasz U. Rutkowski K (2012) Owoce, warzywa, soki – ich kaloryczność i wartość odżywcza na tle zapotrzebowania na energię i składniki odżywcze. Cz. 1. Kaloryczność i mit o wpływie na otyłość. Przem Ferm Owoc-Warz 9:24–27
Markuszewski B, Kopytowski J (2008) Transformations of chemical compounds during apple storage. Scientific Works of the Lithuanian Institute of Horticulture and Lithuanian University of Agriculture. Sodininkyste Ir Daržininkyste 27(2):329–338
Mikołuszko W (2018) Czy stare odmiany jabłek znikną z polskich ogrodów? Polityka, https://www.polityka.pl/tygodnikpolityka/nauka/1730562,1,czy-stare-odmiany-jablek-znikna-z-polskich-ogrodow.read
Min Rol (2018) Produkty zarejestrowane jako Chronione Nazwy Pochodzenia, Chronione Oznaczenia Geograficzne oraz Gwarantowane Tradycyjne Specjalności. Ministerstwo Rolnictwa https://www.gov.pl/web/rolnictwo/produkty-zarejestrowane-jako-chronione-nazwy-pochodzenia-chronione-oznaczenia-geograficzne-oraz-gwarantowane-tradycyjne-specjalnosci
Mitek M, Gasik A (2007) Polifenole w żywności. Właściwości przeciwutleniające. Przem Spoż 9:36–39, 44
Naszkowska K (2017) Jemy coraz mniej jabłek, choć produkujemy ich najwięcej w Europie, Wyborcza.pl, http://wyborcza.pl/7,155287,21998982,jemy-coraz-mniejjablek-choc-produkujemy-ich-najwiecej-w-europie.html
Neri L, Santarelli V, Di Mattia CD, Sacchetti G, Faieta M, Mastrocola D, Pittia P (2019) Effect of dipping and vacuum impregnation pretreatments on the quality of frozen apples: A comparative study on organic and conventional fruits. J Food Sci 84(4):798–806
Nosecka B, Trojanowicz P (2016) Zapotrzebowanie na jabłka zakładów przetwórczych w Polsce. Przem Ferm Owoc-Warz 4(60):10–13
Nosecka B, Mierwiński J (2016) Rynek w UE. Rynek Owoców i Warzyw, 48:8–11
Oszmiański J (2007) Soki owocowe o wysokiej aktywności biologicznej. Przem Ferm Owoc-Warz 4:12–16

Oszmiański J (2009) Nowe trendy w produkcji soków i nektarów jabłkowych. Przem Ferm Owoc-Warz 4:12–15

Oszmiański J, Lachowicz S, Gławdel E, Cebulak T, Ochmian I (2018) Determination of phytochemical composition and antioxidant capacity of 22 old apple cultivars grown in Poland. Eur Food Res Technol, 244(4):647–662

Oszmiański J, Wojdyło A (2006) Soki naturalnie mętne – dobry kierunek w przetwórstwie jabłek. Przem Ferm Owoc-Warz 2:20–22

Panzella L, Petriccione M, Rega P, Scortichini M, Napolitano A (2013) A reappraisal of traditional apple cultivars from Southern Italy as a rich source of phenols with superior antioxidant activity. Food Chem 140(4):672–679

Pissard A, Baeten V, Dardenne P, Dupont P, Lateur M (2018) Use of NIR spectroscopy on fresh apples to determine the phenolic compounds and dry matter content in peel and flesh. Biotechnol Agron Soc Environ 22(1):3–12

Planchon V, Lateur M, Dupont P, Lognay G (2004) Ascorbic acid level of Belgian apple genetic resources. Sci Hortic 100(1):51–61

Poznański G (2014) Tradycyjne sadownictwo na terenie Pogórza Przemyskiego. In: praca zbiorowa. Stare sady owocowe i tradycyjne ogrody wiejskie na terenie Pogórza Przemyskego – dziedzictwem kulturowo-przyrodniczym. Fundacja Dziedzictwo Przyrodnicze, Leszczawa Dolna 2014. https://przyrodnicze.org/wp-content/uploads/2015/10/STARE-SADY-OWOCOWE-NET.pdf

Rembiałkowska E, Hallmann E, Adamczyk M (2004) Porównanie wybranych cech wartości odżywczej jabłek z produkcji ekologicznej i konwencjonalnej. Bromat Chem Toksykol Suppl p 201–207

Rembiałkowska E, Hallmann E, Kapron L, Rusaczonek A (2007) Ocena wartości przeciwutleniającej oraz zawartości związków bioaktywnych w kremogenach wykonanych z owoców starych i nowych odmian jabłoni. Żywn Nauk Technol Jakość 14(1):105–112

Sanoner P, Guyot S, Marnet N, Molle D, Drilleau JF (1999) Polyphenol profiles of French cider apple varieties (Malus domestica sp.). J Agric Food Chem 47(12):4847–4853

Sato Y, Itagaki S, Kurokawa T, Ogura J, Kobayashi M, Hirano T, Iseki K (2011) In vitro and in vivo antioxidant properties of chlorogenic acid and caffeic acid. Int J Pharm 403(1–2):136-138

Sazońska B (2010) Uprawa wybranych starych gatunków roślin uprawnych. Centrum Doradztwa Rolniczego w Brwinowie Oddział w Radomiu. https://www.cdr.gov.pl/images/wydawnictwa/2010/2010-UPRAWA-WYBRANYCH-STARYCH-GATUNKOW-ROSLIN-UPRAWNYCH.pdf

Sieliwanowicz B, Hałasińska AG, Trzcińska M, Jakubowski A, Lipowski J, Skąpska S (2005) Zmiany zawartości związków fenolowych, parametrów barwy i aktywności przeciwutleniającej w czasie przechowywania soków z wybranych odmian jabłek. Acta Sci Pol Technol Aliment 4:83–91

Solomakhin A, Blanke MM (2010) Can coloured hailnets improve taste (sugar, sugar: acid ratio), consumer appeal (colouration) and nutritional value (anthocyanin, vitamin C) of apple fruit? LWT-Food Sci Technol 43(8):1277–1284

Souci SW, Fachmann W, Kraut H (2015) Food composition and nutrition tables. Medpharm Scientific Publishers, Stuttgart. https://www.sfk.online/#/home. Available 28 Jan 2015

Strojewska I (2015) Spożycie owoców, warzyw i ich przetworów w Polsce. Biul Inf ARR 3:2–9

Szajdek A, Borowska J (2004) Właściwości przeciwutleniające żywności pochodzenia roślinnego. Żywn Nauk Technol Jakość 4:5–28

Szewczyk ZP (2010) Chleb nasz powszedni czyli Kuchnia Lachów Sądeckich. Wydawca Biblioteka Gminna w Podegrodziu. Nowy Sącz 2010

Teleszko M, Kolniak J, Oszmiański J (2010) Wpływ odmiany jabłek na zmętnienie i barwę naturalnie mętnych soków. In: Wojtatowicz M, Kawa-Rygielska J (Red.) Jakość i prozdrowotne cechy żywności. Wydawnictwo Uniwersytetu Przyrodniczego we Wrocławiu, p 27–38

Teleszko M, Kolniak J, Wojdyło A, Oszmiański J (2010) Wpływ odmiany jabłek na zawartość polifenoli i aktywność przeciwutleniającą w sokach naturalnie mętnych. Przem Ferm Owoc-Warz 7–8:40-43

Teleszko M, Wojdyło A (2014) Bioactive compounds vs. organoleptic assessment of 'smoothies' type products prepared from selected fruit species. Int J Food Sci Technol 49:98–106

Turek K, Słupski J, Gębczyński P, Skoczeń-Słupska R, Skoczylas Ł, Tomf A 2015 Lactic acid fermentation of apples. In: Staruch L (ed) Laboralim. Zborník vedeckých prác. Fakulta chemickej a potravinárskej technológie STU, Bratislava, p 367–372

Turek K, Słupski J, Tabaszewska T, Skoczylas Ł, Tomf-Sarna A, Skoczeń-Słupska R (2016) Soki jabłkowe naturalnie mętne – produkty bogate w związki biologicznie czynne. In: Duda-Chodak A, Najgebauer-Lejko D, Drożdż I, Tarko T (ed) Rola procesów technologicznych w kształtowaniu jakości żywności. Wyd. Oddział Małopolski Polskiego Towarzystwa Technologów Żywności, Kraków, p 127–134

Varming C, Petersen MA, Toldam-Andersen TB (2013) Ascorbic acid contents in Danish apple cultivars and commercial apple juices. LWT-Food Sci Technol 54(2):597–599

Wojdyło A, Oszmiański J, Laskowski P (2008) Polyphenolic compounds and antioxidant activity of new and old apple varieties. J Agric Food Chem 56(15):6520–6530

Żygała E, Skrzyński J (2008) Wybrane cechy pomologiczne niektórych genotypów jabłoni z kolekcji w Bolestraszycach. Zesz. Nauk. Instytutu Sadownictwa i Kwiaciarstwa, 16:61–68

Żygała E, Piórecki N (2016) Dawne założenia dworsko-ogrodowe jako źródło starych odmian drzew owocowych. Former manor houses and gardens as a source of old cultivars of fruit trees. In: Dolatowski J, Dolatowska A, Dudek-Klimiuk J (ed) Drzewa, parki i ogrody. Dziedzictwo kultury i natury Małopolski Wschodniej. Materiały VIII Zjazdu Polskiego Towarzystwa Dendrologicznego, Bolestraszyce i Lwów, 23–25 września 2016 r. https://www.ptd.pl/ptd/wp-content/download/wydawnictwaPTD/materialy_VIIIzjazd.pdf#page=62

# Chapter 16
# Edible Mushrooms of the Polish Carpathians

**Emilia Bernaś, Jacek Słupski, and Piotr Gębczyński**

**Abstract** For centuries mushrooms have been served on Polish tables, including Carpathian tables. At the turn of the nineteenth and twentieth centuries, 22 species of edible mushrooms were collected in the area of Gorlice and Biecz (Poland), including *Morchella* spp., Agaricus spp., Boletus spp., *Leccinum* spp., Russula spp. (truffles), *Suillus* spp., Lactarius spp., and *Paxillus involutus*. Among the species that were collected were those consumed as raw (Lactarius volemus) or after processing, as well as those that were used for medicinal purposes or in the household, for example, to repel insects. Also, nowadays mushrooms are eaten as raw (sprinkled with salt), fried on the oven stove lid or in a pan, pickled and dried (an addition to dishes), and as tinctures (*Fomitopsis betulina*). One of the old methods of preserving mushrooms is lactic acid fermentation. The Boletus edulis, Leccinum spp., Suillus spp., Xerocomus sp., Lactarius deliciosus, and Tricholoma equestre were the main species used for this.

**Keywords** Carpathian region · Mushrooms · Chemical composition · Regional name

## 16.1 Introduction

In the last few years, there has been increasing worldwide production of edible mushrooms, from 5.9 million tons in 2007 to 10.2 million tons in 2017 (FAOStat 2019). The increase in mushroom production is related to the increase in their consumption

E. Bernaś (✉) · J. Słupski · P. Gębczyński
Department of Plant Products and Nutrition Hygiene, Faculty of Food Technology, University of Agriculture,
122 Balicka Street, 30-149 Krakow, Poland
e-mail: emilia.bernas@urk.edu.pl

J. Słupski
e-mail: jacek.slupski@urk.edu.pl

P. Gębczyński
e-mail: piotr.gebczynski@urk.edu.pl

© Springer Nature Switzerland AG 2022 
J. Hernik et al. (eds.), *Cultural Heritage—Possibilities for Land-Centered Societal Development*, Environmental History 13,
https://doi.org/10.1007/978-3-030-58092-6_16

from about 1 kg/per capita in 1997 to about 5 kg/per capita in 2013. The highest consumption was in China, equaling 22 kg/per capita/per year, while in India it is only 0.08 kg (Royse et al. 2017). As shown in the author's citations, the mushroom market is represented by medicinal (38%), wild (8%), and cultivated (54%) mushrooms. Poland is one of the main producers and exporters of fresh and preserved mushrooms in Europe. Nearly 80% of Poland's mushrooms are destined for the fresh market; the main importers are Russia and Western European countries. The major cultivated species in the world are *Lentinula edodes* (about 22% of the whole world cultivation), *Pleurotus* spp. (19%), *Auricularia* spp. (17%), and *Agaricus bisporus* (15%). In Poland the situation is different because the main cultivated species is *Agaricus bisporus*; the next place is occupied by *Pleurotus ostreatus*. It is estimated that about 35% of imported mushrooms in the world come from Poland; in Europe, it is about 90%. In Poland, a popular form of spending one's free time is collecting wild-growing mushrooms. Nowadays, the main collected species are *Boletus* spp., *Xerocomus* spp., *Suillus* spp., *Leccinum* spp., *Armillaria mellea, Cantharellus cibarius, Lactarius delicious*, and *Tricholoma equestre*. In the past, except for the species listed above, others species were collected, for example, *Morchella* spp., Agaricus spp., *Leccinum* spp., Russula spp., and Tuber spp. (truffles). Some species (*Paxillus involutus* ("olszówka"), Gyromitra esculenta, etc.) were collected, but nowadays we know that they contain poisonous substances and therefore they are no longer collected . In turn, some other species were collected, but now people don't collect them because the collecting tradition has partially disappeared. These species include, among others, *Lactarius piperatus, Lactarius volemus, Lactarius salmonicolor*, and *Lactarius deterrimus*. Poland is one of the main exporters of wild-grown mushrooms in Europe, and the main importers are Germany, France, and Italy. In Europe, only a few species of edible fungi and products obtained from them are important in international trade; they are *Agaricus bisporus, Boletus edulis, Tuber melanosporus, Cantharellus cibarius*, and *Morchella esculenta* (Anonim 1 2019; Domaszewicz 2017; Sanchez and Diaz-Godinez 2004).

## 16.2 Chemical Composition of Edible Mushrooms

Mushrooms can play an important role in human nutrition, but their chemical composition depends, among other things, on the mushroom species, and on the place of picking (Table 16.1). They are rich sources of high-quality protein, including all the essential amino acids, especially lysine. They contain a high amount of vitamins from the B-group (B1, B2, B3, B12, folic acid), ergosterol (provitamin D2), minerals, and antioxidants (phenols, ergothioneine). Because mushrooms are low in fat and sugar, their glycemic index is <3 (Bernaś et al. 2006; Bernaś and Jaworska 2016, 2017; Jasinghe and Perera 2006; Jaworska and Bernaś 2013; Kalac 2013; Table 16.1).

Mushrooms contain many health-promoting substances, like lovastatin (*Pleurotus ostreatus, Agaricus bisporus*), $\gamma$-aminobutyric acid (GABA) (*Flammulina velutipes, Boletus edulis*), ergothioneine (*Pleurotus* spp.), coprine (species of the

**Table 16.1** Minerals in selected wild-growing mushrooms picked in different places of Poland, in 100 g dm (Chudzyński et al. 2007; Falandysz et al. 2007, 2008; Florczak et al. 2014; Kowalewska et al. 2007)

| Place in Poland | Mushroom species | Total nitrogen (g) | Cu (mg) | Zn (mg) | Fe (mg) | Ca (mg) | Mg (mg) | K (mg) |
|---|---|---|---|---|---|---|---|---|
| City of Łódź | *Auricularia auricula-judae* | 2.65 | 0.2 | 4 | 32 | 142 | 147 | * |
| | *Pleurotus ostreatus* | 3.64 | 0.3 | 4 | 8 | 83 | 132 | * |
| | *Flammulina velutipes* | 4.12 | 0.3 | 2 | 14 | 79 | 152 | * |
| Tatra Mountains | *Boletus edulis* | 7.25 | 1.1–8.7 | 8–29 | 1–5 | 2–11 | 69–86 | 2000–2900 |
| Sudety Mountains | *Boletus edulis* | 7.38 | 1.9–8.4 | 14–36 | 2–14 | 4–11 | 53–120 | 1800–3300 |
| Western Beskid | *Suillus grevillei* | * | 0.4–3.9 | 3–26 | 4–160 | 4–86 | 30–120 | 1100–3200 |
| Lubelska Upland | *Leccinum rufum* | * | 2–11 | 2–24 | 3–29 | 3–41 | 89–120 | 2100–4500 |
| Lubelskie Voivodeship | *Macrolepiota procera* | * | 7–32 | 25–71 | 7–27 | 25–420 | 140–270 | 1200–5200 |

*Not analyzed

genus *Coprinus*), purine (*P. ostreatus, Grifola frondosa*), and serotonin (*A. bisporus, Xerocomus badius*). Many medicinal properties have been attributed to mushrooms, including anticancer properties, inhibition of platelet aggregation, reduction of blood cholesterol concentrations, prevention or alleviation of heart disease, reduction of blood glucose levels, prevention or alleviation of infections caused by bacterial, viral, fungal, and parasitic pathogens (Cheung 2008).

## 16.3 Mushrooms Species in Poland and Polish Carpathian Region and Their Processing

The Carpathian Mountains are one of the richest in Europe in terms of species richness and ecological value. Natural or seminatural forest ecosystems are the most valuable ecosystems together with man-made meadows and pastures. One of the elements of this ecosystem is mushrooms. The first documented mention about the consumption of wild-growing mushrooms in Poland comes from the "Passion of the Holy Wojciech Martyr", the so-called "Passion from Tegernsee" in the year 1004. Until the eighteenth century, mushroom picking for food and medicine was not very common in Poland. Only people from rural areas, forestry workers ("budnicy"), people hiding in the woods during the so-called bad air miasmas (outbreaks of dangerous infectious diseases), and those suffering from poverty—especially during periods of famine and wars consumed it. Only at the turn of the eighteenth and nineteenth centuries, when scientific research in the fields of botany, systematics, and geography in Europe and Poland developed, interest in mushrooms and their use increased (Grzywacz 2015). In Poland, including the Carpathians region, there are many species of edible mushrooms. *Boletus* spp., *Xerocomus* spp., *Leccinum* spp., *Lactarius* spp., *Cantharellus cibarius*, *Morchella* spp., and *Suillus* spp. are some of the most frequently collected. Chachuła et al. (2016) in four nature reserves of Polish Carpathians in 2006–2007 and 2011–2014 found some species of edible mushrooms, including *Lactarius salmonicolor, Tuber puberulum, Russula nigricans, Lactarius piperatus, Boletus edulis, Cantharellus cibarius, Armillaria gallica, Cordyceps militaris, Agaricus sylvaticus,* and *Boletus reticulatus.*

The impoverishment of forests in Poland, observed over the centuries, forced the necessity to protect the species of edible mushrooms, and thus to determine the number of species allowed for marketing on the market. In Polish law in the thirteenth and fourteenth centuries, there were no legal regulations regarding the picking of mushrooms. At that time the forests were available to everyone. The "Forest Act" of 1567 changed this situation and allowed the picking of *Boletus* spp., *Lactarius* spp., and mushrooms with a hymenophore of the so-called "bedłki". At the beginning of the nineteenth century, the first official lists of edible mushrooms allowed for trade at the marketplaces appeared. The following species were allowed to be traded at that time: *Boletus* spp., *Agaricus* spp., *Lactarius deliciosus, Russula spp., Xerocomus* spp., *Morchella* spp., *Sparassis crisp*, *Tuber* spp., *Tricholoma equestre*, but we have limited knowledge about other species (Grzywacz 2015). Nowadays, in Poland, 47 species of edible mushrooms are permitted for trading in the market, including several from *Suillus* spp., *Agaricus* spp., *Leccinum* spp., *Lactarius* spp., *Tuber* spp., and *Xerocomus* spp. The actual number of edible species is definitely larger, but due to their rare occurrence, they are on the national list of endangered species. Such species include, for example, *Morchella esculentae.*

The impoverishment of forests in mountainous regions has been observed since about the eighteenth century when their exploitation increased and their area was limited. This phenomenon has been largely influenced by the so-called "wolnizny"—relief, that was granted to settlers for the use of the land available due to deforestation. In the case of the Pieniny Mountains, the most intensive period of destruction of forests is in the nineteenth and twentieth centuries. At that time, the felling of individual trees was often replaced with more extensive logging, chosen mainly by landowners, in order to quickly improve their profits. This phenomenon was connected with the almost complete liquidation of forest biocenoses and intensification of erosion, which in extreme cases leads to the creation of wasteland. Many places were reforested, but often they were not local species of trees (beech, fir), but foreign species such as spruce, which naturally did not occur in the area of the Pieniny Mountains. This phenomenon had an impact on the fauna and flora of these areas (Karwowski 2014).

An especially difficult period for the Polish nation, including those inhabiting the mountain areas, was the period of the First and Second World Wars as well as the post-war periods when the nation had to face hunger. This phenomenon had a very large impact on diet and was associated with the necessity to obtain new sources of food and income. This increased the collection of mushrooms from the forest. The more valuable species were sold, while others were consumed, sometimes even with the risk to health. The following mushroom species can be mentioned as sold: Boletus spp. (prawdziwki), *Leccinum* spp. (kozaki) (*L. scabrum, L. rufum,* L. *pseudoscabrum*), *Cantharellus cibarius* (kurki), and *Lactarius deliciosus* (rydze). Inferior species included: Xe*rocomus* spp., *Macrolepiota* spp., *Armillaria* spp., *Suillus* spp., *Russula virescens, Tricholoma equestre,* and *Paxillus involutus* (currently classified as poisoned). According to Orłoś (1946), in Poland in the period before Second World War, the picked mushroom species, taking into account their commercial value, were as follows, in the order from the most valuable to the least valuable: Boletus edulis, *Cantharellus cibarius,* Agaricus spp., *Lactarius deliciosus, Gyromitra esculenta, Tricholoma equestre, Leccinum scabrum, Suillus bovinus, Morchella esculenta, Suillus luteus, Armillaria mellea, Marasmius oreades,* edible species from *Russula* spp., *Paxillus involutus, Leccinum aurantiacum, Suillus granulatus, Gyroporus cyanescens, Macrolepiota procera, Sarcodon imbricatus,* and *Tuber aestivum.* On this list, attention is drawn to the presence of mushrooms nowadays regarded as belonging to poisonous varieties: *Gyromitra esculenta, Paxillus involutus,* as well as the absence of some currently picked mushrooms, like Pleurotus ostreatus.

In the nineteenth and twentieth centuries, 22 species of edible mushrooms were collected in the area of Gorlice and Biecz (Poland), including *Morchella* spp., Agaricus spp., Boletus spp., *Leccinum* spp., Russula spp. truffles, *Suillus* spp., Lactarius spp., and *Paxillus involutus.* The number of mushroom species used from the nineteenth century onward has not been changed significantly and includes about 20, in contrast to the observed disappearance of wild plants. However, it should be emphasized that the use of some species is disappearing, including *Paxillus involutus* ((Batsch: Fr.) Fr.) ("olszówka") and Gyromitra esculenta ((Pers.: Fr.) Fr.) as a result of a campaign carried out for several decades, associated with the occurrence of poisoning with this mushroom (Bartnicka-Dąbkowska 1964; Burda 1998; Łuczaj 2011).

Edible mushrooms, due to the high popularity of consumption, gained many common names, which were often associated with their appearance, or place of occurrence. In many cases, these names are characteristic only for a given Polish region and therefore they can be treated as a part of its cultural heritage. Examples of names commonly used in the Carpathian region were listed in Table 16.2, e.g., Polish female's names like "Marysia", "Zuzia", "Zosia" for mushrooms from Russula spp. or "pociece" for Neoboletus erythropus. Unfortunately, these names, the customs associated with picking mushrooms and old ways of preparing dishes from them, are being slowly forgotten. On the one hand, this phenomenon is associated with the increasing number of protected species; on the other hand, with a change in lifestyle and civilizational progress. Many species of mushrooms, formerly harvested, nowadays are not known; for example, *Catathelasma imperiale, Sarcodon* spp., *Hydnellum* spp., *Lactarius* spp., *Hygrocybe* spp. According to surveys conducted by the Department of Mycology and Forest Phytopathology of the Warsaw University of Life

**Table 16.2** Regional name of mushrooms eaten and picked in Polish Carpathian region (Bartnicka-Dąbkowska 1964; Grzywacz 2015; Referowska-Chodak 2015)

| Regional name | Polish name | Latin name |
|---|---|---|
| Koziabródka | Szmaciak gałęzisty (siedzuń sosnowy) | *Sparassis crispa* |
| Grzyb | Borowik szlachetny | *Boletus edulis* |
| Hubka | Hubiak pospolity | *Fomes fomentarius* |
| Gąbka modrzewiowa | Pniarek lekarski (modrzewiowy) | *Laricifomes officinalis* |
| Maślak, pępek | Maślak zwyczajny | *Suillus luteus* |
| Bedłki | Opieńki, grzyby blaszkowe | *Armillaria spp.*, lamellar mushrooms |
| Rydz | Mleczaj rydz | *Lactarius deliciosus* |
| Lisówka | Pieprznik jadalny (kurka) | *Cantharellus cibarius* |
| Babie uszy | Piestrzenica jadalna (kasztanowata) | *Gyromitra esculenta* |
| Gąska żółta | Gąska zielonka | *Tricholoma equestre* |
| Podbrzeźniak | Koźlarz babka | *Leccinum scabrum* |
| Sitarek | Maślak sitarz | *Suillus bovinus* |
| Maślacz, ślimak | Maślak zwyczajny | *Suillus luteus* |
| Smardz | Smardz jadalny | *Morchella esculenta* |
| Podpieńka | Opieńka miodowa | *Armillaria mellea* |
| Surojadki | Gołąbki jadalne | *Russula* spp. *(only edible)* |
| Marysia<br>Zuzia<br>Zosia | Gołąbek wyborny<br>Gołąbek zielonawy<br>Gołąbek zielonofioletowy | *Russula vesca*<br>*Russula virescens*<br>*Russula cyanoxantha* |
| Olszówka | Krowiak podwinięty | *Paxillus involutus* |
| Podosiniak, osiczan | Koźlarz czerwony | *Leccinum aurantiacum* |
| Truflica | Trufla letnia | *Tuber aestivum* |
| Sarna, sarniak, cygany | Kolczak dachówkowaty (sarniak dachówkowaty) | *Sarcodon imbricatus* |
| Siniak | Piaskowiec modrzak | *Gyroporus cyanescens* |
| Pępek | Maślak ziarnisty | *Suillus granulatus* |
| Chrząszcz, biel | Mleczaj biel | *Lactarius vellereus* |
| Mleczaj krówka | Mleczaj smaczny | *Lactarius volemus* |
| Pociece | Borowik ceglastopory | *Neoboletus erythropus* |
| Czartopłochy | Gołąbek czerwony (wymiotny), zasłonaki, wieruszki, lejkówki, krowiaki (grzyby trujące) | *Russula emetica, Paxillus spp., Clitocybe* spp., *Entoloma spp.* (poisonous mushrooms) |
| Surojadki | Gołąbki jadalne | Edible species from *Russula* spp. |

Sciences (SGGW) (Poland), the average high school graduate can usually name 5–8 mushroom species. Nowadays, picking mushrooms is one of the ways to spend one's free time. According to expert estimates, about 90% of picked mushrooms are eaten by pickers. The mushrooms season lasts, depending on the weather conditions, from several weeks to 4 or 5 months (Grzywacz 2015).

The sale of mushrooms picked in forests may also be a way to improve the home's budget, because the market value of the annual harvest of edible mushrooms in Polish forests has been estimated at 740 million PLN. 60% of this value is used by pickers, 30% is processed or exported (mainly to Germany, France, and Belgium), while 10% is sold by the road and at marketplaces. It is estimated that about 40–60 species of mushrooms are collected in Poland, with species suitable for consumption being 1100–1400. In other European countries, the legislation regarding harvesting mushrooms from natural sites is often more restrictive than in Poland. Examples are Germany, Austria, or Switzerland, where the method of picking and the maximum amount of picked mushrooms at one time is regulated, up to 1–5 kg of fresh mushrooms, depending on the country. In some countries, for example, in Austria, the hours of gathering mushrooms are regulated (7.00–19.00), and the period during the year—from June to October; the law prohibits the organization of collective mushroom picking. In contrast, the inhabitants of Scandinavia do not have the habit of picking mushrooms because they usually do not eat them (Grzywacz 2008, 2015).

Besides for purposes of consumption, mushrooms were also used to treat humans and animals (Table 16.3). St. Hildegard of Bingen, the prioress of the Benedictine monastery, used mushrooms as a cure for various diseases. At present, the German Society for the Knowledge of Vital Mushrooms is operating in Bingen, on the Rhine (Germany), referring to the old traditions. Similarly, Marcin from Urzędów (Poland), who is a botanist, physician, priest, professor of the Krakow Academy in "Herbarz Polski", mentions medicinal mushrooms, among others larch sponge—*Laricifomes officinalis* (Grzywacz 2015).

## 16.4 Summary

In the Polish Carpathian regions, mushrooms have been playing an important role, but over time the spectrum of their applications has gradually narrowed. They were used as food, in medicine for treating people and animals, and in folk religious ceremonies. Initially, their use was a feature of rural circles, later in higher society too (mushroom picking has been a way of spending free time). The main sources of information on traditions related to the collection and use of fungi are oral communications, which according to the UNESCO Convention (2003) can be included in the intangible cultural heritage of Poland.

**Table 16.3** Old ways of using mushrooms in Poland for medicinal purposes (Grzywacz 2015; Referowska-Chodak 2015; Szczepkowski 2012; Trojanowska 2001)

| The way of application | Application | Mushroom species |
|---|---|---|
| Alcoholic tincture | Fever, indigestion, diarrhea | *Fomes* spp. from a pear tree (Carpathian highlanders) |
| Harvested from live trees, dried, ground, flooded with boiled water for the night and drunk in the morning | Diseases of the stomach, liver, kidneys, bladder, malignant tumors | *Inonotus obliquus* |
| Decoction or infusion | Tumors of the stomach and intestines, strengthening the body | *Inonotus obliquus* |
| Alcohol tincture from fresh or dried fruiting bodies | Rheumatism | *Amanita muscaria* |
| Boiled or alcoholic tincture | Shigellosis , heartburn | *Amanita muscaria* |
| Young fruiting bodies after cutting or spores from old fruiting bodies | Wraps for bleeding wounds, poorly healing | *Lycoperdon* spp. |
| Chopped | Ulcers on the skin | *Lactarius deliciosus* |
| Consumed after cooking | Treatment of the alcohol addiction | *Coprinus comatus* |
| Dried, powdered | Treatment of the old wounds | *Sparassis crispa* |
| Dressings—the internal part of the fruiting body cut into strips or shredded | Stopping the bleeding | *Fomes fomentarius* |
| Infusion | Treatment of the skin problems (pimples, ulcers) and rheumatism; for wounds and bleeding | *Inonotus obliquus*, *Fomitopsis betulina* |
| Alcohol tincture from young fruiting bodies | Pouring on scalded places | *Amanita phalloides* |
| Dried mushrooms soaked in milk—wraps | Ulcer of the skin | *Boletus* spp., *Leccinum* spp., *Suillus* spp. |
| Dried mushrooms boiled in milk | Alleviation of breast inflammation in lactating women | *Boletus* spp. |
| Agaryk medicine—the inner part of the fruiting body, dried, powdered: extracts, elixirs, pillsn | Laxative, emetic | *Agaricum officinale* |

## References


Anonym 1 (2019) The mushrooms market in Poland—import, export, perspectives. www.polskibiznes.info (Download 2 July 2019) (in Polish)

Bartnicka-Dabkowska B (1964) Polish folk names of mushrooms. Linguistic work PAN No. 42. Publisher National Institute Ossolińskich, Wroclaw-Warsaw-Krakow (in Polish)

Bernaś E, Jaworska G (2016) Vitamins profile as an indicator of the quality of frozen *Agaricus bisporus* mushrooms. J Food Comp Anal 49:1–8

Bernaś E, Jaworska G (2017) Culinary-medicinal mushroom products as a potential source of vitamin D. Int J Medic Mushrooms 19(10):925–935

Bernaś E, Jaworska G, Lisiewska Z (2006) Edible mushrooms as a source of valuable nutritive constituents. Acta Scientiarum Polonorum, Technologia Alimentaria 5(1):5–20

Burda PR (1998) Poisonings with higher mushrooms and plants. PWN, Warsaw, Poland (in Polish)

Chachuła P, Bodziarczyk J, Kozubek R, Widlak M, Siwy M (2016) Macrofungi in fir-beech forests with admixture of common yew *Taxus baccata* L. in the Polish Carpathians. Roczniki Bieszczadzkie 24:53–85 (in Polish)

Cheung PCK (ed) (2008) Mushrooms as functional foods. Wiley, Hoboken, NJ

Chudzyński K, Bielawski L, Falandysz J (2007) Mineral elements and their bioconcentration factors in the fruiting bodies of larch bolete (*Suillus grevillei*) from the area of Beskid Zachodni. Bromat Chem Toksykol 40(2):159–166 (in Polish)

Domaszewicz B (2017) Forestry. Central Statistical Office, Warsaw, Poland, pp 270–288 (in Polish)

Dz.U. 2014 poz. 1408 (2014) Regulation of the Polish Minister of the Environment of 9 October 2014. On the protection of species of fungi (in Polish)

Dz.U. 2018 poz. 1281 (2018) Regulation of the Polish Minister of Health of June 12, 2018. Amending the regulation on mushrooms authorized for the marketing or production of mushroom products, foods containing mushrooms and the classifier of mushrooms and mushroom specialists (in Polish)

Falandysz J, Gucia M, Mazur A (2007) Some mineral elements and their bioconcentration factors in parasol mushroom (*Macrolepiota procera*) from the region of Poniatowa in the Lubelskie Voievodeship. Bromat Chem Toksykol 40(3):249–255 [in Polish]

Falandysz J, Kunito T, Kubota R, Bielawski L, Frankowska A, Falandysz JJ, Tanabe S (2008) Multivariate characterization of elements accumulated in King Bolete *Boletus edulis* mushroom at lowland and high mountain regions. J Environ Sci Health A Tox Hazard Subst Environ Eng 43(14):1692–1699

FAOStat (2019) Mushrooms and truffles. Food and Agriculture Organization of the United Nations. http://faostat3.fao.org/. Accessed 2 July 2019

Florczak J, Chudy J, Barasińska M, Karwowski B (2014) Contents of selected nutrients in wild-grown *Hirneola auricula judae*, *Pleurotus ostreatus* and *Flammulina velutipes* mushrooms Bromat Chem Toksykol 47(4):876–882 (in Polish)

Grzywacz A (2008) Biodiversity of fungi in forests. In: Natural resources of Polish forests. PTL publishing house, Cedzyna near Kielce, pp 23–37 (in Polish)

Grzywacz A (2015) Traditions of the collections of wild mushroom in Poland. Studies and Materials CEPL in Rogów 17, 44(3):189–199 (in Polish)

Jasinghe VJ, Perera CO (2006) Ultraviolet irradiation: the generator of vitamin $D_2$ in edible mushrooms. Food Chem 95(4):638–643

Jaworska G, Bernaś E (2013) Amino acid content of frozen *Agaricus bisporus* and *Boletus edulis* mushrooms: effects of pretreatments. Int J Food Prop 16:139–153

Kalac P (2013) A review of chemical composition and nutritional value of wild-growing and cultivated mushrooms. J Sci Food Agric 93:209–218

Karwowski K, Koper K, Wawrzczak M (2014) In book: The folk culture of the Pieniny Mountain. Traditional use of the forest in Pieniny and surroundings (Ceklarz K, Janicka-Krzywda U, eds). COTG PTTK, Krakow (in Polish)

Kowalewska I, Bielawski L,Falandysz J (2007) Some metals, phosphorus, and their bioconcentration factors in red aspen bolete (*Leccinum rufum*) from the Lubelska Upland. Bromat Chem Toksykol 40(2):153-158 (in Polish)

Łuczaj Ł (2011) Wild food plants used in Poland from the mid-19th century to the present. Etnobiologia Polska 1:57–125 (in Polish)

Orłoś H (1946) Production and trade of mushrooms in Poland. Forest Research Institute, Krakow, B:8 (in Polish)

Referowska-Chodak E (2015) Folk customs related to mushrooms in Poland. Studies and Materials CEPL in Rogów 17, 44 (3):200–2017 (in Polish)

Royse DJ, Baars JJP, Qui Tan (2017) Current overview of mushroom production in the world: technology and applications. In Zield DC (ed) Edible and medicinal mushrooms: technology and application. Wiley, New York

Sanchez C, Diaz-Godinez G (2004) Agro-FOOD Industry Hi-tech. 15(3):44–45

Szczepkowski A (2012) Arboreal fungi in a different light—the use of sporocarps. Studies and Materials CEPL in Rogów 14, 32(3):171–189 (in Polish)

Trojanowska A (2001) On the medicinal uses of fungi in the 19th century. Analecta, 10/2(20):111–12 (in Polish)

UNESCO Convention (2003) Convention for the safeguarding of the intangible cultural heritage, 17 October 2003, Paris

# Chapter 17
# Usage of wild-Growing Plants as Foodstuff

**Piotr Gębczyński, Emilia Bernaś, and Jacek Słupski**

**Abstract** Wild edible plants (WEP) have long been an inherent ingredient in human life. They grow spontaneously in self-maintaining populations in natural ecosystems and exist independently of direct human action. The meaning of WEP can be referred to as wild vegetables. They are usually cheap, readily and simple available food. Thanks to their proximity to humans they are known to local communities. These plants make it possible to ensure food security in areas or periods of food shortages. Food security should be dealt with to satisfy hunger or malnutrition and supply nutritional components. The particular role of WEP is in the poorly developed regions and where the agricultural economy is difficult to sustain. This situation is typical of mountain and foothill areas. It is worth emphasising that the use of WEP as a food source promotes biodiversity. Nowadays, the share of plants obtained from nature for food has decreased significantly because of the dynamic development of horticulture and the possibility of supplying mountain areas with cheap fruit and vegetables; law regulations regarding wildlife and nature protection; and labour-intensive harvesting. However, in areas with no food shortage, WEP can still be a valuable food supplement. They can also be a factor that distinguishes locally prepared food. Particular attention should be paid to those plants commonly found in nature in a given area. Such plants include, among others, bear's garlic (*Allium ursinum* L.), white goosefoot (*Chenopodium album* L.), and garden orache (*Atriplex hortensis* L.). The above species are common and well adopted in mountain and foothill areas from Asia to Europe. Thanks to this, they can be obtained from their natural state and protected plants from local crops.

---

P. Gębczyński (✉) · E. Bernaś · J. Słupski
Department of Plant Products and Nutrition Hygiene, Faculty of Food Technology, University of Agriculture, Krakow, Poland
e-mail: piotr.gebczynski@urk.edu.pl

E. Bernaś
e-mail: emilia.bernas@urk.edu.pl

J. Słupski
e-mail: jacek.slupski@urk.edu.pl

© Springer Nature Switzerland AG 2022

J. Hernik et al. (eds.), *Cultural Heritage—Possibilities for Land-Centered Societal Development*, Environmental History 13,
https://doi.org/10.1007/978-3-030-58092-6_17

**Keywords** Wild edible plants · Food · Bear's garlic · White goosefoot · Garden orache

## 17.1 Introduction

Wild edible plants (WEP) have long been an integral ingredient in human food. The most cited definition about WEP is the one by the FAO, which states that the plants grow spontaneously in self-maintaining populations, in natural or semi-natural ecosystems and can exist independently of human action. WEP, are contrasted with cultivated plants that have arisen through human activity, such as selection or breeding, and that depends on management for their continued existence (Heywood 1999). The meaning of WEP sometimes referred to as wild vegetables is manifold to man. They are usually a cheap and readily available food source, which generally occurs in the immediate/imminent human environment. Thanks to this direct proximity the properties of these types of raw materials are often known to local communities; hence, their use as food usually does not pose a threat to well-being (Shaheen et al. 2017). These plants make it possible to ensure food security in areas with food shortages or during periods of food shortage. Food security should be dealt with in two ways, both in terms of satisfying hunger or avoiding malnutrition, as well as supplying many valuable components of dietary well-being (when appropriately consumed)—proteins, fat, carbohydrates, fibre, minerals, vitamins, and some other active constituents. The use of WEP as a source of food allows the preservation of the biodiversity of the natural environment (Heywood 1999; Pinela et al. 2017). Thanks to this biodiversity, in areas where plant-based agricultural production of a monoculture, mainly cereal, is typical, it is also possible to obtain a variety of food, especially in the poorest population (Mavengahama et al. 2013). So, the primary role of WEP is in the poorly developed regions of the world, and where the agricultural economy is challenging to organise (Bvenura and Afolayan 2015; Bvenura and Sivakumar 2017).

This situation is characteristic for the mountain and sub-mountain areas of southern Poland, including the southern part of the Małopolska region. It is caused by a large proportion of the soils being weak, adverse of physiographic and climatic conditions, ground fragmentation, as well as poorly developed agricultural infrastructure. These factors contribute to the outcome in addition to the reduced profitability of farms (Dmochowska 2003; Musiał 2013). In Małopolska, the share of mountainous farmland is unusually large, above 50%, while in the remaining provinces at the intersection with the Carpathians, Sudeten, and the Świętokrzyskie Mountains it does not exceed 20% (Dmochowska 2003). These unfavourable conditions cause a worrying phenomenon of simplification of plant production tending towards increased cultivation of cereals, which already occupy more than 70% of crops in Małopolska (Apryjas et al. 2017; Musiał 2013). At the same time, the vegetable growing area in these areas is reduced to not more than 1%. With this background, only fruit production is different. In Małopolska, fruit-growing has been present for a long time, and in the early twentieth century, significantly increased development took place, also in some of the sub-mountain areas. In Małopolska, in the Carpathian foothills, there are three active orchard districts—Raciechowice, Limanowa, and Łącko. Currently, only fruit production is expected to develop further in the above regions (Sroka and Wojewodzic 2015).

Nowadays, the share of plants obtained from wild habitats for food needs has decreased significantly. The reason for this is above all the dynamic development of fruit-growing and horticulture in lowland areas in Poland and the opportunity to supply mountain areas with cheap fruit and vegetables. Another factor is the legal regulations regarding nature protection. Such law protects natural sites against uncontrolled and excessive exploitation, but at the same time limits people's access to wild plant sources (Baričevič et al. 2015; Łuczaj 2010). Besides, the gathering of wild-growing materials is done by hand, so it is labour-intensive and is usually treated only as an additional source of income, mainly for the rural population (Piwowar 2015).

However, in areas where food shortages are not recorded, edible wild plants can still be a valuable component of the diet. They can also be a distinguishing element of food prepared in mountainous areas. Among these plants we should pay particular attention to those that are commonly found as the part of wild nature in a given area. The ongoing aim of the paper is the study of plants characteristic of mountain areas—bear garlic, goosefoot, and orache.

## 17.2 Bear Garlic (*Allium ursinum* L.)

Bear garlic belongs to the large and economically important genus *Allium* of the *Amaryllidaceae* family. It is also called ramson, wild garlic, or broad-leaved garlic. It belongs to the subgenus *Allium* of which almost all species are present in temperate Europe, the Mediterranean region, Near and the Middle East, the Caucasus, northern Asia, with only a few species occurring outside this area (Odeny and Narina 2011; Khassanov 2018). In a natural state, it occurs on a floor of wet deciduous or mixed forests, with dominant beech, hornbeam, or oaks. In Poland, it inhabits mainly the south of the country, in the mountains. Even the so-called garlic subassembly stands out in the wild habitats of mountain forest formations. In the Carpathians, lush beech is the main component of the landscape (Herden et al. 2012; Matuszkiewicz et al. 2012; Rola 2012; Kaplan 2017; Peinado et al. 2017; Kolosova et al. 2017). It is a perennial herb plant with underground onion, growing up to heights of 30–50 cm, and usually forms colonies for up to several hundred square metres. The stem is soft and short. The leaves are mostly oval lanceolate. White flowers form umbels that bloom from April to June (Ghedira and Goetz 2016).

The plant is usually obtained from a natural state, but it can also be obtained from cultivation (Ernst 1979; Amagova et al. 2018). In Poland, it is on the list of protected species (Regulation MS 2014), and plants that go to market should (according to law) come from cultivation. Verifying these conditions is difficult, and it is suspected that plants found on local markets in the southern part of the country are growing wild but illegally harvested (Kasper-Pakosz et al. 2016). Legality in obtaining the plant need not be demonstrated and many of the habitats of bear garlic are parks, cemeteries, and abandoned gardens (Rola 2012). Moreover, the authors of the present paper encountered a habitat of bear garlic in a neglected garden in the centre of the city (Krakow, Poland).

*Allium* is the largest and an essential representative genus of the *Alliaceae* family. Besides the well-known onion and common garlic, several other species are widely used in everyday culinary and herbal medicine. Its consumption is attributed to several factors, mainly heavy promotion that links flavour and health. Most *Allium* species are usually cultivated for the edible bulb, and only a few for their leaves—chives, scallions. All of them are used mainly as a spice in foods. Their characteristic and dominant flavour and possible nutritional effects and medical applications have attracted the attention of food and medical researchers (Hanen et al. 2012).

Bear garlic properties and their peculiar sensory features are very similar to aspects of common cultivated garlic. All parts of the plant, leaves, bulbs, flowers, and can be used for consumption, but it is obtained primarily for its large, green leaves. Bear garlic is considered to be one of the sentinel plants of ancient forests. It should be assumed that it has accompanied man for a very long time (Honnay et al. 1998; Rose 1999). However, there is relatively little information on the use of garlic leaves as food in antiquity and the Middle Ages (Sobolewska et al. 2015). Only the works of botanists from the nineteenth and twentieth centuries mention this plant as a food ingredient, but mainly in the context of use during periods of food shortage (Łuczaj 2008).

There has been a growing interest in use of bear garlic, mostly as a functional food and dietary supplement. Such re-discovery, in a sense, re-invention of plants from a close human environment is mainly associated with the dynamic development of chemical analytics and medical sciences, which enable interested persons to find in those plants many active biological substances potentially useful for humans and other living organisms. Hunger now disappearing in many places and the threat of obesity is rising globally. It is estimated that the number of persons who are obese or at risk of obesity may soon exceed the number of persons suffering from hunger (Konarzewski 2015). Obesity is associated with some civilizational or diet-dependent diseases, associated with the functioning of the cardiovascular system. Because obesity is mainly associated with excessive intake of fats, and carbohydrates, there are attempts to prevent it by replacing its sources, cereals, and animal foods, with plants especially containing low amount of carbohydrates, mainly vegetables (Yahia et al. 2019). Besides, many plants have been shown to have many substances in their composition, mainly antioxidants, which are capable of counteracting disease-related oxidative stress (Krivokapic et al. 2017; Petkova et al. 2019). Vegetables of the genus *Allium*, including bear garlic, contain such substances including, above all, numerous sulphur compounds (Scheffler et al. 2018; Putnik et al. 2019) and the polyphenols (Lachowicz et al. 2017; Skrovanokva et al. 2018). Many of these substances can act on humans and animals directly as well as indirectly, for example, by inhibiting the development of parasites and pathogenic microorganisms (Grzesiak et al. 2018; Krstin et al. 2018).

An essential feature of bear garlic as a food product is its occurrence as early vegetation. It is one of the earliest plants on the forest floor, as it appears already at the turn of winter and spring. It is known and used in local cuisine, as a fresh or processed herb or vegetable, in Eastern Europe, from Ukraine to the Caucasus and Romania, as well as in Poland, the Czech Republic, and Germany. Primarily the leaves of the plant are used, mainly for salads and hot dishes like soups, and as a component of fillings in different pies (Łuczaj 2004; Barla et al. 2014). Bear garlic, in addition to its culinary uses, is also increasingly used as an additive to processed food products. For example, in Poland and Turkey, it is added to local rennet cheeses (Apak 2019; MRiRW 2019), and in the Czech Republic to camembert cheese for barbecue.

## 17.3 White Goosefoot (*Chenopodium album* L.)

The white goosefoot is an annual plant belonging to the *Amaranthaceae* (*Chenopodiaceae*) family. *Amaranthaceae* contains almost two hundred genera and over two thousand of species and is a widespread cosmopolitan family spread from the tropics to cool temperate regions (Müller and Borsch 2005). The *Chenopodium* genus includes herbaceous, annual plants, and is widespread worldwide, mainly in the moderate and subtropical zones, and contains about 250 species (Pandey and Gupta 2014).

White goosefoot, also known as pigweed, fat hen, and lamb-quarters, is an aggressive, fast-growing weed of many crop plantations and is considered one of the leading representatives of weedy species in Europe. In Poland, there are currently found about 30 species of *Chenopodium* (Nowak et al. 2016). It comes to Europe from Central and Western Asia accompanying the westward spread of man and agriculture in the Neolithic period (Krak et al. 2019). Nowadays, the goosefoot is a cosmopolitan species occurring within a broad area from 50° south to 70° north latitude. It can sprout under high water stress conditions as well as in a wide range of pH soils from acid to alkaline. It is a troublesome weed in many crops, and it can be most commonly found in sugar cane, potato, corn, and spring cereals. The primary action of the weed is reflected in the reduction of yield (Brijacak et al. 2018; Habibi et al. 2018).

White goosefoot is highly variable in all morphological characteristics and possesses a high level of phenotypic plasticity because it represents a highly diversified polyploid species complex. For example, its hight as a single plant could be from 0.10 to 3.50 m, and its natural habitats cover both lowland areas and mountain slopes, even up to an altitude of over 4500 m. The plant is usually much-branched. Leaves are distinctly longer rather than broad and usually ovate-deltoid, sometimes rhombic or ovate-lanceolate, entire or toothed. The texture of the plant is mealy in all parts (Batsatsashvili et al. 2017; Krak et al. 2019; Pandey and Gupta 2014).

## 17.4 Food Use of White Goosefoot

It should be noted that the *Amaranthaceae* family includes several species of beneficial plants, including quinoa, beet, and grain amaranth (Müller and Borch 2005) and also plants with edible green parts—spinach and vegetable amaranth. White goosefoot can be classified as useful plants because in many parts of the world it functions as food, whether obtained from the natural state or from cultivation. It is cultivated as a pot-herb and usually grown in gardens (Sikarwar et al. 2017). Green leafy vegetables (GLV) are, in general, good sources of many nutrients, especially vitamins, minerals, essential amino acids (EAA), and polyunsaturated fatty acids. Leafy vegetables are especially rich in vitamins and compared with the essential plant foods, grain or pulses, GLV are more abundant in vitamins A, B1, E, C, and K. Although fresh plant leaves usually contain more than 80% of the water they are a valuable source of protein, mainly because of high content of EAA. The content of methionine and cysteine is the most crucial because, in the other plant foods, the lack of one or the other of them is observed. GLV are abundant in methionine and cysteine, and thanks to this their protein value is not much less than animal protein (Edelman and Colt 2016). GLV exceed the essential plants in the content of some minerals such as calcium, potassium and magnesium, zinc, and iron. GLV are an essential source of polyunsaturated fatty acids with proper nutritional relationship of omega-6 to omega-3 fatty acids. Thanks to their diverse components, GLV have many positive health outcomes, among other benefits improvements with regards to cardiovascular diseases (Edelman and Colt 2016; Pollock 2016). A factor that may support the broader use of rowing wild vegetables is their much greater wealth in some nutrients than cultivated plants. Creation of new varieties with breeding can yield benefits if aimed primarily at improving varieties in terms of the essential nutrients—proteins, carbohydrates, or fats. Some modern varieties of plants are lower in some nutrients than older varieties due to the dilution effect of increased yield associated with accumulation mainly of carbohydrates, without a proportional increase of other nutrients (Marles 2017).

Consumption of white goosefoot leaves alone, or whole young tender shoots is widespread in Central and Western Asia—India, China, Georgia. Today, its use is minor in Europe. However, until the early twentieth century, the seeds of the goosefoot in Estonia produced flour for bread and porridge. At present, Estonians sometimes use whole young shoots as a green addition to an omelette, as well as (from the late twentieth century on) more often, as a supplement to vegetable salads (Kalle and Sõukand 2016). In Central Europe (Poland, Hungary, Ukraine) the white goosefoot was, alongside nettle, a vital spice plant for poor people, in times of hunger or wartime. However, also in Central Europe its use has disappeared, although in the middle of the twentieth century it was still possible to meet with its use (Łuczaj 2010; Pieroni and Sõukand 2017). However, up to the present day white goosefoot could often be found in many Mediterranean countries. In Croatia, it is a component of ready-made salad blends, which can be bought at local markets (Łuczaj et al. 2013). It is known and used from Turkey to Spain (Dogan 2012; Nedelcheva 2013; Pieroni et al. 2015a; Rigat et al. 2009). In some areas of the world, a critical botanical feature of this plant is halophility. It can survive and complete its life cycle in the presence of soil salinity (Tug and Yaprak 2017).

It is customary to eat the white goosefoot raw in vegetable salads alone or with curd (Poonia and Upadhayay 2015). More often it is consumed after boiling or stewing. In many meals, it is the main component, but frequently it is mixed with various other herbs or vegetables, and as extra greens are used: amaranthus, nettle, sorrel, purslane, bittercress, dill, onion, even leaves of the potato or tomato. Other additives, nuts, flour, curd, cheese, eggs, are also used in dishes that are based on such plant mixtures. Such foods are either separate dishes or are an addition to other meals, usually as a filling. The plant, especially eaten raw, is used most often in the spring. During this period, they are not excessively mature in growth, the leaves are soft, and the stems do not grow woody (Batsatsashvili et al. 2017). Simple preservation methods are also practiced. For winter use goosefoot is dried or pickled. Dried leaves are then used as a spice for dishes but also as a medicinal agent (Poonia and Upadhayay 2015).

## 17.5 Medicinal Use of White Goosefoot

Man has lived with plants which were healers and health-rejuvenators since ancient times. The use of plants and plant products in the form of medicines for the treatment of humans and animals has its roots in prehistory. Even today, plants play an essential role in the health care of about 80% of the world population (Lone 2017). The white goosefoot is used as a folk medicine agent in different parts of the world. It is used as diuretic, laxative, sedative, hepato-protective, and antiparasitic (Sikarwar et al. 2013; Zheng 2017; Arora and Itankar 2018). In traditional system of medicine, plant infusion was used to treat stomachache, scurvy, and diarrhea, and the leaves were used as a poultice over wounds and bites. The leaves possess anthelmintic, antibacterial, antioxidative, antiphlogistic, antirheumatic, mildly laxative and odontalgic properties, applied as a wash, infusion or poultice to bug bites, pathogenic microorganisms, sunstroke, rheumatic joints, and swollen feet (Kokanova-Nedialkova et al. 2009; Aman et al. 2016; Lone et al. 2017). Studies on male rats have also shown that white goosefoot acts as a potent aphrodisiac (Sewani-Rusike et al. 2015).

The medicinal importance of a plant is due to the presence of some biologically active substances (Bayram et al. 2018). The most important are alkaloids, carotenoids, flavonoids, glycosides, gums, resins, tannins, and volatile oils (Arora and Itankar 2018; Kaur and Kaur 2018). Most of the antioxidant potential of the plants results from the redox properties of polyphenols. They act as reducers and can react with free radicals and metal ions (Gawlik-Dziki 2008).

Although at present, the medicinal uses of *Chenopodium* are not widely known, some reports try to propagate its health-enhancing properties. Unusual properties of white goosefoot include action against urolithiasis. Their leaves in ethnomedical practices are used for the treatment of kidney stones and urinary tract disorders. This effect was confirmed by in vitro and in vivo studies in animals (Sharma et al. 2016; Sikarwar et al. 2017). These are apparently odd outcomes because many GLV contain oxalic acid, which is responsible for the formation of kidney stones by binding into insoluble salts of calcium and magnesium ions (Savage and Vanhanen 2018). However, it is believed that limits on the formation of insoluble calcium oxalate result from the participation of saponins included in the leaves of goosefoot (Sharma et al. 2016).

## 17.6 Garden Orache (*Atriplex hortensis* L.)

The garden orache is an annual plant of the *Atriplex* genus. *Atriplex* species, and as described above, white goosefoot, is a member of the *Amaranthaceae* family. The genus contains over two hundred or even more of species (Benzarti et al. 2013; The Plant List 2013). The garden orache is grown as a cultivated plant. In Europe, the plant was already cultivated in ancient Rome and in mediaeval cloister (Boseva and Bosseva 2016; Bosi et al. 2017). Garden orache is found as a relic of former cultivation that has returned to a semi-wild state as a result of displacement by more popular crops (Bacchetta et al. 2016). It is also met as a wild plant widespread as a ruderal in waste ground. It is also called mountain spinach, French spinach, sea purslane, and saltbush (Greszczuk et al. 2010). It is a robust and hard herb, up to 150 cm tall. Stems are erect and branched. Leaves are alternate, triangular, and often broadly hastate at the base. For food or feed young stems and leaves or whole plants can be used. Orache is rich in vitamins and biologically active compounds with antioxidant activity (Zeipina et al. 2015). It also has positive dietetic and medicinal properties (Greszczuk et al. 2010).

Orache is a plant that is used as the world's most popular leafy vegetable—spinach (*Spinacia oleracea* L.), which belongs to the same family. That is why its names contained the word 'spinach'—French spinach, mountain spinach, similarly to a few other leafy vegetables, for example, New Zealand spinach (*Tetragonia tetragonioides* Pall./Kuntze), a species from the *Aizoaceae* family. Garden orache is commonly grown as a warm-weather alternative to *S. oleracea*. In a warm climate, it is more tolerant of heat and slower to break into seed shoots (Mehideyeva et al. 2017). Additionally, the plant grows naturally in arid and semi-arid regions of the world, where most such plants are highly tolerant of drought and salt. Usually, the naturally grown plants require very much less water for survival and to complete their life cycle. The name 'saltbush' indicates that some species are halophytes; a few are widespread ruderals (Rinchen and Nachendra 2015).

## 17.7 Food Uses of Garden Orache

Garden orache, like the other wild vegetables, is traditionally consumed fresh. Therefore, harvesting and consumption times relatively closely coincide. Sometimes simple preservation techniques were used to make this food available out of the vegetation season. One of the oldest methods is drying or pickling and brining. Actually, in Spain, freezing is used for the preservation of orache leaves (Tardío and Pardo-de-Santayana 2016).

Leaves alone or with stems can be used raw or after culinary heat processing. Young shoots and leaves are used raw and fresh in salads, and after cooking or frying are a component of traditional meals. The most popular are soups in Armenia, Azerbaijan, Belarus, and Romania usually with other vegetables and meat, prepared on broth or yoghurt basis (Mehideyeva et al. 2017; Soukand et al. 2017). In other regions, it is used for vegetable spread, pies, sarmale, tarts, appetiser cakes, omelettes, puree, soufflé, rolls, pilaf, pasta (Mehideyeva et al. 2017; Sengeanu et al. 2017). In many regions one of the most common ways to use leaf vegetables, aside from soups, is adding as a filling for a variety of dishes, mainly farinaceous meals. Garden orache, as well as the previously described white goosefoot, individually or with other plants, with meat, curd, cheese is added to pita, dumplings, pies (Pieroni et al. 2015a, b).

As shown above, the majority of applications of the leaves and young shoots of garden orache and white goosefoot are similar. However, there is one such product, in which white goosefoot does not apply. That is a Turkish origin dish called sarma. The term sarma defines leaves rolled around a filling made of rice, bulgur and, minced meat, possibly with vegetables and seasoning plants, and gently cooked and generally consumed warm, with meat, or cold, without meat. It is an iconic meal of the traditional cuisine of the former Ottoman Empire. Even today, it is still a prevalent dish in Balkan countries. Leaves from almost one hundred of an annual or perennial plant can be used raw or more often shortly cooked, or kept in salt brine. Garden orache leaves are sufficiently large, durable, and readily available. They are often used mainly in Bulgaria (Dogan et al. 2015). Nowadays, in Poland there is a product similar to sarma, but prepared with white or savoy cabbage leaves, cooked or fermented with lactic acid bacteria. In addition orache is used as an ornamental plant and source of natural red and blue dyes for wool (Mehideyeva et al. 2017).

## 17.8 Medicinal Uses of Garden Orache

Currently, an increasing interest is granted to these species because of their high content in bioactive compounds (Ksouri et al. 2011). Various *Atriplex* species have medicinal values. The aerial parts of orache were used in folk medicine against diseases of the respiratory tract, and digestive and urinary systems, and due to their analgesic properties, in rheumatism. Leaves are diuretic, emetic, and purgative. It is also a stimulant of metabolism; an infusion is used as a spring tonic and remedy for tiredness and nervous exhaustion. Seeds mixed with wine are said to cure yellow jaundice. They also excite vomiting. Heated with vinegar, honey, and salt, *Atriplex* is used for gout. Liniments and emollients prepared from the whole plant, like the juice of the plant, are said to be remedies for indurations and tumours, especially of the throat (Rinchen and Narendra 2015). *Atriplex* species may contain phytochemical compounds with anti-fungal properties (Boughalleb et al. 2009). Several studies also attributed the anti-carcinogenic, anti-inflammatory, and antioxidant activities potential of plant extracts to their bioactive compounds compositions (Ksouri et al. 2011). Chemical investigation of the species of the genus *Atriplex* showed the presence of saponin glycosides, alkaloids, and phytosterols (Keckeis et al. 2000). Wild or cultivated orache is used as food mostly, either fresh or stewed, but sometimes it is used as medicinal plant, for example, as a remedy for anaemia for its depurative, detoxifying, antioxidant, regeneration, immunomodulant, cardiovascular, digestive properties (Savića et al. 2019; Segneanu et al. 2017).

## 17.9 Conclusions

Food for modern consumers is not only a source of essential nutrients but also a means of improving health and quality of life. This is why new, unconventional foods are sought to meet these requirements and not necessarily come from typical, large-scale agricultural production. Such specific expectations can be meet by wild edible plants. This type of food can provided by unconventional sources—forests, wastelands, infield areas, and even farmland, where weeds grow alongside arable crops.

Wild edible plants, including green leafy vegetables, have been known and used by humans for thousands of years. Until today they are practically unchanged in a natural state, since they were not subjected to breeding. Today they are no longer used as a primary component of human diet because food shortages are rather rare. They are, however, regarded as a spice and diversifying additive, to enhance attractiveness, or give local character to food products. They are also expected to increase the value of food through features related to the improvement of health and well-being. As described above, GLV could be an interesting substitute of typical, well-known leafy vegetables. The species mentioned above are commonly found in sub-mountain areas and are well adapted to local environmental conditions. As a result, they can be obtained not only from natural sites but also from local crops, the latter especially necessary in the case of plants that are growing in a wild state. The knowledge of WEP and GLV is primarily known to the rural population, who lives in direct contact with their place of occurrence. However, also in this environment this awareness disappears due to changes in people's living conditions and lifestyles. The knowledge passed between generations in the most common oral form. It has only been systematically documented for just over a hundred years, and it needs to be sustained and further promoted.

## References


Amagova ZA, Matsadze VH, Golubkina NA et al (2018) Fortification of wild garlic with selenium. Veg Crops Rus 4:76–80. https://doi.org/10.18619/2072-9146-2018-4-76-80

Aman S, Mazumder A, Gupta AK et al (2016) Pharmacological activities of *Chenopodium album* Linn—a review. World J Pharm Res 5:361–371

Apak R (2019) Current issues in antioxidant measurement. J Agri Food Chem. https://doi.org/10.1021/acs.jafc.9b03657

Apryjas A, Chadzińska B, Dzień M et al (2017) Rolnictwo. In: Dobrzańska J (ed) Województwo Małopolskie 2017 (Agriculture. Małopolska Voivodeship 2017). Urząd Marszałkowski Województwa Małopolskiego, Kraków, pp 193–197

Arora S, Itankar P (2018) Extraction, isolation and identification of flavonoid from *Chenopodium album* aerial parts. J Trad Comp Med 8:476–482. https://doi.org/10.1016/j.jtcme.2017.10.002

Bacchetta L, Visioli F, Cappelli G et al (2016) A manifesto for the valorization of wild edible plants. J Ethnopharm 191:180–187. https://doi.org/10.1016/j.jep.2016.05.061

Baričevič D, Máthé Á, Bartol T (2015) Conservation of wild crafted medicinal and aromatic plants and their habitats. In: Máthé Á (ed) Medicinal and aromatic plants of the world, vol 1. Springer, Dordrecht, pp 131–144

Barla G, Poroch-Seritan M, Sanduleac (Tudosi) E et al (2014) Antioxidant activity and total phenolic content in Allium ursinum and *Ranunculus ficaria*. Food Environ Saf 13:349–353

Batsatsashvili K, Kikvidze Z, Khutsishvili M et al (2017) *Chenopodium album* L. *Chenopodium foliosum* L. Amaranthaceae. In: Bussmann RW (ed) Ethnobotany of the Caucasus. European Ethnobotany. Springer International Publishing AG, pp 191–198. https://doi.org/10.1007/978-3-319-50009-6_126-1

Bayram B, González-Sarrías A, Istas G et al (2018) Breakthroughs in the health effects of plant food bioactives: a perspective on microbiomics, nutri(epi)genomics, and metabolomics. J Agric Food Chem 66:10686–10692. https://doi.org/10.1021/acs.jafc.8b03385

Benzarti M, Rejeb KB, Debez A et al (2013) Environmental and economical opportunities for the valorisation of the genus Atriplex: new insights. In: Hakeem RK et al (eds) Crop improvement. Springer Science+Business Media LLC, pp 441–457. https://doi.org/10.1007/978-1-4614-7028-1_16

Boseva KY, Bosseva YZ (2016) Edible and medicinal plants in the cloister gardens of West Europe (800s–900s AD). Phytol Balcan 22:161–166

Bosi G, Bandini Mazzanti M, Montecchi MC et al (2017) The life of a Roman colony in northern Italy: ethnobotanical information from archaeobotanical analysis. Quat Int 460:135–156. https://doi.org/10.1016/j.quaint.2016.08.008

Boughalleb N, Trabelsi L, Harzallah-Skhiri F (2009) Antifungal activity from polar and non-polar extracts of some *Chenopodiaceae* wild species growing in Tunisia. Nat Prod Res 23:988–997. https://doi.org/10.1080/14786410802168494

Brijacak E, Starostis V, Scepanovic M (2018) Biology and ecology of common lambsquarters (*Chenopodium album* L.). Agron Glas 1:19–34

Bvenura C, Afolayan AJ (2015) The role of wild vegetables in household food security in South Africa: a review. Food Res Int 76:1001–1011

Bvenura C, Sivakumar D (2017) The role of wild fruits and vegetables in delivering a balanced and healthy diet. Food Res Int 99:15–30. https://doi.org/10.1016/j.foodres.2017.06.046

Dogan Y (2012) Traditionally used wild edible greens in the Aegean Region of Turkey. Acta Soc Bot Pol 81:329–342

Dogan Y, Nedelcheva A, Łuczaj Ł et al (2015) Of the importance of a leaf: the ethnobotany of sarma in Turkey and the Balkan. J Ethnobiol Ethnomed 11:26. https://doi.org/10.1186/s13002-015-0002-x

Edelman M, Colt M (2016) Nutrient value of leaf vs. seed. Front Chem 4:32. https://doi.org/10.3389/fchem.2016.00032

Ernst R (1979) Population biology of Allium ursinum in Northern Germany. J Ecol 67:347–362

Gawlik-Dziki U (2008) Effect of hydrothermal treatment on the antioxidant properties of broccoli (*Brassica oleracea* var. *botrytis italic*) florets. Food Chem 109:393–401. https://doi.org/10.1016/j.foodchem.2007.12.058
Ghedira K, Goetz P (2016) Ail des ours: *Allium ursinum* L. (Amaryllidaceae) (Bear garlic: *Allium ursinum* L.). Phytothérapie 14:165–169. https://doi.org/10.1007/s10298-016-1042-7
Greszczuk M, Jadczak D, Kawecka A (2010) Effect of sowing date on biological value of garden orache. Acta Sci Pol Hort Cult 9:163–169
Grzesiak B, Kołodziej B, Głowacka A et al (2018) The effect of some natural essential oils against bovine mastitis caused by *Pyrototheca zopfii* isolates in vitro. Mycopathologia 183:541. https://doi.org/10.1007/s11046-018-0246-9
Habibi F, Vít P, Rahiminejad M et al (2018) Towards a better understanding of the *Chenopodium album* aggregate (*Amaranthaceae*) in the Middle East: a karyological, cytometric and morphometric investigation. J Sys Evol 56:231–242
Hanen N, Fattouch S, Ammar E et al. (2012) *Allium* species, ancient health food for the future? In: Valdez B (ed) Scientific, health and social aspects of the food industry. IntechOpen. https://doi.org/10.5772/1869
Dmochowska H (ed) (2003) Rolnictwo na terenach górskich i terenach o słabych warunkach glebowych (Agriculture in mountainous areas and areas with poor soil conditions ). GUS, Warszawa
Herden T, Neuffer B, Friesen N (2012) *Allium ursinum* L. in Germany—surprisingly low genetic variability. Feddes Repertorium 123:81–95
Heywood HV (1999) Use and potential of wild plants in farm households. FAO Farm Systems Management Series 15, Rome
Honnay O, Degroote B, Hermy M (1998) Ancient-forest plant species in western belgium: a species list and possible ecological mechanisms. Belg J Bot 130:139–154
Kalle R, Sõukand R (2016) Current and remembered past uses of wild food plants in Saaremaa, Estonia: changes in the context of unlearning. Debt Econ Bot 70:235. https://doi.org/10.1007/s12231-016-9355-x
Kaplan Z (2017) Flora and phytogeography of the Czech Republic. In: Chytrý M, Danihelka J, Kaplan Z et al (eds) Flora and vegetation of the Czech Republic. Plant and Vegetation, vol 14. Springer, Cham, pp 89–163
Kasper-Pakosz R, Pietres M, Łuczaj Ł (2016) Wild and native plants and mushrooms sold in the open-air markets of south-eastern Poland. J Ethnobiol Ethnomed 12:45. https://doi.org/10.1186/s13002-016-0117-8
Kaur G, Kaur N (2018) Effect of processing on nutritional and antinutritional composition of bathua (*Chenopodium album*) leaves. J App Nat Sci 10:1149–1155. https://doi.org/10.31018/jans.v10i4.1856
Keckeis K, Sarker SD, Dinan L (2000) Phytoecdysteroids from *Atriplex nummularia*. Fitoterapia 71:456–458
Khassanov FO (2018) Taxonomical and ethnobotanical aspects of Allium species from Middle Asia with particular reference to subgenus *Allium*. In: Shigyo M, Khar A, Abdelrahman M (eds) The *Allium* genomes. Compendium of plant genomes. Springer, Cham, pp 11–21
Kokanova-Nedialkova Z, Nedialkov PT, Nikolov SD (2009) The genus *Chenopodium*: Phytochemistry, ethnopharmacology and pharmacology. Pharmacogn Rev 3:280–306
Kolosova V, Svanberg I, Kalle R et al (2017) The bear in Eurasian plant names: motivations and models. J Ethnobiol Ethnomed 13:14. https://doi.org/10.1186/s13002-016-0132-9
Konarzewski M (2015) Na początku był głód (At the start there was a hunger). PIW, Warszawa
Krak K, Habibi F, Douda J et al (2019) Human-mediated dispersal of weed species during the Holocene: a case study of *Chenopodium album* agg. J Biogeo 46:1007–1019. https://doi.org/10.1111/jbi.13545
Krivokapic M, Bradic J, Petkovic A (2017) Phytochemical and pharmacological properties of Allium ursinum. Ser J Exp Clin Res 1. https://doi.org/10.2478/SJECR-2018-0003

Krstin S, Sobeh M, Braun MS et al (2018) *Tulbaghia violacea* and *Allium ursinum* extracts exhibit anti-parasitic and antimicrobial activities. Molecules 23:313. https://doi.org/10.3390/molecules 23020313

Ksouri R, Ksouri WM, Jallali I et al (2011) Medicinal halophytes: potent source of health promoting biomolecules with medical, nutraceutical and food applications. Crit Rev Biotechnol 32:289–326. https://doi.org/10.3109/07388551.2011.630647

Lachowicz S, Kolniak-Ostek J, Oszmianski J et al (2017) Influence of maturity on the content of phenolic compounds of *Allium ursinum* L. J Food Process Pres 41:e12921. https://doi.org/10.1111/jfpp.12921

Lone BA, Chishti MZ, Bhat FA et al (2017) Evaluation of anthelmintic antimicrobial and antioxidant activity of *Chenopodium album*. Trop Anim Health Prod 49:1597–1605. https://doi.org/10.1007/s11250-017-1364-y

Łuczaj Ł (2004) Dzikie rośliny jadalne Polski: przewodnik survivalowy (Wild Polish edible plants: survival guide). Chemigraf, Krosno

Łuczaj Ł (2008) Dziko rosnące rośliny jadalne w ankiecie Józefa Rostafińskiego z 1883 (Wild growing edible plants in the poll of Józef Rostafiński from 1883). Wiad Bot 51(1/2):39–50

Łuczaj Ł (2010) Changes in the utilization of wild green vegetables in Poland since the 19th century: a comparison of four ethnobotanical surveys. J Ethnopharmacol 128:295–404. https://doi.org/10.1016/j.jep.2010.01.0

Łuczaj Ł, Zovko Končić M, Miličević T et al (2013) Wild vegetable mixes sold in the markets of Dalmatia (southern Croatia). J Ethnobiol Ethnomed 9:2–13. https://doi.org/10.1186/1746-4269-9-2

Marles RJ (2017) Mineral nutrients comparison of vegetables, fruits, and grains: the context of reports of apparent historical declines. J Food Comp Anal 56:93–103. https://doi.org/10.1016/j.jfca.2016.11.012

Matuszkiewicz W, Sikorski P, Szwed W et al (2012) Przegląd zespołów leśnych występujących w Polsce. In: Matuszkiewicz W, Sikorski P, Szwed W et al (eds) Lasy i zarośla: zbiorowiska roślinne Polski (Overview of forestry groups present in Poland. Forests and scrubs: plant congeries of Poland). PWN, Warszawa, pp 136–197

Mavengahama S, McLachlan M, de Clercq W (2013) The role of wild vegetable species in household food security in maize based subsistence cropping systems. Food Sec 5:227–233. https://doi.org/10.1007/s12571-013-0243-2

Mehdiyeva N, Fayvush G, Aleksanyan A et al (2017) *Atriplex hortensis* L., Atriplex *tatarica* L. Amaranthaceae. In: Bussmann RW (ed), Ethnobotany of the Caucasus. European Ethnobotany, Springer International Publishing AG, pp 133–136. https://doi.org/10.1007/978-3-319-50009-6_126-1

MRiRW (2019) Produkty regionalne i tradycyjne (Polish Ministry of Agriculture and Rural Development, Local and traditional products database). https://www.gov.pl/web/rolnictwo/produkty-regionalne-i-tradycyjne. Accessed 11 May 2019

Müller K, Borsch T (2005) Phylogenetics of *Amaranthaceae* based on matK/trnK sequence data: evidence from parsimony, likelihood, and Bayesian analyses. Ann Mo Bot Gar 92:66–102

Musiał W (2013) Charakterystyka specyfiki rozwoju rolnictwa na obszarach górskich i podgórskich. In: Matyka M (ed) Rolnictwo na obszarach specyficznych (Characteristics of agricultural development in mountainous and submountainous areas. Agriculture in specific areas). GUS, Warszawa, pp 169–182

Nedelcheva A (2013) An ethnobotanical study of wild edible plants in Bulgaria. Eurasia J Biosci 7:77–94. https://doi.org/10.5053/ejobios.2013.7.0.10

Nowak R, Szewczyk K, Gawlik-Dziki U et al (2016) Antioxidative and cytotoxic potential of some *Chenopodium* L. species growing in Poland. Saudi J Biol Sci 23:15–23. https://doi.org/10.1016/j.sjbs.2015.01.017

Odeny DA, Narina SS (2011) Allium. In: Kole C (ed) Wild crop relatives: genomic and breeding resources. Springer, Berlin, Heidelberg, pp 1–10

Pandey S, Gupta RK (2014) Screening of nutritional, phytochemical, antioxidant and antibacterial activity of *Chenopodium album* (Bathua). J Pharmacogn Phytochem 3:1–9

Peinado M, Aguirre JL, Aparicio A (2017) The Iberian ranges and highlands. In: Loidi J (ed) The vegetation of the Iberian peninsula. Plant and vegetation, vol 12. Springer, Cham, pp 439–512

Petkova NT, Ivanov IG, Raeva M et al (2019) Fructans and antioxidants in leaves of culinary herbs from Asteraceae and Amaryllidaceae families. Food Res 3:407–415

Pieroni A, Sõukand R (2017) Are borders more important than geographical distance? The wild food ethnobotany of the Boykos and its overlap with that of the Bukovinian Hutsuls in Western Ukraine. J Ethnobiol 37:326–345. https://doi.org/10.2993/0278-0771-37.2.326

Pieroni A, Ibraliu A, Abbasi AM et al (2015a) An ethnobotanical study among Albanians and Aromanians living in the Rraice and Mokra areas of Eastern Albania. Genet Resour Crop Evol 62:477–500. https://doi.org/10.1007/s10722-014-0174-6

Pieroni A, Nedelcheva A, Dogan Y (2015b) Local knowledge of medicinal plants and wild food plants among Tatars and Romanians in Dobruja (South-East Romania). Genet Resour Crop Evol 62:605–620. https://doi.org/10.1007/s10722-014-0185-3

Pinela J, Carvahlo AM, Ferreira ICFR (2017) Wild edible plants: nutritional and toxicological characteristics, retrieval strategies and importance for today's society. Food Chem Tox 110:165–188. https://doi.org/10.1016/j.fct.2017.10.020

Piwowar A (2015) Zasoby leśne—baza surowcowa i znaczenie w agrobiznesie (Forestry resources—raw material base and agribusiness significance). J Agribus Rur Dev 35:101–106. https://doi.org/10.17306/JARD.2015.11

Pollock RL (2016) The effect of green leafy and cruciferous vegetables intake on the incidence of cardiovascular disease: a meta-analysis. JRSM Cardiovasc Dis 5. https://doi.org/10.1177/2048004016661435

Poonia A, Upadhayay A (2015) Chenopodium album Linn: Review of nutritive value and biological properties. J Food Sci Technol 52:3977–3985. https://doi.org/10.1007/s13197-014-1553-x

Putnik P, Gabrić D, Roohinejad S et al (2019) An overview of organosulfur compounds from Allium spp.: from processing and preservation to evaluation of their bioavailability, antimicrobial, and anti-inflammatory properties. Food Chem 276:680–691. https://doi.org/10.1016/j.foodchem.2018.10.068

Regulation MS (2014) Rozporządzenie Ministra Środowiska z dnia 9 października 2014 w sprawie ochrony gatunkowej roślin (Regulation of the Minister of Environment of 9 October 2014 on the conservation of plant species). Dz. U. 2014, poz. 1408

Rigat M, Bonet MA, García-Giménez S et al (2009) Ethnobotany of food plants in the High River Ter Valley (Pyrenees, Catalonia, Iberian Peninsula): non-crop food vascular plants and crop food plants with medicinal properties. Ecol Food Nutr 48:303–326

Rinchen T, Narendra S (2015) Exploring nutritional potential of *Atriplex hortensis*. Indian Hort 60(2):16–17

Rola K (2012) Taxonomy and distribution of Allium *ursinum* (*Liliaceae*) in Poland and adjacent countries. Biologia, Botany 67:1080–1087. https://doi.org/10.2478/s11756-012-0101-2

Rose F (1999) Indicators of ancient woodland. The use of vascular plants in evaluating ancient woods for nature conservation. Brit Widlife 10:241–247

Savage G, Vanhanen L (2018) Oxalate contents of raw, boiled, wok-fried and pesto and juice made from fat hen (*Chenopodium album*) leaves. Food 8:2. https://doi.org/10.3390/foods8010002

Savića J, Mačukanović-Jocića M, Jarić S (2019) Medical ethnobotany on the Javor Mountain (Bosnia and Herzegovina). Eur J Integr Med 27:52–64. https://doi.org/10.1016/j.eujim.2019.02.007

Scheffler L, Sharapa C, Amar T et al (2018) Identification and quantification of volatile ramson-derived metabolites in humans. Front Chem 6:410. https://doi.org/10.3389/fchem.2018.00410

Segneanu AE, Damian D, Velciov S et al (2017) Some less unknown application of perennial plants from Romania. Food Nutr J 3:1–5. https://doi.org/10.29011/2575-7091.100037

Sewani-Rusike CR, Iputo JE, Ndebia EJ et al (2015) A comparative study on the aphrodisiac activity of food plants *Mondia whitei*, *Chenopodium album*, *Cucurbita pepo* and *Sclerocarya birrea* extracts in male wistar rats. Afr J Tradit Complem 12:22–26. https://doi.org/10.4314/ajtcam.v12i2.5

Shaheen S, Ahmad M, Haroon M (2017) Edible wild plants: a solution to overcome food insecurity. Springer Cham, pp 41–58

Sharma D, Dey YN, Sikarwar I et al (2016) In vitro study of aqueous leaf extract of Chenopodium album for inhibition of calcium oxalate and brushite crystallization. Egypt J Basic Appl Sci 3(2):164–171. https://doi.org/10.1016/j.ejbas.2016.02.001

Sikarwar I, Wanjari MM, Baghel SS et al (2013) A review on phytopharmacological studies on *Chenopodium album* Linn. Indo Amer J Pharm Res 3:3089–3098

Sikarwar I, Dey YN, Wanjari MM et al (2017) *Chenopodium album* Linn. Leaves prevent ethylene glycol-induced urolithiasis in rats. J Ethnopharm 195:275–282. https://doi.org/10.1016/j.jep.2016.11.031

Sobolewska D, Podolak I, Makowska-Wąs J (2015) *Allium ursinum*: botanical, phytochemical and pharmacological overview. Phytoch Rev 14:81. https://doi.org/10.1007/s11101-013-9334-0

Soukand R, Hrynevich Y, Vasilyeva I et al (2017) Multi-functionality of the few: current and past uses of wild plants for food and healing in Liubań region. Belarus. J Ethnobiol Ethnomed 13:10. https://doi.org/10.1186/s13002-017-0139-x

Sroka W, Wojewodzic T (2015) Stan i perspektywy rozwoju rolnictwa w województwie małopolskim ze szczególnym uwzględnieniem wsparcia środkami Unii Europejskiej (State and prospects of agricultural development in Małopolska voivedeship with particular emphasis on support for European Union funds). Urząd Marszałkowski Województwa Małopolskiego, Kraków

Tardío J, Pardo-de-Santayana M (2016) Ethnobotanical analysis of wild fruits and vegetables traditionally consumed in Spain. In: Sánchez-Mata MC, Tardío J (eds) Mediterranean wild edible plants. Springer Science+Business Media New York, pp 57–79. https://doi.org/10.1007/978-1-4939-3329-7_4

The Plant List (Version 1.1) (2013). http://www.theplantlist.org/ Accessed 21 May 2019

Tug GN, Yaprak AE (2017) Halophytes as a potential food source. Anad J AARI 27:78–81

Yahia EM, Garica-Solis P, Celis MEM (2019) Contribution of fruits and vegetables to human nutrition and health. In: Yahia EM (ed) Postharvest physiology and biochemistry of fruits and vegetables. Elsevier-Woodhead Publishing, pp 19–45. https://doi.org/10.1016/B978-0-12-813278-4.00002-6

Zeipina S, Alsina I, Lepsie L et al (2015) Antioxidant activity in nettle (*Urtica dioica* L.) and garden orache (*Atriplex hortensis* L.) leaves during vegetation period. Chem Technol 66:29–33. https://doi.org/10.5755/j01.ct.66.1.12055

Zheng W (2017) Chemical compounds from *Chenopodium album* Linn. ACMM, IOP Conf Ser. Mat Sci Eng 207:012009. https://doi.org/10.1088/1757-899X/207/1/012009

# Chapter 18
# Landscape Ecological Structure in a Suburban Area: Case Study

**Renata Różycka-Czas, Barbara Czesak, and Wojciech Sroka**

**Abstract** Suburban areas of Polish agglomerations are areas of dynamic structural and spatial changes, which results in adopting some new functions such as residential functions, and in some cases abandoning previous agricultural functions. These changes are mostly turbulent, uncontrolled and with little reversibility, often as a result of lack of coherent comprehensive spatial planning in suburban areas. A high volume of anthropogenic processes affects existing spatial and environmental structures. As a result environmentally valuable areas often shrink which may introduce local ecological imbalance. The results of such imbalance are difficult to measure. However, results can usually be predicted to be twofold: heterogenic mosaic landscapes, commonly chaotic in structure or recurring single function landscapes. Landscape shaping based on nature conservation is a valuable concept as it is not only a way of protecting environmental structures but also helps building better designed living spaces for the residents of suburbs. This paper aims at characterising and analysing the types of ecological structures of landscape (ecological-spatial models) in suburban communities. The research was conducted in selected suburb communities of the Krakow Metropolitan area which are associated with prevailing agricultural areas. Recommendations for landscape shaping have been given for each ecological-spatial model of landscape.

**Keywords** Suburbanisation · Agricultural land loss · Landscape changes · Landscape units

R. Różycka-Czas (✉) · B. Czesak
Department of Land management and Landscape Architecture, Faculty of Environmental Engineering and Land Surveying, University of Agriculture in Krakow, Krakow, Poland
e-mail: renata.rozycka-czas@urk.edu.pl

B. Czesak
e-mail: barbara.czesak@urk.edu.pl

W. Sroka
Department of Economics and Food Economy, University of Agriculture in Krakow, Krakow, Poland
e-mail: wojciech.sroka@urk.edu.pl

© Springer Nature Switzerland AG 2022

J. Hernik et al. (eds.), *Cultural Heritage—Possibilities for Land-Centered Societal Development*, Environmental History 13,
https://doi.org/10.1007/978-3-030-58092-6_18

## 18.1 Introduction

Spatial changes in the urban–rural fringe, especially visible in changes in land use can be uncontrolled and rapid (Prus et al. 2018). Suburban areas change their functions which in turn leads to changes in landscape character. Changes are particularly noticeable in municipalities where the share of agricultural land (AL) in the total area is high, because it is relatively easy to change land status designation in such areas, as well as because changes in functions in those areas will have a larger, more pronounced impact.

One of the most important negative consequences of the process of (sub)urbanisation is the consumption of agricultural land (Busko and Szafrańska 2018). The disappearance of AL is especially pronounced in developing countries but is occurring in Europe as well (Piorr et al. 2011). In Poland, around 0.30% of AL is converted to non-agricultural purposes on an annual average, while in suburban areas this rate is up to five times faster (Sroka 2018). These processes are widely regarded as adverse because agriculture in peri-urban areas (PUA) is perceived as a "counterbalance" to urbanised and densely populated areas. The literature often emphasises many benefits of maintaining and restoring agriculture (Lovell 2010). They include, in particular, creation of green infrastructure, maintenance of biodiversity (preserving habitats of animals and plants) and climate protection (Specht et al. 2016). Social and culture-forming functions are also mentioned, including social integration as part of common initiatives, propagation of traditions and customs, etc. Urban agriculture also contributes to increased food security (Mok et al. 2014).

Suburban areas are suitable for urban agriculture as the conditions for agricultural development do not differ much from those in promising areas elsewhere. The Index of Adjustment of Agricultural Production Space (IAAPS) has been developed by the Institute of Soil Science and Plant Cultivation in Pulawy (The higher the index is the better is the area for agriculture. Index values are between 20 and 125 points). The Index is 67.4 points for suburban areas while the average for Poland is 66.8 points. It is farm fragmentation that can limit urban agriculture in suburbs. Good conditions for agriculture, urbanising development pressure and resulting loss of agricultural land make protection and maintenance of landscapes desirable but difficult in suburban areas.

All those factors are reflected in the landscapes of suburbs, which according to the Landscape Convention should be identified and adequately managed (Council of Europe 2000). Landscapes are diverse and they should be managed sustainably. But landscape variety makes this task difficult. Therefore, there is a need to identify and designate significant landscape to facilitate their monitoring and planning. There are multiple methods of landscape characterisation (Chuman and Romportl 2010; Simensen Halvorsen and Erikstad 2018). All methods of landscape characterisation inevitably involve simplifications and require identifying spatial units (landscape units) (Campos-Campos et al. 2018; Ostaszewska 2006; Soto and Pintó 2010; Verhoeven et al. 2008). Defining and good understanding of landscapes is crucial for their protection, management and planning (Mücher et al. 2010). Moreover, as Antrop and

Van Eetveld (2017) pointed out it is easier to proceed when we classify landscapes as to types and spatial units. In the multitude of multidisciplinary approaches to landscape classification there are no incorrect classifications; these can only be divided according to the aim they would be used for and that would make them either appropriate or unsuitable (Hettner 1928). The approach depends on the scale, available data and the needs of the researchers. There are three main approaches to landscape identification. Researchers analyse the perception of landscape, its aesthetics or its physiography (Cosgrove 2008). Landscape classification should be adequate for the aim of the classification.

One of the best-known Polish landscape classifications is based on the physico-geographical regions delimited by Kondracki (1955, 1959). This classification covers the whole of Poland. It distinguishes various landscapes according to their elevation, soil type and physiognomy. It relates to the physiography of landscape so it reflects one of the three main approaches (Simensen et al. 2018). The European Landscape Convention (ELC) promoted an increased interest in recognising and identifying landscapes. The ELC (Council of Europe 2000) left it to the countries themselves to classify landscapes. It took over a decade to implement the ELC in Poland, where the significant changes in landscape protection law started in 2015. But during those years the ELC induced research on various landscape classification methods such as the Basic Landscape Units (BLU) (Chmielewski and Kułak 2014) approach used by Michalik-Śnieżek et al. (2019) to identify landscapes on the local level of one landscape park. Another approach (Chmielewski et al. 2015) was built on previous approaches (Bogdanowski 1990; Kondracki 1955, 1959; Ostaszewska 2002; Richling 2005; Solecka et al. 2018) and is rooted in the European Landscape Convention which prompted the effort to characterise Polish landscapes. The approach though is not detailed enough to be used at the municipality level. Apart from considering only the physiognomy of the landscape also ecological types, the ecological value of landscapes (Chmielewski and Chmielewski 2015; Chmielewski and Kułak 2016; Richling 2005) and sustainable development (Solecka et al. 2019) are is discussed in the literature, especially in the context of anthropogenisation of the landscapes (Michalik-Śnieżek et al. 2019). One of the classifications that include natural aspects of the landscape was created by Matuszkiewicz (2008) who delimited geobotanical regions based on land cover in Poland. Also Żarska (2006) attempts to identify landscapes according to their ecological features.

Landscape classification should be useful for spatial planning at the level of a municipality (Chmielewski et al. 2015) which could facilitate recognition of valuable natural and anthropogenic landscapes. Polish spatial planning already recognises the need to protect landscapes. The Act on Spatial Planning and Development from 2003 specifies that local spatial plans should have mandatory provisions that should include rules for environmental and landscape protection, rules for landscape shaping, rules for cultural landscape protection, boundaries and rules for priority landscapes development. The Żarska approach allows for landscape identification at the municipality level, which is also the main level of spatial planning in Poland, and thus it was chosen as a test approach in the Kraków metropolitan area.

With the introduction of the ELC there is an increasing need for planning tools and quick methods, which will help to create strategies that combine preservation of landscape diversity with sustainable use of an area. Such methods could help municipalities to control and manage landscapes, protect spatial order and support municipality spatial planning. Existing landscape audits (Chmielewski et al. 2015) based on the ELC are time consuming, obligatory only for provinces and expensive which makes it difficult to do such audits at the local level. The aim of this paper is to use Żarska's method (2006) to identify landscapes in the test areas and to create a list of recommendations that could support a local spatial planning process. We assumed that there is a relationship between the ecological structure of a landscape and the fitness of agricultural production space (measured by IAAPS) as well as the distance from the core of the metropolitan area.

## 18.2 Method

*Delimitation of Research Area*

First the Functional Urban Area (FUA) of Krakow, located in Malopolska voivodeship was delimited. The FUA was delimited based on Urban Atlas (UA) data. Out of 43 municipalities those without national parks, or nature parks, as well as without landscape protection parks were chosen. Then the municipalities in which the share of agricultural land in the total area is at least 50% were chosen. The score of agricultural production space indicator (IAAPS[1]) was used as the last criterion. We chose municipalities for which the IAAPS score was the highest and the lowest in a previously limited group. Figure 18.1 shows six municipalities in which the study was conducted. Three of them are located in the direct neighbourhood of Kraków in the suburban area and the other three in the second ring of the municipalities surrounding the city. The quality of agricultural production space (IAAPS) differed among the municipalities. The municipalities located north from Kraków have a higher IAAPS score. In two out of three municipalities located south from Kraków the quality of agricultural production space is average and in one of them it is poor in the PUA.

The highest score of IAAPS was calculated for Proszowice municipality and Igołomia-Wawrzeńczyce municipality, and the lowest: Świątniki Górne municipality and Mogilany municipality located in the suburban area and Myślenice and Wiśniowa municipalities located in the PUA (Fig. 18.1)

*Classification of a Landscape Type*

Landscape types are identified based on land cover forms. In this paper the method proposed by Żarska (2006) was used. First landscapes were classified into two groups:

[1] Index of adjustment of agricultural production space (IAAPS), developed by IUNG in Puławy.

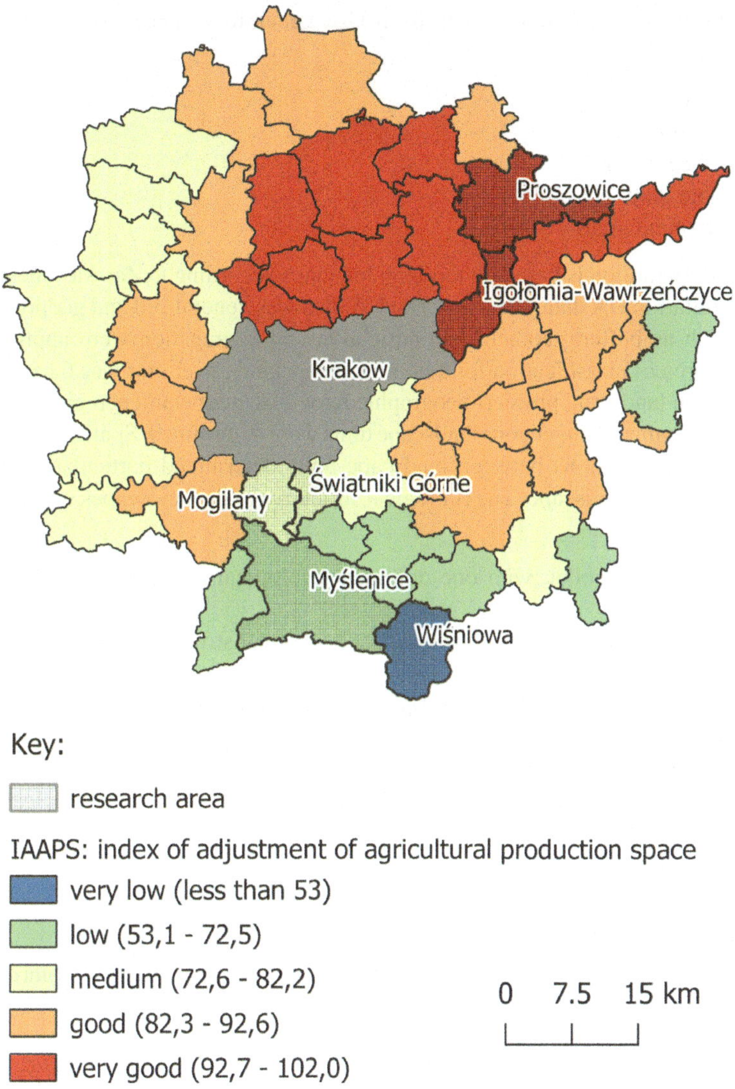


**Fig. 18.1** The score of IAAPS for FUA of Krakow

upland landscape and valley landscape. Classification is based on topographic maps and a digital elevation model from Web Map Service (WMS) by Polish Head Office of Geodesy and Cartography.

In the second step, the physico-geographical regions proposed by Kondracki (1998), as well as geobotanical regions according to Matuszkiewicz (1993) were delimited.

In the third step, based on data from UA, generalised land cover classes were determined:

A. forests
B. meadows
C. surface water
D. arable land
E. settlement areas

This delimitation is based on topographic maps according to Żarska's method. In our delimitation UA data was used. The UA data was generalised and geoprocessing tools, such as buffer, geometry simplification and aggregation were applied. The minimal area of classification for a homogeneous land cover unit was 6 ha.

Based on landform, physico-geographical regions, geobotanical regions and land cover classes the homogeneous landscape units were determined. Spatial distribution, shapes and proportions between landscape units determined particular ecological structures of the landscape. The ecological structures of the landscapes were grouped into four types:

Type I. Axial model, with one dominant ecological structure in the elongated shape.
Type II. Knot model, with one dominant ecological structure of which the area is at least 500 ha.
Type III. Mosaic model, lack of one dominant ecological structure, a mosaic of small or medium size ecological/natural units surrounded by highly transformed landscapes (e.g. settlements).
Type IV. Compound model, with at least two big, distinctive ecological structures or with one dominant ecological structure surrounded by a mosaic of ecological/natural units.

*Classification of a Landscape Subtype*

Based on qualitative criterion, spatial criterion, as well as quantitative criterion subtypes for each classified type were delimited.

## 18.3 Results

The dominant elements of ecological structures of analysed landscapes relevant for wild species preservation and natural valuable plants conservation are forests, river valleys, as well as land depressions.

According to Kondracki (1998) Proszowice municipality is located in the Niecka Nidziańska macroregion (physico-geographical regions). It is classified as landscape ecological-spatial axial model (type I). In this type a dominant ecological structure should have an elongated shape. Above-mentioned criterion is fulfilled by meadows located in the central part of the municipality. The meadows are the main ecological structure of this research area (Fig. 18.2). There is a lack of distinctive forest; forestation rate is only 3% (Statistical Office in Krakow 2018). A main meadows ecosystem is located in the valley, which can be considered as an ecological axis. According to qualitative criterion Proszowice municipality is a valley landscape. According to spatial criterion, it is a central (classic) subtype of the axial model. According to quantitative criterion which is based on the share of natural and seminatural ecosystems in the total area, it is middlingly affluent (a total share of surface area of meadows, forests, and surface water is 30%).

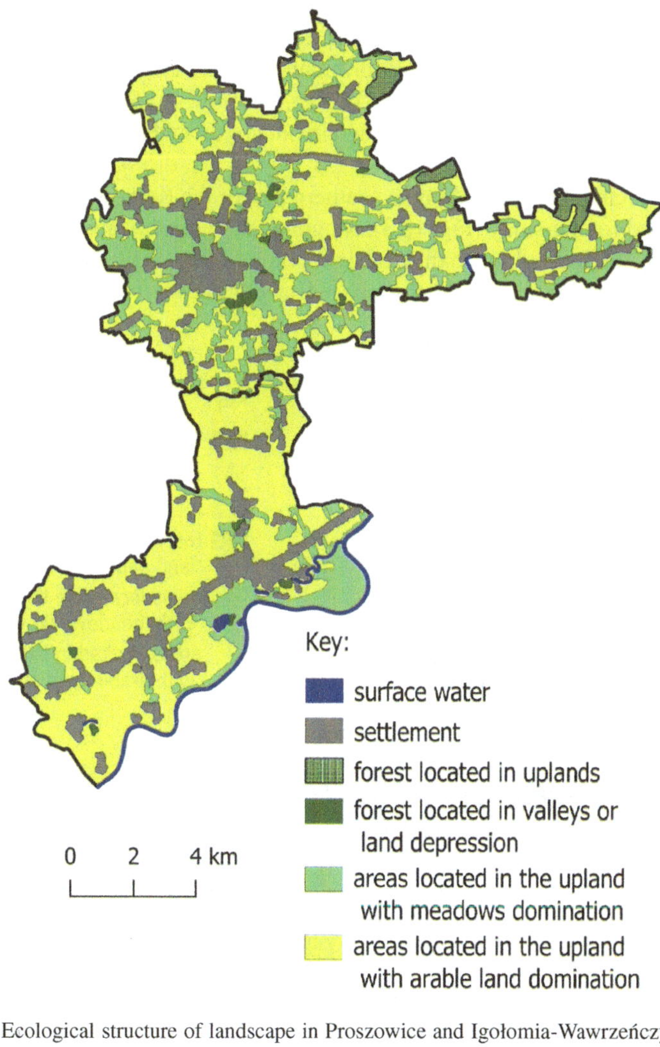


**Fig. 18.2** Ecological structure of landscape in Proszowice and Igołomia-Wawrzeńczyce municipalities

Igołomia-Wawrzeńczyce municipality is located in two physico-geographical macroregions: Niecka Nidziańska and Kotlina Sandomierska (Kondracki 1998). It is classified as landscape ecological-spatial axial model (type I). A dominant ecological structure is constituted by the Vistula River valley located on the eastern border of the analysed municipality (Fig. 18.2). A considerable disparity in quality of ecological structure occurs between areas located in the valley and those located in the uplands. The area in the direct neighbourhood of the Vistula is particularly ecologically valuable as in these areas there occur old river beds and swampy land depressions with meadow as well as marshland vegetation. The upland is rich in fertile arable lands. According to spatial criterion it is a border type of axial model. The landscape of Igołomia-Wawrzeńczyce is less affluent. A total share of surface area of meadows, forests and surface water is 23%. There are almost no forests, and the forestation rate is only 0.1%. Both municipalities, Proszowice and Igołomia-Wawrzeńczyce are located in Kraina Wyżyn Miechowsko-Sandomierskich geobotanical region (Matuszkiewicz 1993) and in both of them the quality of agricultural space production is high.

Four municipalities located in the southern part of the Krakow metropolitan area were selected for our study (Fig. 18.3). Mogilany municipality and Świątniki Górne municipality are located in the direct neighborhood of Krakow, and in both of them the quality of agricultural production space is average.

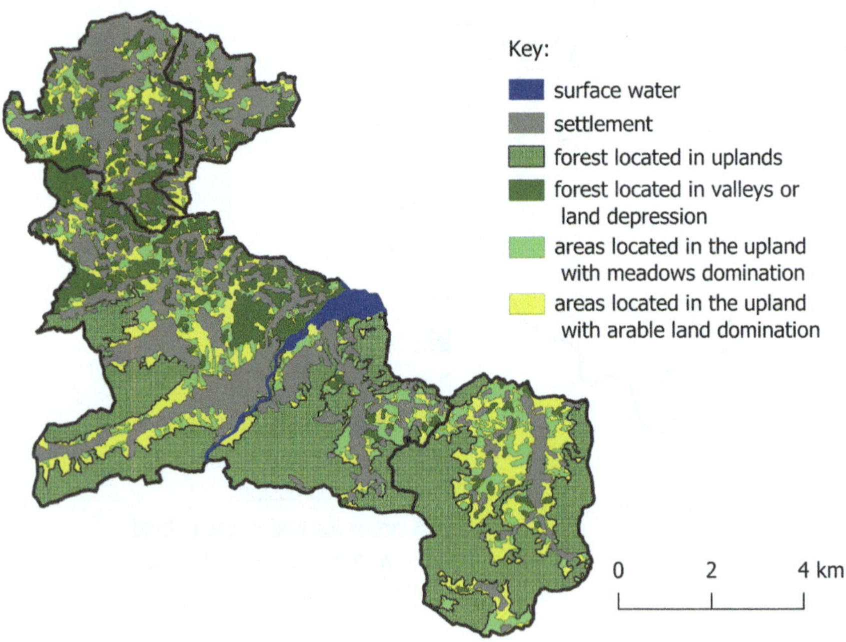


**Fig. 18.3** Ecological structure of landscape in Mogilany, Świątniki Górne, Myślenice and Wiśniowa municipalities

According to Kondracki (1998) classification of the northwestern part of Mogilany municipality implies that it belongs to Brama Krakowska physico-geographical macroregion and the rest of the municipality's area is located in the Pogórze Zachodniobeskidzkie physico-geographical macroregion. Mogilany municipality was identified as a mosaic model (type III) and a forest-valley subtype. Forests cover 13% of the municipality (Statistical Office in Krakow 2018). Natural and semi-natural areas cover 60% of the municipality area, which means, according to quantitative criterion, that it belongs to the affluent subtype. There is no dominant or distinctive natural or semi-natural structure. The landscape is constituted by small and a medium size homogeneous, natural and semi-natural landscape units (mostly meadows and coppices), surrounded by agricultural landscape with domination by arable lands.

Świątniki Górne municipality belongs to the same landscape ecological-spatial model as Mogilany municipality. Forests cover 13.7% of total municipality area (Statistical Office in Krakow 2018). Świątnik Górne and almost the entire area of Myślenice municipality belong to Pogórze Zachodniobeskidzkie. The main and dominant natural structure in Myślenice municipality is the Raba River valley (axis). It is an important ecological corridor by local and regional scale. Located in the southern part of the municipality, mixed forests (knot) are the second important natural structure. Occurrence of two dominant natural structures resulted in assigning type IV of the ecological structure of landscape to Myślenice municipality. Because of the relative location of axes and ecological knots the landscape was identified as within an integrated compound axial-knot model subtype.

The last researched municipality, Wiśniowa, is located in the PUA. It has the most hilly ground among the entire set of researched municipalities. Wiśniowa municipality has rich resources of animate and inanimate elements of nature. A dominant role is played by upland forests located in the southern part of the municipality. It is an important regional and local ecological structure. The forests belong to the Beskidy Zachodnie national ecological corridor which covers the southern part of Myślenice and Wiśniowa municipalities. Geobotanicaly both municipalities are located in Kraina Kotliny Oświęcimskiej and Kraina Karpat Zachodnich. Based on the Żarska (2006) method Wiśniowa municipality is identified as a knot model (type II). A dominant ecological structure is big areas of woodland, located along the southern part of the municipality's border (forestal, border landscape subtype). Natural and semi-natural areas cover more than 35% of the municipality area, which means, according to quantitative criterion, that it belongs to the affluent subtype.

## 18.4 Conclusion and Recommendations

The spatial data used in this research is freely and easily accessible, which can increase the ease of use. In our delimitation UA data instead of data from topographic maps was used to minimise time needed for classification, as well as required expenditure of labour.

The landscape ecological-spatial models identified for suburban and peri-urban municipalities vary. All four types of landscape ecological-spatial model were identified in the research area. These include type I (axial) in Proszowice and Igołomia-Wawrzeńczyce municipalities, type II (knot) in Wiśniowa municipality, type III (mosaic) in Mogilany and Świątniki Górne municipalities, and type IV (compound) in Myślenice municipality. For each of the main types a subtype, based on qualitative, quantitative and spatial criterion was determined. No relationship was observed between the distance from the metropolitan core and the type of landscape ecological-spatial model. There are no relationships as well, between score of IAAPS and distance from metropolitan core. On the other hand, there is a strong connection between score of IAAPS and the type of land use. In the northern part of the studied area, especially in Proszowice and Igołomia-Wawrzeńczyce municipalities the surface of arable land is bigger than in the southern municipalities. Both IAAPS index and the landscape ecological-spatial model should be taken into account in spatial planning policy.

The main recommendations for type I of landscape ecological–spatial model is to preserve the main ecological axis, as well as link other natural structures with them. Creation of dense ecological structure (meadows in this case) should be the most important action for Proszowice municipality of which the landscape is middlingly affluent in natural elements. That kind of action should result in biological strengthening of the main ecological axis. Moreover Żarska (2006) recommends forestation as the main action for municipalities with low forestation index. She proposed creation of at least one big (more than 500 ha) forest anchor. But, as she mentioned, this can be difficult for municipalities with upland, in terms of agricultural usefulness, which is the case for the analysed municipality. Igołomia-Wawrzeńczyce municipality, the second municipality with the axial model, should focus its spatial actions on Vistula River valley preservation, as it is the most important and dominant ecological structure. It can be restored and its role reasserted by forest or bushes, especially since the forestation rate for this municipality is extremely low. and that municipality has its productive lands in the uplands.

Mogilany and Świątniki Górne municipalities have the most complicated landscape ecological-spatial model, a mosaic one. Municipalities with that type of landscape should strive for creation of one significant natural anchor, as well as some ecological corridor formation. The municipalities studied are highly developed, with significant and rapidly changing investment processes, so the abovementioned actions can be extremely hard. In that case preservation of existing small, but ecologically valuable units is the proper action.

Myślenice and Wiśniowa municipalities, which are more affluent in natural elements should focus on dominant ecological structures in terms of spatial integration (axes and knots). Żarska (2006) recommends creation of an ecological corridor parallel with the dominant axis and radiating around knots. That action can be appropriate for the studied municipalities.

Further research should be done to investigate the landscape ecological–spatial models of external ecological connections, which are important for the long-term survival of the living resources of the municipality and its surroundings.

**Acknowledgements** We thank the editor and the reviewers for their valuable comments and detailed suggestions to improve the paper. Further, we also acknowledge the support by the National Science Centre, Poland under project no. 2016/21/D/HS4/00264.

## References

Antrop M, Van Eetvelde V (2017) Landscape perspectives: the holistic nature of landscape. Springer, Dordrecht. https://doi.org/10.1007/978-94-024-1183-6

Bogdanowski J (1990) Metoda jednostek i wnętrz architektoniczno-krajobrazowych (JARK WAK) w studiach i projektowaniu. Kraków: Politechnika Krakowska

Busko M, Szafranska B (2018) Analysis of changes in land use patterns pursuant to the conversion of agricultural land to non-agricultural use in the context of the sustainable development of the malopolska region. Sustainability (Switzerland) 10(1). https://doi.org/10.3390/su10010136

Campos-Campos O, Cruz-Cárdenas G, Aquino RJC, Moncayo-Estrada R, Machuca MAV, Meléndez LAÁ (2018) Historical delineation of landscape units using physical geographic characteristics and land use/cover Change. Open Geosci 10(1):45–57. https://doi.org/10.1515/geo-2018-0004

Chmielewski TJ, Chmielewski S (2015) Podstawowe przyrodnicze jednostki przestrzenne, a spójność i stabilność ekologiczna systemów krajobrazowych [The basic natural spatial units, versus ecological connectivity and stability of landscape systems] (3):145–160

Chmielewski TJ, Kułak A (2014) Struktura fizjonomiczna krajobrazu. In Struktura środowiska przyrodniczego, a fizjonomia krajobrazu. Instytut Geografii i Gospodarki Przestrzennej; Uniwersytet Jagielloński, pp 33–52

Chmielewski TJ, Kułak A (2016) Ekotony w krajobrazie i krajobraz ekotonów: nowe wyzwania dla uznanej koncepcji. Prace Komisji Krajobrazu Kulturowego 31:25–42

Chmielewski TJ, Myga-Piątek U, Solon J (2015) Typologia Aktualnych Krajobrazów Polski. Przeglad Geograficzny 87(3):377–408. https://doi.org/10.7163/PrzG.2015.3.0

Chuman T, Romportl D (2010) Multivariate classification analysis of cultural landscapes: an example from the Czech Republic. Landsc Urban Plan 98(3–4):200–209. https://doi.org/10.1016/j.landurbplan.2010.08.003

Cosgrove DE (2008) Geography and vision seeing, imagining and represeninc the world. Match Studio Projects

Council of Europe (2000) European Landscape Convention, Florence 2000. European Treaty Ser (176):1–7. https://rm.coe.int/1680080621

Hettner A (1928) Die Oberflächenformen des Festlandes: Probleme und Methoden der Morphologie. Teubner, Leipzig

Kondracki J (1955) Problematyka Fizyczno-Geograficznej Regionalizacji Polski. Przegląd Geograficzny 27:298–309

Kondracki J (1959) Typy środowiska i podział regionalny. In: Z badań środowiska geograficznego w powiecie mrągowskim. Pr. Geogr. Inst. Geogr. PAN, pp 91–102

Kondracki J (1998) Geografia regionalna Polski. Wydaw. Naukowe PWN, Warszawa

Lovell ST (2010) Multifunctional urban agriculture for sustainable land use planning in the United States. Sustainability 2(8):2499–2522. https://doi.org/10.3390/su2082499

Matuszkiewicz JM (1993) Krajobrazy roślinne i regiony geobotaniczne Polski. Prace Geograficzne IGiPZ PAN 158:107

Matuszkiewicz JM (2008) Geobotanical regionalization of Poland. https://www.igipz.pan.pl/Regiony-geobotaniczne-zgik.html

Michalik-Śnieżek M, Chmielewski S, Chmielewski TJ (2019) An introduction to the classification of the physiognomic landscape types: methodology and results of testing in the area of Kazimierz Landscape Park (Poland). Phys Geogr 40(4):384–404. https://doi.org/10.1080/02723646.2018.1551009

Mok HF, Williamson VG, Grove JR, Burry K, Barker SF, Hamilton AJ (2014) Strawberry fields forever? Urban agriculture in developed countries: a review. Agron Sustain Dev 34(1):21–43. https://doi.org/10.1007/s13593-013-0156-7

Mücher CA, Klijn JA, Wascher DM, Schaminéecd JHJ (2010) A new European Landscape Classification (LANMAP): a transparent, flexible and user-oriented methodology to distinguish landscapes. Ecol Indicators 10(1):87–103. https://doi.org/10.1016/j.ecolind.2009.03.018

Ostaszewska K (2002) Geografia Krajobrazu. Wydawnictwo Naukowe PWN, Warszawa

Ostaszewska K (2006) Models of Landscape units—utopia or necessity. Miscellanea Geographica 12:5–11

Piorr A, Ravetz J, Tosics I (2011) Peri-Urbanisation in Europe: towards European policies to sustain urban-rural futures. In: Forest & landscape. University of Copenhagen, Copenhagen, pp 30–42. http://forskningsbasen.deff.dk/Share.external?sp=Sac92f938-1fd0-4556-b96e-2697a0e87707&sp=Sku

Prus B, Bacior S, Dudzinska M (2018) Identification of spatial-settling status of rural areas of southern Poland a case study. Eng Rural Dev 17:439–444. https://doi.org/10.22616/ERDev2018.17.N020

Richling A (2005) Krajobraz naturalny, pierwotny, kulturowy i potencjalny. In: Geografia fizyczna Polski. Wydawnictwo Naukowe PWN, Warszawa, pp. 294–296

Simensen T, Halvorsen R, Erikstad L (2018) Methods for landscape characterisation and mapping: a systematic review. Land Use Policy 75(October 2017):557–569. https://doi.org/10.1016/j.landusepol.2018.04.022

Solecka I, Bothmer D, Głogowski A (2019) Recognizing landscapes for the purpose of sustainable development—experiences from Poland. Sustainability 11(12):3429. https://doi.org/10.3390/su11123429

Solecka I, Raszka B, Krajewski P (2018) Landscape analysis for sustainable land use policy: a case study in the municipality of Popielów, Poland. Land Use Policy 75(September 2016):116–126. https://doi.org/10.1016/j.landusepol.2018.01.021

Soto S, Pintó J (2010) Delineation of natural landscape units for Puerto Rico. Appl Geogr 30(4):720–730. https://doi.org/10.1016/j.apgeog.2010.01.010

Specht K, Siebert R, Thomaier S (2016) Perception and acceptance of agricultural production in and on urban buildings (ZFarming): a qualitative study from Berlin, Germany. Agricult Hum Values 33(4):753–769. https://doi.org/10.1007/s10460-015-9658-z

Sroka W (2018) Conversion of agricultural land to non-agricultural purposes in selected Polish Metropolitan Areas. Acta Scientarum Polonorum 17(2):97–107

Statistical Office in Krakow (2018) Statistical Vademecum of Regional Civil Servant

Verhoeven JTA, Soons MB, Janssen R, Omtzigt N (2008) An operational landscape unit approach for identifying key landscape connections in wetland restoration. J Appl Ecol 45(5):1496–1503. https://doi.org/10.1111/j.1365-2664.2008.01534.x

Żarska B (2006) Modele ekologiczno przestrzenne i zasady kształtowania krajobrazu gmin wiejskich. Wydawnictwo SGGW, Warszawa

# Chapter 19
# South African Agricultural Oenology, Viticulture, Land Ownership, and Sustainable Development

**Betty J. Harris and Edward Sankowski**

**Abstract** We interpret aspects of cultural heritage, land use, and sustainable development in South Africa in relation to agriculture and, most specifically, viniculture and the wine industry. We review relevant South African history, including colonialism, land expropriation, and the formation of a mixed-race community and identity within South Africa. We then proceed to an examination of key features of the political economy and culture since 1994 and the South African transformation. We provide a research commentary on land rights and societal plans to transform farmworkers into farmers. We note the absence of government programs to facilitate the goal that farmworkers would become farmers. We note the population group of "colored" South Africans making efforts to be recognized as indigenous people who thereby would be granted more land rights. We discuss features of the situation of black middle-class vintners, including the requirements to establish a wine estate. Black middle-class starting points pertain to wine-making essentials, as well as acquiring wine labels, and obtaining other resources to generate an income to buy and operate a farm. We proceed to a discussion of further contemporary land issues in South Africa, relating this to issues about sustainable development.

**Keywords** South Africa · Land · Apartheid · Agriculture · Wine industry · Sustainability

B. J. Harris (✉)
Department of Anthropology, College of Arts and Sciences, University of Oklahoma, Dale Hall Tower, Norman, OK 73019, USA
e-mail: bharris@ou.edu

E. Sankowski
Department of Philosophy, College of Arts and Sciences, University of Oklahoma, Dale Hall Tower, Norman, OK 73019, USA
e-mail: esankowski@ou.edu

© Springer Nature Switzerland AG 2022

J. Hernik et al. (eds.), *Cultural Heritage—Possibilities for Land-Centered Societal Development*, Environmental History 13,
https://doi.org/10.1007/978-3-030-58092-6_19

## 19.1 Contemporary Land Issues in South Africa

In the 25 years after the 1994 emergence of the New South Africa, land issues are increasingly contentious. In 1994, the black majority only controlled 13% of the land in South Africa. In the constitutional negotiations leading up to the first all-race elections, white commercial farmers, who own most of South Africa's most productive farmland, organized a strong lobby to influence policy on the status of farmworkers and continued control of farmland. As a consequence, the South African Constitution oriented its land policies to norms emerging from the 1913 or 1936 Land Acts to the present. In the 1990s, significantly, Zimbabwe's land issues were becoming more violent because little land had been acquired by black Zimbabweans.

Entering into "independence" during a period when neoliberalism was becoming more ubiquitous, the South African government was discouraged from developing various social programs to facilitate a transformation in which the previously disadvantaged would have greater opportunities for social mobility. South Africa was saddled with a large International Monetary Fund loan from the apartheid era. The Eastern Bloc countries that had supported the anti-apartheid movement provided refuge to African National Congress leaders, and had educated South African doctors and other professionals, had experienced the collapse of the Soviet Union, and underwent their own transformations.

## 19.2 History of Land Alienation from Indigenous South Africans of the Western Cape

Cape Town, established in 1652, became the site of a refreshment station for the Dutch East India Company along the Spice Route between Amsterdam and Batavia (currently Jakarta, Indonesia). Company crews represented a variety of European nationalities. Those who were assigned to Cape Town had the opportunity to remain there after completion of their contracts with the Company thereby infringing on the rights of Khoisan foragers and cattle-herders.

While the marriage between a Khoisan woman, Eva, and a member of the Dutch East India Company, Van Meerhof, is often highlighted as initiating the development of a mixed-race or colored population in the Western Cape, this narrative minimizes the impact of genocide and land alienation on the Khoisan people.

As more crew members completed their contracts, the Company sought land for them in the environs around Cape Town, including major wine-growing areas of Stellenbosch and Drakenstein, in the late 1670s through 1680s. White settlement in those areas precipitated a series of wars with indigenous populations. Elphick (1977) indicates that the first Khoikhoi–Dutch war occurred in 1659 and the second in 1673.

In the formation of a mixed-race identity, by 1701, Company authorities rarely referred to Khoikhoi leaders whose polities were no longer viable or on the verge of collapse.

Slavery persisted in the Cape from 1658 to 1833, with slaves being imported into the Cape from parts of Asia under Dutch East India Company rule and southeast Africa. Even though landless and dependent Khoikhoi were reduced to working on white-owned farms, they have often been described as indentured servants. When slavery was abolished, many recently emancipated people moved to missions, such as Genadendal and Mamre, where they were taught agricultural skills and established two-parent, domestic-type families. A substantial free colored population lived in Cape Town.

## 19.3 Segregationist and Apartheid Legislation

The 1913 Natives Land Act prohibited black South Africans from renting or owning land outside of designated areas. Only 7–8% of land was designated for black South Africans, and such land was less fertile than land occupied by whites. The 1936 Native Land and Trust Act increased the potential for reserves to 13%, which was never achieved during Apartheid. It prohibited black South Africans from purchasing land outside of designated areas. The Prohibition of Mixed Marriages Act of 1949 made it illegal for whites and non-whites to marry. The Group Areas Act of 1950 aimed to eliminate mixed neighborhoods by prohibiting occupation in, and ownership of land in areas designated for other racial groups. Some neighborhoods were reclassified requiring relocation of entire groups of non-white people.

## 19.4 Farmer-Initiated Wine Projects After 1994

After the activism that led to the end of legal apartheid, the new constitution, and all-race elections in 1994, there were notable initiatives in the wine sector. These included the Nelson's Creek Wine Estate's New Beginnings worker land allocation in 1997; and Solms-Delta's Wine Estate's establishment of the Wijn de Caab Trust in 2007 with 50% worker ownership proposed.

## 19.5 Strategies for Breaking into the South African Wine Industry

Since 1994, South Africa had had a willing-buyer/willing seller land transfer policy. Therefore, for black entrepreneurs, land acquisition has been the most difficult task in establishing a wine estate. Some black vintners buy wine grapes from estates that produce a surplus and produce on their own labels. Without a farm, one must also contract with a winemaker and cellar to produce wine. Black entrepreneurs also need to contract with marketers. If one owns a wine estate, one can have these functions under one roof. However, by starting with acquiring grapes and creating labels, vintners can seek to acquire capital to buy land.

The South African wine industry ranks 6th in the world, generates more revenue than the mining industry, has created skyrocketing land prices, and expanded at least ninefold during 1994–2013, after international economic sanctions were lifted against South Africa.

## 19.6 Some Especially Noteworthy Facts

One often reads in the literature that black South Africans only controlled 13% of the land during apartheid. This would include the various homelands demarcated by the South African government.

During the negotiations to end apartheid from 1990 to 1994, the farmers' lobby had the clout to negotiate to keep land acquired prior to the 1913 Land Act. Of particular interest is the apartheid-era Cape Province, which today includes the Western, Eastern, and Northern Cape, where most of the land was alienated, at least by the end of slavery in 1833. With the exception of government land, there was little probability that the previously disadvantaged would acquire substantial amounts of land.

The colored population would have difficulty acquiring land as a racial group, having been proletarianized in the eighteenth century, having lost tribal leadership and Khoikhoi ethnic identity. In the post-apartheid period, efforts are being made to resurrect tribal structures. The Center for Human Origins at the University of the Witwatersrand is offering free DNA tests and analyzing the DNA of the South African population. While the Western Cape served as a Coloured Labor Preference Area, coloreds did not have a designated land area as indigenous people.

Wine estates were typically located in areas with higher average rainfall with vineyards on flat land and on mountain slopes where they benefit from ocean breezes. However, in the post-apartheid period, wine estates have been established in low-rainfall areas, such as the Swartland, where irrigated wheat is produced, and other areas of South Africa. There are now eighteen official wine routes in South Africa to which tourists travel by car and bus. Wine estates often have restaurants and tasting rooms.

***Two Policy Proposals***

1. Government incentives need to be developed to encourage entrepreneurs of color to invest in the wine industry to break with neoliberal policies.
2. There should be a global movement to give all Khoisan descendants "first nations status."

## 19.7 Khoisan Land Alienation and the Establishment of White Commercial Agriculture

Globalization has had a tremendous impact in shaping the South African political economy and culture from its Dutch mercantilist founding as a temporary refreshment station at the Cape of Good Hope to the emergence of the post-apartheid state into a neoliberal global context. It is significant that South Africa's colonization was initiated by a company rather than a nation-state. At the time, the Northern Provinces were just developing into a nation-state. The Dutch East India Company (VOC), which colonized South Africa, had been formed in 1602 to consolidate competing shipping lines, in order to minimize competition on the Old Spice route to Batavia.

In 1652, when Jan van Riebeeck and members of the Dutch East India Company set up a refreshment station at Table Bay, they intended to be there temporarily because they didn't consider the venture economical (Theal 1897). The 90 members, predominately male crew represented different European nationalities although the majority were Dutch or German Protestants. During this period, the Company governed the entire white population of Table Bay.

The Dutch East India Company demarcated an area along the shore of Table Bay in which to establish a fort which included a hospital, workshops, a storage area for agricultural produce, and stables for stock animals. Thus, they were equipped to supply passing ships with meat and fruit and vegetables—the last two of which prevented scurvy among crew members. Although the Europeans grew their own vegetables and fruit, they had to bargain for cattle and sheep from the Khoikhoi. Nevertheless, the Khoikhoi began to resist this arrangement when they saw that their supply was being depleted.

The first healthy vine cuttings were imported from Europe for wine production in 1656 and the first wine pressed in 1659 (Seely 1997: 2). The grapes were likely of French origin and could be related to Muscat d'Alexandrie or Haneport or Chenin Blanc or Steen. The first wine estate was established in Constantia in 1685.

By 1657, the Company had allowed several of its freed servants—men who had completed their service to the Company—to have freehold rights on farmland some distance from the settlement. The settlers encountered resistance from the Khoikhoi with whom they fought the First Khoikhoi War (1658–1660). The settlers had encroached on their pastureland thus inhibiting their migratory pattern. Continued white expansion in this area resulted in the Second Khoikhoi War (1673–1677) after which white settlement was temporarily confined to the Cape flats.

Since the Cape was rather sparsely populated and Khoikhoi were not readily available as laborers, slaves were imported from Angola in 1658 and Guinea in 1659. The slaves were leased out to the farmers. However, most fled and eventually returned to the direct ownership of the Company. Later, slaves were to be imported from Mozambique and Madagascar as well as from parts of Asia that were under Dutch influence.

By the time of the first British takeover of the Cape in 1795, the Khoikhoi were virtually landless. However, there were a few groups who managed to maintain, to some extent, their traditional social organization. When the British arrived to administer the Cape, they saw the necessity of formulating legislation to regulate black labor. However, implementation of that legislation was interrupted by the restoration of the Cape to Dutch rule as provided by the Treaty of Amiens in 1803. Nevertheless, three years later when the British recaptured the Cape, they began to institute South Africa's first labor legislation.

The Cape Colony, upon incorporation into the British Empire, was subject to laws passed by the British Parliament. The first such legislation was the prohibition of the slave trade in 1808. The Cape colonial government also passed legislation applicable to the economic conditions prevailing there. Of course, the prohibition of the slave trade created a labor shortage on the farms although the supply of slaves was not completely cut off.

To alleviate the situation, the Cape colonial government passed legislation that would make the landless Khoikhoi more accessible to farmers as laborers. In the plantation setting, there is little data on relations between Khoikhoi laborers and slaves.

Caledon's Law stipulated that all labor contracts between farmers and Khoikhoi must be registered with a field cornet. This law, along with existing pass laws, made Khoikhoi labor more stable and depressed wages because laborers could not move freely from one plantation to another to bargain about wages. Slavery was abolished in the Cape Colony in 1835. While many former slaves remained on the farms of their former masters, some moved to mission stations where they were able to acquire land and learn new agricultural techniques.

Wilson (1971: 23) views the Mineral Revolution, beginning in 1867 with the discovery of diamonds in Kimberly and continuing in 1886 with the discovery of gold in the Witwatersrand, as coinciding with the closing of the South African frontier. The process of white farm subdivision had intensified by that time.

## 19.8 Wine Estates in the Post-Apartheid Period

Historically, the agricultural sector has been viewed as the most repressive in the South African economy. Farmers have been characterized as being bigoted, autocratic, and patriarchal. The farm is the domain over which the farm owner has complete control. He controls his own family labor as well as that of farmworkers. He controls labor relations, technological innovation, ecological degradation, wages, alcohol use, domestic violence, housing, schooling, shopping, medical care, recreation, unionization, etc. Thus, farm workers have been very dependent on farmers for their total livelihood. Farmers in the Western Cape, among the most progressive in apartheid South Africa, have been anti-union, anti- "progressive" politics, and anti-protective labor legislation. For the most part, they supported the National Party in the 1994 election and the Democratic Alliance in the 1999 election.

Farms vary considerably in how well farmworkers are treated. However, wine farms tend to have better labor relations than wheat farms, higher wages, and better housing. On the more benevolent farms, farmworkers may be treated as extensions of the farmer's family with the farmer's wife acting as nurse to the farmworkers' families.

"Colored" men have engaged in basic farm labor and are most likely to be chosen for leadership positions and to be sent for further training. Farmworkers' wives may be employed as domestics and as seasonal labor on farms and their children may be employed during school holidays. The summer holidays coincide with the grape harvest.

Since 1994, South Africa has opened its doors to international tourism. In response to hordes of tourists from Western Europe, the US, and elsewhere, entrepreneurs and government entities have invested in new hotels, shopping malls, game reserves, and museums, while upgrading existing infrastructure. Cape Town is the number one tourist destination in South Africa. The wine routes established throughout the Western Cape are well trodden by tourists exploring rural areas in the vicinity of Cape Town.

## 19.9 Wine Estate Survey Results for Farmers and Farmworkers

In 1997–1998, Betty J. Harris conducted a survey of the South African wine industry in the Western Cape towns of Stellenbosch, Paarl, and Robertson. Her survey included forced answer and open-ended questions. As discussed here, the present chapter compares the responses of wine farmers and farmworkers to major open-ended questions about the status and future of the wine industry for their perspectives. With regard to the impact of tourism on the wine industry, farmers were generally enthusiastic. The views of two farmers reflect those of most farmers:

> A hell of an impact! Very big! We have a system of cooperative wine cellars. Wine estates cultivate tourism and make lots of money from it. Overseas marketing is important as most wine is sold overseas. Wine goes out. Brandy stays here.
>
> Very big! The exposure through communication with foreigners broadens your vision and thinking.

One less enthusiastic farmer stated:

> There are too many tourists in December and January. We can't cope with parking, toilet facilities, selling and keeping stock. We either have to expand or limit admission by charging an entrance fee. It would provide job creation and a valued-added product-wine. We have to project the right image through advertising and publicity companies.

Many farmworkers, however, viewed tourism in relation to their own remuneration:

> It will increase wine sales. If we sell more wine, it means we might get a raise. It's a good influence of the industry in general.
>
> I feel that it's unfair because money is coming in and tourism has had a positive impact on farming but it has not benefited the farmworkers in any way.

The following reflects the pride that some farmworkers expressed about the farm and the industry as a whole:

> It is very good for the wine industry. People from overseas and elsewhere in our land can taste our wine and buy it elsewhere. We all believe in it—all the workers. We need more domestic tourism from Natal and the Transvaal so that they will know what we've busy doing here. We know what they are doing there.

Harris viewed government support as having been essential to a globalizing wine industry that was re-entering the international system relatively late. Nevertheless, many wine farmers did not want government support or thought it should be minimal:

> The South African government should leave the wine industry alone because it doesn't need money. Ten years from now, it will be down in surplus. Four years later, it will be full. The wine industry has got its ups and downs. I don't ask for any money.
>
> We only PAY, PAY!! The wine industry is a milking cow. We want privatization and globalization but are not used to it. We had 50 years of protection. Free trade? I suppose it should stay. I don't know how good it is for Africa. Free trade is coming to the wine industry.

Those farmers who found the government supportive were in the minority. Here is one comment:

> Yes, they have been supportive. Traditionally, they have provided too much support. Incentive money is available to develop farms for export…for the first 4–5 years. The government can always do more. First and foremost, it should provide a good education. There should be taxation after a farm's success—no tax for 5 years and extra tax for 10 years.

Finally, Harris asked farmers and farmworkers if they feared violence on farms given the situation in Zimbabwe and violence on farms in other South Africa provinces. Responses were varied and enlightening:

> No, not at all. I have very good relations with farmworkers. Definitely not.
>
> I do not fear violence of the staff, but criminal violence, given our proximity to Khayelitsha (+/−2 km).
>
> Yes, to some extent. There is domestic violence, although rare. There are very few violent interactions among laborers. I had a break-in at the foreman's house about a week ago. It is a tightly knit community.

Farm workers were asked if the farmer supported their personal goals:

> He gives us lots of information and new skills.
>
> There is a good relationship between owners and workers. We have good salaries and our basic needs are addressed.

Negative responses were few. Here is one:

> He will not help me because he thinks that if he helps me, he stands the chance of losing me.

To many farmers, the post-apartheid transition has provided them new entrepreneurial opportunities. However, they are aware that good labor relations are an important aspect of their international image. Although affirmative action is not viewed very positively by the white population, some wine farmers are making efforts to recruit and train wine-making apprentices of color for the first time and to recruit the most enterprising farmworkers into positions of responsibility.

Most farmers and farmworkers in the survey came from farming backgrounds. The farm often houses its inhabitants from birth to death. The South African Parliament passed security of land tenure legislation to enable aging farm workers to remain on farms where they have worked after their retirement. One farmer had been critical of the legislation and said that there would be a backlash from farmers who would terminate workers to avoid such an obligation. He thought that farmers had been benevolent about this matter in the past and would continue to do so if it were not legislated. Subsequent to the legislation's enactment, farmers have begun to expel workers who have no alternative means of support in a tight South African economy. However, some workers have been able to procure the government housing subsidy granted to victims of apartheid to move into housing off-farm while continuing to work there.

In order for the South African wine industry to be more competitive in the global market, it must reach a high level of efficiency. That will place more pressure on farmworkers to produce more without commensurate remuneration and farmers to engage in greater technological innovation.

## 19.10 Black Entrepreneurs: Backgrounds and Strategies to Acquire Wine Estates

The first black winemaker, Carmen Stevens, graduated from Elsenburg College in Stellenbosch in 1995. She was finally admitted to Elsenburg after her third application having been rejected twice due to South Africa's racial laws, not having military service and not having an agricultural background. Before being admitted to Elsenburg, she earned money to attend the college by doing factory work and selling shoes as an informal trader. Stevens worked her way up in the industry by starting as an assistant winemaker, winning a variety of wine-making awards, including the Decanter 2008 International Red Bordeaux Varietals Trophy for her Amani Cabernet Franc/Merlot 2006. Becoming the winemaker at Naked Wines, she was the Naked Wines' Winemaker of the Year in 2015 with a cash prize of R 6 million for wine projects. Her wines, launched in 2011, have been among Naked Wine's bestsellers in the UK and the US ("Carmen Stevens: Breaking Barriers as South Africa's First Black Winemaker," The Buyer (by Su Birch); July 13, 2016. http://www.the-buyer.net/author/su-birch/) (Richard Kershaw Wines).

Jabulani Ntshangase, another advocate of black participation in South Africa's wine industry in the post-apartheid period, gained exposure to the US wine industry during the time he was a student in New York. A Johannesburger, he settled in Cape Town in 1996 to implement a plan to support black participation in the wine industry at all levels, as wine estate owners, winemakers, vineyard scientists, marketers, etc. Through collaboration with the Wine Industry Trust, the Ministry of Agriculture and Land Affairs and South African Airways, he organized a scholarship program at Stellenbosch University in viticulture and oenology in 1998. Student recruitment was conducted around the country.

As an entrepreneur, Jabulani Ntshangase was able to form a partnership with three prominent South African vintners, Charles Back, John Platter, and Gyles Webb in 1998. Each of the four partners borrowed $250,000 to develop a winery, called the Old Spice Route Wine Company, on a choice section of a 1000 acre farm purchased by Charles Back previously. After two years, Ntshangase sold his share in the Old Spice Route Company and purchased two estates, Thabani Wines and later Highberry Wines with other partners.

Beverly Farmer is the CEO of Women in Wine which she co-founded with 19 other women in 2006. She and the other women are shareholders in a wine company which has its own label, exports wine to the US, China, Ireland, Spain, Sweden, and Denmark, and is available at Makro stores throughout South Africa. The company was established without "a traditional vineyard with rows of vines stretching into the distance, and a vast cellar" ("Our Story"—Women in Wine—http://www.womeninwine.co.za—November 25, 2019) due to more limited funds. Women in Wine buys grapes from farmers in the Stellenbosch–Paarl area and uses the cellar of a progressive wine farmer. Women in Wine has established a Women Worker's Trust that has shares in the company and an ongoing concern about improving the lives and livelihoods of farmworkers. Beverly Farmer has served as the vice-chairperson of the South Africa Wine and Brandy Transformation Unit and is a member of the Hortfin board, a parastatal company funded by the South African Land Bank, the Jobs Fund, and several industries to provide loans and expertise to the deciduous fruit industry.

Vivian Kleynhans founded Seven Sisters Vineyards in 2007. It is one of few wineries owned by a woman of color in South Africa. The winery's name reflects her family background in which there are seven sisters and one brother from the Atlantic coastal village of Paternoster. As children, the Brutus family was separated, going to live with different family members in groups of two, when their father lost his job and company housing. Kleynhans was sixteen years old. She views Seven Sisters Vineyards as a way of symbolically reuniting her family, husband, children, and grandchildren. Having her own family, she decided to integrate the cost of a 1 year marketing and management course at Stellenbosch Business School into the family budget.

Kleynhans entered the wine market in 2003 without land, and she had to access wine resources from an established white-owned winery. Among other discriminatory practices, the established winery marketed poor-quality wine under the label she used at the time. She later established the Seven Sisters label, more cautious about her collaborations. In 2007, Seven Sisters made a deal with Walmart (in the US) to market several of its wines, which are sold in 42 US states. Furthermore, in 2009,

American Airlines selected Vivian's The Sauvignon Blanc to be served in its first and business classes. In the next 5–10 years, Kleynhans plans to market her wines in other parts of Africa and in China.

Mzokhona Mvemve was a chemical engineering major who switched to a Bachelor of Science in Agriculture at Stellenbosch University after receiving an Indaba Scholarship, an award created by the wine distributor Cape Classics for black students to study aspects of the wine industry (South Africa Encourages Black Vintners, BBC News, http://news.bbc.co.uk/2/hi/business/3687615.stm, May 6, 2004). Having completed his degree in 1997, he became a winemaker's assistant at Cape Classics 4 years later. In that position, he selected and managed the vineyard and decided how and when the grapes should be harvested and processed. Since the integration of the wine industry had been slow, Mvemve noted that a mining charter developed in 2002 had set goals for black equity at future dates. He was hopeful that such a charter would be developed in the wine industry.

Mvemve has been a director at Mvemve Raats since January of 2004 and also, appointed operations director at FarmSol Africa in 2017. The maiden line MR de Composella vintage of high-scoring Bourdeaux blends was created in the same year ("MR de Compostella 2017—An Iconic Wine that Flirts with Perfection," Greg Sherwood, MW, August 23, 2019, http://gregsherwoodmw.com/2019/08/23/mr-de-compostella-2017-an-iconic-wine-that-flirts-with-perfection). FarmSol Africa assists emerging black farmers with loans and expertise in sustainable agricultural development. Bruwer Raats, of which Mvemve is affiliated, has received a five-star rating for eight wines in *Platt's 2018 South African Wine Guide,* the highest number of wines for any South African wine estate in 1 year. Bruwer Raats was also designated Platter's Winery of the Year in 2018 (Van Zyl 2018). Mvemve also received an MBA from Stellenbosh.

Ntsiki Biyela, from her home in rural KwaZulu Natal, became one of the early beneficiaries of the Wine Industry Trust 's SAA Wine Education Bursary, which Jabulani Ntshangase had participated in creating, in 1998 ("Winemaker Ntsiki Biyela: Like Climbing Table Mountain," City Press, http://city-press.news24.com/News/winemaker-ntsiki-biyela-like-climbing-table-mountain20181226). She lived with her grandmother while her mother worked as a domestic in Durban, did not engage in agricultural production, had never tasted wine, and did not speak Afrikaans. In 1999, she entered the Afrikaans-medium Stellenbosch University to study oenology and viticulture, with linguistic, cultural, and residential challenges. Despite those challenges, Biyela, received her Bachelor of Science degree and continued to work on Delheim wine estate where she had developed in interest in wine-making while learning the entire wine production process. She was the second black woman to qualify as a winemaker. With Ntshangase's assistance, she became a winemaker at Stelleyaka, where she worked for 13 years. Biyela won the Michelangelo International Wine Award in 2006 for her Cape Cross 2004. She received South Africa's 2009 Woman Winemaker of the Year award. She started her own brand Aslina, named for her grandmother, in 2012, and left Stellekaya in 2016. Like most incipient black entrepreneurs, she began by buying grapes from selected farms, whose growth she monitored.

Some members of the black elite have owned or currently own wine estates, such as former Minister of Human Settlements Tokyo Sekwale, Makaziwe Mandela, and Tukwini Mandela, daughter and grand-daughter of President Nelson Mandela, and investment executive Raymond Ndlovu. There are other ventures in which there is one black partner and one or more white partners. Clearly, the black impact on the wine industry has been minimal thus far despite the efforts, education and training, and awards achieved by black entrepreneurs.

Current South African President Cyril Ramaphosa supports amending current land law to make it easier to cease white-owned land without compensation. However, the proposal of such a strategy for the equitable distribution of land has met with considerable controversy in a potentially explosive situation in which black farmworkers and entrepreneurs and counterposed to white commercial farmers and conservative politicians.

Ben Cousins, an academic expert on South African land and agricultural reform, has been consulted by President Ramaphosa for policy formulation on these issues. In an article on land reform, Cousins (2017:135–140) argues that land reform as formulated in the 1994 South African Constitution and under policies formulated by three presidents has suffered a variety of failures. He suggests that land reform in the present day should not just be tinkered with but should be reformulated given 350 years of land-grabbing under settler colonialism. With regard to that reformulation, intersectionalities of race, gender, and class should be integrated into the process so as not to replicate previous apartheid structures. During the Mandela era, neoliberal strategies for deregulation and liberalization resulted in white commercial farmers no longer receiving loans and subsidies with the most competitive 20% of commercial farmers integrated into the agri-business value chain. In the Mbeki era, despite the ruling African National Congress (ANC) support for integrated rural development, there was some purchase of farms by the state, which were leased to rural applicants for 3–5 years. During the Zuma era, despite discussion of food security and land reform, there was little policy formulation and implementation. Due to backlog of land claims, the Restitution of Land Rights Amendment of 2014, was extended for 5 years.

Cousins (2017: 142–144) indicates, with regard to overall impacts that <10% of farmland was transferred to black South Africans from restitution and redistribution between 1994 and 2016 and that the majority of urban land restitution claims were settled through cash settlements rather than land reclamation. No reclaimed farms have been sub-divided for smallholders. With water shortages, there is also a need for irrigation for smallholders in rural and peri-urban areas. Cousins also suggests the creation of peri-urban markets for smallholder produce. He observes that there has been a reduction in civil society functioning in the post-apartheid period with many civil society participants receiving government and consultancy positions.

## 19.11 Some Connections with Sustainable Development

Section 24 of the South African Constitution reads as follows:

*24. Environment*

*Everyone has the right*:

a. *to an environment that is not harmful to their health or well-being; and*
b. to have the environment protected, for the benefit of present and future generations, through reasonable legislative and other measures that:
    i. *prevent pollution and ecological degradation;*
    ii. *promote conservation; and*
    iii. *secure ecologically sustainable development and use of natural resources while promoting justifiable economic and social development.* (https://www.gov.za/documents/constitution/chapter-2-bill-rights#24).

Thus, the South African Constitution includes "sustainable development" as a right. How is "sustainable development" to be construed? It is plausible that the constitutional meaning has some affinities with the Brundtland Report (1987), Our Common Future, issued in 1987. "Brundtland" condemns South African apartheid, referring specifically to inequity in land distribution, and to conditions in rural and urban areas. The Brundtland Report is the canonical formulation of sustainable development, which must also be understood in relation to the 2015 Sustainable Development Goals issued by the UN. The 2015 SDG 2 says: *"End hunger, achieve food security and improved nutrition and promote sustainable agriculture," and adds various "targets." Target 2.3 reads*:

> By 2030, double the agricultural productivity and incomes of small-scale food producers, in particular women, indigenous peoples, family farmers, pastoralists and fishers, including through secure and equal access to land, other productive resources and inputs, knowledge, financial services, markets and opportunities for value addition and non-farm employment.

One cannot be too literal or "originalist"; legal (or other official, e.g., UN publicized) meanings require interpretation, and can change. Also, some of the advocacy may seem abstractly utopian. Nonetheless, it is clear that the South African Constitution implies support for both "ecologically sustainable development" and the promotion of "justifiable economic and social development." Equitable access to land ownership and other uses of land (among other "natural resources") do imply racial and ethnic equity in respect to land, and "justifiable economic and social development." Applying the advocacy of the 2015 SDGs to the South African wine industry would (to be plausible) require active governmental interventions of a magnitude not at present in the works. As with so many policy questions, the devil is in the details, in measures affecting the lives of not only entrepreneurs, but farmworkers and many others. In light of the bracketing of SDG 2, Target 2.3 of food security and nutrition with sustainable development in agriculture, it might seem a case of misplaced priorities to refer to the South African wine industry as a topic of high human welfare concern. But that industry plays a significant socioeconomic role in South Africa, and there would be broader consequences of relevant sustainability factors that could be anticipated if there were more general land reforms bearing on the wine industry, or if reforms directly impacting the wine industry spread to other parts of South African societal organization.

## 19.12 Conclusion

There is a long and complex history of cultural heritage, land use issues, and sustainable societal development issues. Globalization is certainly basic in understanding that history, and in understanding post-1994 events. The wine industry is an important part of the political economy and culture of South Africa. The wine industry can be used to illustrate features of South African agriculture, and local and regional economic development. Moreover, the ideology of sustainable development is enshrined in the South African Constitution, and may well play a notable role (explicitly or implicitly) in South African governmental policy.

## References


Brundtland GH (1987) Our common future. The Brundland Report. World Commission on Environment and Development.Oxford University Press, Oxford

Cousins B (2017) Land reform in South Africa is failing: can it be saved? Transf: Crit Perspect Southern Africa 92:135–157

Elphick R (1977) Kraal and Castle: Khoikhoi and the founding of White South Africa. Yale University Press, New Haven

Seely J (1997) The wines of South Africa. Faber wine series. Faber & Faber, London

Theal GM (1897) History of South Africa under the administration of the Dutch East India Company, 1652–1795. S. Sonneschein & Company, London

Van Zyl P (2018) Platter's 2018 South African wine guide. Jean-Pierre Rossouw, Hermanus, South Africa

Van Zyl JK, Binswanger HP (eds) (1996) Agricultural reform in South Africa. Oxford University Press, Cape Town

Wilson F (1971) Farming 1866–1966. In: Wilson M, Thompson L (eds) The Oxford history of South Africa, 1870–1966, vol II. Oxford University Press, Oxford, UK

# Chapter 20
# Metamorphosis of the Polish Village as a Result of Semi-Urbanisation

**Magdalena Wilkosz-Mamcarczyk and Barbara Olczak**

**Abstract** The last decade of the twentieth and the beginning of the twenty-first centuries brought many changes into the landscape of the Polish countryside. They have resulted from changes in its functioning and various processes initiated after World War II, which became apparent or intensified after the political and economic transformation in 1989. The ongoing globalisation and intensifying suburbanisation contributed to the dynamic development of suburban zones. The penetration of rural areas by the city structure and urban lifestyle is most evident in the areas of impact of the largest cities in the country. The changes are intensified by the slowdown in the development of rural areas, migration of the rural population to cities and residents of urban areas to the countryside and the development of non-agricultural economic activity in these areas. These changes are conditioned by the semi-urbanisation of rural settlements, which is defined as the socio-economic and morphological changes of the countryside, which has included contemporary country gardening along with quasi-urbanisation but do not lead to total urbanisation. It does not result in rural areas being included in the administrative boundaries of cities or transforming into a developed city. These changes are visible both in the changing spatial development of rural areas and in the social sphere. Spatial conflicts are widespread in places where the traditional, rural layout of the village is permeated by modern urban housing and urban lifestyles contrasting with the traditional rural culture of the region. In the light of the extensive problems of the village discussed in this article, the focus is on the analysis of the consequences of spatial and landscape processes related to rural building traditions, regionalism, the way of shaping the landscape, home gardens and

M. Wilkosz-Mamcarczyk (✉)
Department of Land management and Landscape Architecture, Faculty of Environmental Engineering and Land Surveying, University of Agriculture in Krakow, ul. Balicka 253c, 30-149, Krakow, Poland
e-mail: magdalena.wilkosz-mamcarczyk@urk.edu.pl

B. Olczak
Department of Land management and Landscape Architecture, Faculty of Environmental Engineering and Land Surveying, University of Agriculture in Krakow, ul. Balicka 253c, 30-149, Krakow, Poland
e-mail: barbara.olczak@urk.edu.pl

© Springer Nature Switzerland AG 2022

J. Hernik et al. (eds.), *Cultural Heritage—Possibilities for Land-Centered Societal Development*, Environmental History 13,
https://doi.org/10.1007/978-3-030-58092-6_20

the surrounding countryside. The issue of social behaviour related to the migration of the urban population to the countryside is also presented. The selected research area is located on the border of urban–rural and rural municipalities located along state road 44 in the Kraków impact zone. The conclusions can be used to develop guidelines for the protection of the cultural landscape of rural areas undergoing intensified semi-urbanisation resulting from their location within the city's impact zone, which may apply to the emerging local zoning plan and zoning conditions and directions.

**Keywords** Semi-urbanisation processes (semi-urbanisation) · Deruralisation · Kraków suburban zone

## 20.1 Introduction

The landscape, spatial management and environment have become key study areas in the context of sustainable development in recent years. The urbanisation of rural areas is a remarkable landscape change. It is a global phenomenon particularly common in very developed regions. On the one hand, it is a positive fact related to the growth of non-agricultural functions in rural areas, which improves their competitive capabilities, but on the other hand, the demographic, social and economic growth transform how space is used and its value (Brańka 2015). Semi-urbanisation can be linked to the common issue of suburbanisation, although by definition, rural areas do not necessarily need to be wholly transformed into urban areas (Golachowski 1969).

The study focuses on analysing changes in spatial management along state road 44 from the City of Kraków to and including the Municipality of Skawina located in its suburban zone. The study area near road 44 was selected because of the significant impact of road infrastructure on functional and spatial relationships and its decisive influence on the dynamics of suburbanisation processes (Zimnicka and Czernik 2007, p. 26). The investigated area exhibits a linear growth of suburbanisation (Fig. 20.1). The goal of the paper is to identify the degree and nature of the transformation of buildings, spatial changes and landscape changes over the stretch of the road. Subsequent conclusions will help determine the progress of semi-urbanisation, and develop further guidelines that can be used to amend the local zoning plan.

## 20.2 Study Area and Methods

The study area is located in the impact zone of the City of Kraków, along a section of one of the key traffic routes, state road 44 from the north-east to the south-west towards Oświęcim. The route crosses the Kraków Metropolitan Area (KrMA) at the interface of the south-west part of Kraków and the Municipality of Skawina. It encompasses a part of the Dębniki District of Kraków up to the city limits, the Town of Skawina and its villages, Borek Szlachecki, Zelczyna, Facimiech, Wielkie Drogi and Jaśkowice (Fig. 20.2).

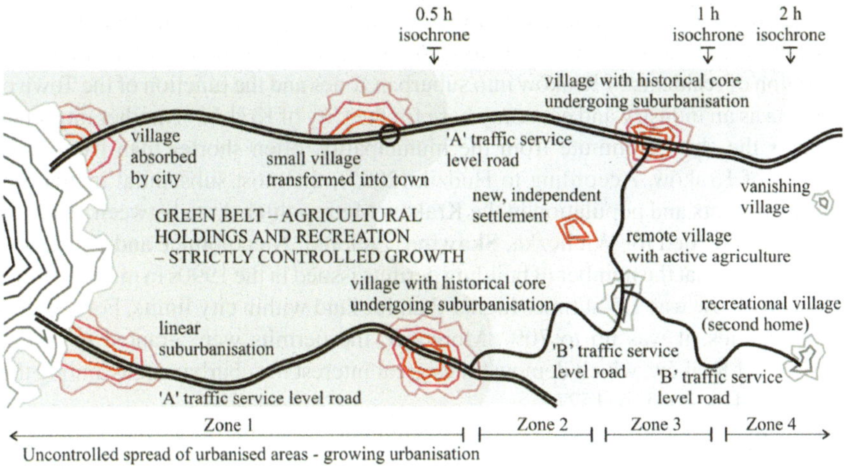


**Fig. 20.1** Suburbanisation zones, Zimnicka, Czernik 2007 in LC 1995 in accordance with Waugh D. Geography, an integrated approach

**Fig. 20.2** The study area in the context of locations and transport links, by B. Olczak

The area was selected because of the interactions between the metropolitan city of Kraków and the settlements mentioned above. Some of the keys factors are the migration of residents of Kraków into suburban zones and the function of the Town of Skawina as an industry and economy base for the City of Kraków. Another important factor is the short commute from the municipality, often shorter than from some districts of Kraków. According to Hudzik (2005), the most substantial increase of new apartments and population in the Kraków Metropolitan Area between 1989 and 2002 was recorded for Wieliczka, **Skawina**, Zielonki, Niepołomice and Zabierzów. Studies show that the number of building permits issued in the 1990s in municipalities around Kraków was 1.5–4 times higher than for land within city limits. For adjacent municipalities, it was up to 70%. Moreover, the permits were granted to former residents of Kraków, which demonstrates great interest in suburban areas among this group (Hudzik 2005, p. 152).

Note that being an industrial satellite community, Skawina grows more important thanks to its thriving plants and developed transport network (the A4 motorway), which attracts new residents. It is considered an alternative settlement location for residents of Kraków because of its small-town character contrasting with the large-city atmosphere of the second largest city in Poland. The villages in the municipality have a rural landscape, which invites new residents who seek calm locations not far from a large city.

Based on the toolbox and knowledge of an architect and landscape architect, a research project was conducted to analyse changes associated with spatial and landscape transformations related to such parameters as the size of the plot, buildings and their surroundings or the interweaving of functions as the distance to the limits of Kraków grows. The study involved an analysis of the 2006 local zoning plan for the Town and Municipality of Skawina, 2014 zoning plan for the Town of Skawina and the 2016 zoning plan for the Municipality of Skawina. Land intended to be developed according to the 2016 zoning plan was characterised (Fig. 20.3).

The state of the space was diagnosed by interpreting the map documentation for the administrative boundaries of the Municipality of Skawina regarding the presentation of spatial development after 2006, verification of the existing planning documentation and site visits. Conclusions may contribute to guidelines that could direct the growth of residential developments, thus preventing negative consequences of suburbanisation in the suburban zone of Kraków and the semi-urbanisation it entails. Such guidelines will help protect local values, regional architecture, culture and landscape as well.

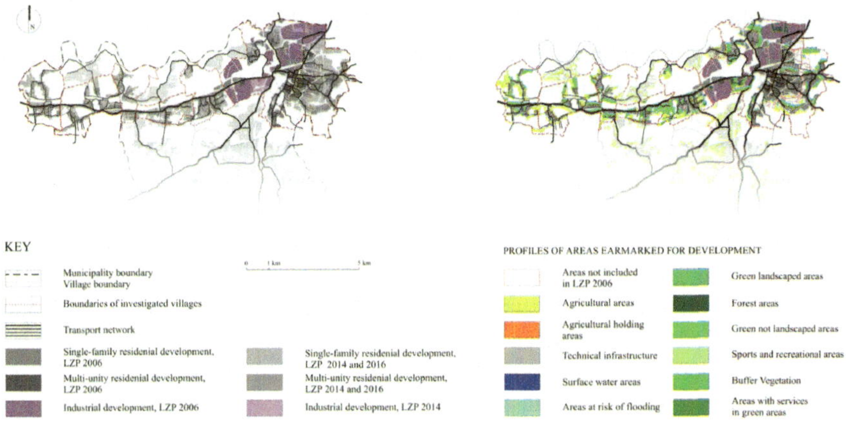


**Fig. 20.3** The growth of areas in the municipality intended to be developed according to local zoning plans from 2006 and 2016 (2014 for the Town of Skawina) by B. Olczak

## 20.3 Results

*Profile of the Municipality of Skawina focusing on the cultural context and rural traditions of the region.*

The Municipality of Skawina is a mixed urban–rural municipality. Its surface area is 9984 ha providing space for 43,137 residents, of which the Town of Skawina takes up 2050 ha and has 24,203 residents and the rural areas of the municipality (16 villages) occupy 7934 ha in total and have 18,934 residents (Municipal Programme…).

Skawina has a long history and abundant traditions. The town was founded by King Casimir III the Great in 1364 in the place of existing villages, Babice Nowe, Babice Stare and Pisary. Over centuries, it grew around the main square in the form of a rectangle with buildings adjoining it on all sides. They were tenement houses, initially wooden and then brick after the nineteenth and twentieth centuries, most of which remain till this day. The core of the town is very urbanised today. It has several-storey high buildings.

Skawina grew mostly as a town of artisans and residents engaged in trade and farming. In the nineteenth and twentieth centuries, the town went through substantial changes related to the construction of factories and establishment of manufacturing companies. The opening of a railway line in 1884 contributed to the establishment of a kerosene refinery, stoneware and chamotte products factory, a coffee products factory and a brewery.

An important factor affecting the economic growth of the town was the establishment of an aluminium smelter after World War II. This stimulated more investment projects, a power station and residential housing schemes both in the town and in neighbouring villages. The population grew as the nearby plants promised job opportunities. This trend continues today. The agricultural industry is withering; farmland is unused or left idle (Monograph of Skawina 2014). Locals have mostly abandoned agriculture in favour of jobs offered by numerous businesses in the area.

The detailed study area is the rural zone of the municipality. Note that the village, rural hamlet or rural settlement are defined in the literature as a functional and spatial unit consisting of people, single or multiple houses, farmyards, i.e. yards with residential and functional buildings, roads and other infrastructure. In a broader sense, the notion of the village encompasses land being part of the village together with functional relationships of the whole area, mainly relationships among residential locations and fields, which are an integral part of human settlement. The village is a link in the settlement network of the country. It has formal limits, separating it from other settlement units. (Szymańska 2009, p. 77). Types of spatial arrangements of Polish villages depend to a large extent on the time when a specific settlement unit was established and its geographical and environmental surroundings. The shape of the village today is a result of dynamic transformations of rural areas. The first stage was after World War II when an agrarian reform caused a significant dispersion of developments due to breaking-up of large estates. The subdivision of estates prevented well-thoughtout planning of developments or necessary functions, which led to increased spatial chaos in rural areas (Kowicki 2010, p. 37).

It was not possible to make long-term and consistent plans for rural areas because of many agrarian reforms motivated politically; there were nine major agrarian reforms in the period of 60 years from 1933 to 1999. Political transformations towards a free-market economy in 1989 were reflected in exceptionally dynamic changes in the spatial and functional structure of rural settlements. As a consequence, the social and economic structure of the then-agricultural village changed. The changes are evident in the increase of the share of single-family housing owned by persons not affiliated with agricultural production. Moreover, spontaneous growth of services and expansion of the transport network and road and technical infrastructure have been consolidated as part of the image of a modern village (Krzyk 2010 p. 26).

The plans for rural urbanisation strove mostly to present it as an overwhelming process of changes that would replace old structures. In the post-war communism period in Poland, the idea of rural development was, to a large extent, a negation of rurality understood as a specific cultural tradition, sense of identity, social, cultural, economic and landscape otherness (Wójcik 2015, p. 44). The modern rural layout is considered a vitally important component of the heritage and thus local resources of the village. It is defined by law as a 'rural planning scheme with building complexes, individual buildings and landscaped green areas arranged according to historical ownership and functional structure, including that of streets or road network' (Article 3(12) of the Act on the protection and care of historical monuments of 23 July 2003, Polish Journal of Laws of 2003 No. 162, item 1568 as amended).

The dynamics of changes in the settlement structure depend mostly on local and regional conditions related to the features of the city, its nature, location, traditions, cultural heritage, environmental conditions and demographic variations. The population of the Municipality of Skawina increased by 2.8% in the period from 2009 to 2014. Simultaneously, rural areas of the municipality exhibited a greater dynamics of population growth (Municipality Programme of Revitalisation…, p. 20). An analysis of the pace of changes in the population size is a factor to identify areas where suburbanisation transformations can be expected. The growth of the population directly increased the population density in the same period resulting in 250–350 people per km$^2$ in the Municipality of Skawina in 2016 (Górniak and Kusek 2018, pp. 291–292). Migration dynamics in the suburban zone of Kraków indicate an advanced stage of suburbanisation in 2014–2016 in the Małopolskie Voivodeship in four municipalities bordering on Kraków, including in Skawina (Górniak and Kusek 2018, p. 303). The development in the direction towards Kraków has intensified in Skawina in recent years. The attractiveness of new settlement areas in terms of location, accessibility and access to the motorway may bring spatial chaos between Kraków and Skawina and a merger of their urban structures in future (Bajwoluk 2015, s. 89). The dynamic growth of the settlement structure in agricultural areas of suburban zones of large cities leads to a disharmonious landscape. The natural environment is being excessively developed and used (Fig. 20.3), which leads to its dynamic dwindling and unnecessary devastation. Suburban villages lose their environmental and cultural qualities related to farming. New forms are being introduced that are urban in nature. They disturb the traditional cultural image as dissonant elements.

## 20.4 Analysis of Development Along State Road 44 by Villages

The villages of the Municipality of Skawina under the study, Borek Szlachecki, Zelczyna, Wielkie Drogi and Jaśkowice that make up the rural areas of the municipality were formed mostly in the fourteenth and fifteenth centuries. They feature rather compact development. Most houses are located on road 44 or near it.

The first village bordering on Skawina from the west is **Borek Szlachecki** with 1400 residents. It dates back to the fourteenth century when it was first mentioned (as the Village of Werner) in documents held by the Benedictines from Tyniec. It had been called Borek since about 1400, which is indicative of its forest affiliations as 'bór' in Polish means coniferous forest. Since, the nineteenth century, it has been called Borek Szlachecki, most likely to commemorate its owners who belonged to the 'szlachta', which is Polish for the gentry.

The locality is situated on the state road and stretches south as well. Streets perpendicular to it have single-family houses. The village has a church, a chapel and around 30 businesses (http://www.borekszlachecki.pl/). The four analysed single-family houses, three residential and one commercial, along the road illustrate the

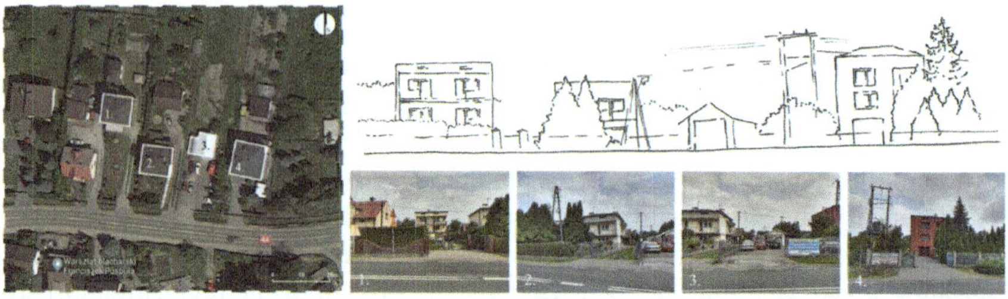

**Fig. 20.4** An example of the development of Borek Szlachecki. Disharmony caused by the form of the building, its height, colour of the façade and garden: 1–2. a 'Polish cube house' 3. a fire station 4. a 'Polish cube house', by B. Olczak, M. Wilkosz-Mamcarczyk

general developmental problem of the village. The frontage in Fig. 20.4 exhibits significant variability in terms of building height and form. Vegetation is shaped variously according to the preferences of the owners.

Zelczyna, a village of 1000 residents stretches over 3 km of state road 44. Just as Borek Szlachecki, it dates back to the fourteenth century. Its owners frequently changed until the interwar period when estates were broken up. The development is rather compact both along the main road and parallel roads. The central point of the village is a school, a bus stop area and a shop (Zelczyna Redevelopment Plan for 2010–2017, pp. 5–6). The area selected for detailed investigation shows a gap between traditional rural development north to state road 44 and buildings from the 1970s and later on the south side. Larger distances of fronts of buildings to the road make possible what is there, including gardens with modern landscaping trends where large lawns are surrounded by thuja hedges. The spatial chaos is exacerbated by buildings with flat roofs, gable roofs, 45°-pitched roofs and bright colours of façades (Fig. 20.5, 20.6).

**Fig. 20.5** Example of the development in Zelczyna. Disharmony caused by the form of the building, its height, colour of the façade and garden: 1. 1970s semi-detached housing with disharmonious façade colour, 2. no garden; just lawn and hedge, 3. a neutral colour 'Polish cube house', 4. a semi-detached 'Polish cube house' with eye-catching façade colour, by B. Olczak, M. Wilkosz-Mamcarczyk

**Fig. 20.6** Example of the development in **Wielkie Drogi**. 1. a single-family house, brick, early twentieth century 2. a single-family house, brick, early twentieth century with expansion, a porch not consistent with the form of the building 3. preserved traditional wooden building from the early nineteenth century 4. a 'Polish cube house'; by B. Olczak, M. Wilkosz-Mamcarczyk

Another village bordering on Zelczyna over a short section of state road 44 is **Facimiech**. It was first mentioned in the thirteenth century. It was reported to host twelve farmsteads then. In the eighteenth century, Facimiech became the property of the Haller family. Its descendants lived in a local manor until the end of World War II. After the war, the estate was broken up.

Today, the developed part of the village is located north to the state road and gravitates towards the unused manor, a community centre and a church that make up the centre of the village. The village development on state road 44 was not analysed as the village is crossed at a place that is of little importance for the structure of the settlement.

The next settlement on the state road is **Wielkie Drogi**. The name 'Wielga Droga' was first recorded in the late sixteenth century, and assumed its modern form in the eighteenth century. The settlement established here in the twelfth century was the cradle for Wielkie Drogi and several other villages nearby. Privileges granted by Bolesław V the Chaste in 1274 stimulated an intensive growth of settlements in the area. Individual hamlets with their own names got separated already in the first half of the fourteenth century. It was a period in which the settlement on 'great roads' ('wielkie drogi' in Polish) grew slowly. Before 1795, Wielkie Drogi was purchased by a Kraków merchant Marcin Alojzy Haller de Hallenburg. In 1879, Wielkie Drogi became the property of the Brandys family from Brody. It was then that the construction of a grand neoclassical palace commenced. The end of World War II was also the end of the glory of the palace. In 1941, it was taken over by German invaders who set it on fire together with other estate buildings in January 1945 during evacuation. After 1945, the estate was expropriated and broken up under the 1944 Polish Committee of National Liberation's decree on agrarian reform (Wielkie Drogi Redevelopment Plan 2016, pp. 14–15). The detailed study area covers the intersection of state road 44 with Dańkowskiego Street heading south to Paszkówka and further to Kalwaria Zebrzydowska. There are commercial units south to the state road near the intersection such as a pharmacy, grocery stores, a motel and a confectionery. There is residential housing to the north. It features traditional wooden buildings from the early nineteenth century surrounded by single-family houses from the early twentieth

**Fig. 20.7** Example of the development in **Jaśkowice**. 1. a single-family house, brick, early twentieth century with visible lack of a garden 2. a single-family house, brick 3. a 'Podhale-style' house of disproportional size compared to neighbouring buildings 4. a new modern building, by B. Olczak, M. Wilkosz-Mamcarczyk

century and the 1970s. Variable heights of neighbouring buildings and different roof pitches with ridge lines parallel to the road were considered disharmonious. The colours of façades and roofs do not contribute to the sense of spatial chaos.

The last village on state road 44 in the Municipality of Skawina is **Jaśkowice**. The village was part of the parish in Pobiedr, first mentioned in 1325. In 1448, the place was called Jaskowycze according to Liber beneficiorum dioecesis Cracoviensis by Jan Długosz. In 1857, the village had 86 houses with 535 residents, parishioners of the Pobiedr parish. In 1927, the population of Jaśkowice grew to 722, in 2000, to 1036. A neo-Gothic resort chapel was built in Jaśkowice in 983. In 1991, it was replaced with a new, large Blessed Virgin Mary, Mother of Church chapel (Kucharczyk 2004, pp. 3, 10). The detailed study area was selected because of the disharmony in the building-road distances and significant variability of forms (Fig. 20.7). In the case of the Podhale-style house (Fig. 7.3), the increased distance to the road was considered a positive trait as it reduced the visual exposure of the building with its disharmonious form and colour. Modern evidence of a lawn garden consistent with modern trends were noticed. The effect is enhanced by a significant distance of neighbouring buildings to the road (Figs. 7.2 and 7.3).

A characteristic feature of the investigated villages is the absence of a clearly defined centre. The function is often served by areas of bus stops, schools, churches or fire stations. An analysis of historical cartographic documentation for the investigated area determined the historical layout of the village to be a linear one, expanded over time so that its initial arrangement has become lost. The construction of the Łączany-Skawina Canal consistent with the present process of abandonment of agricultural activities and subdivision of plots into smaller, building plots contributes to the state of affairs.

The area is where the suburban zone and urban zone of the Town of Skawina and that of Kraków merge. Some features of this trend are spatially continuous built-up areas and blurred boundaries between settlement structures.

## 20.5 The Residential Building and Its Vicinity

The traditional, regional, rural architecture, which developed over thousands of years in Europe, including Poland, threatens to be annihilated entirely as a result of a change of lifestyle in the twentieth century. Urban–rural migration also affected imagination about how residential buildings should look, which is related to life style and quality of life (Uruszczak 2013). It is evident in the forms of the buildings investigated in villages near Skawina. The general principle is, the closer it is to the town, the less apparent the specific 'rural' developments. The past form of development was wooden, then there were put up brick, low, single-storey buildings with gable roof and the ridgeline parallel to the road (Fig. 20.8).

This type of development with gable roof about 20 m from the road centreline is the most popular also today (only old houses are located about 4 m from the road centreline). The landscape of the investigated villages features a lot of 1970 and 1980s developments in the form of high, usually three-storey buildings with flat roofs known as 'Polish cube houses' (Fig. 20.9). The attitude towards the space around the house and the distance to the road have changed as well. The developments are not spatially continuous and residential housing includes commercial functions on large-area plots (Fig. 20.10). The traditional forms of buildings are being replaced by new houses designed as kinds of quasi-manors or homes from catalogues of ready-made house plans that fail to manifest appreciation of the context of building location. The colour of the façade became an essential part of the rural landscape. The paints are usually pastel hues, but some buildings are intense red or orange, which increases the disharmony (Fig. 20.11).

The analysed layouts of gardens and ways of arranging the space around residential buildings differ significantly from traditional rural approaches. Most of them lack the front garden, usually close to the building wall, full of hollyhocks, sunflowers, zinnias, rudbeckias and asters; the entrance garden, a mixed private–public space with night-scented stocks, hollyhocks, larkspurs, pot marigolds and zinnias. It used

**Fig. 20.8** Traditional rural wooden nineteenth/early twentieth -century development: 1–2. Zelczyna 3–4. Wielkie Drogi, brick, twentieth century: 5. Zelczyna, 6–7. Borek Szlachecki, 8. Jaśkowice; by B. Olczak, M. Wilkosz-Mamcarczyk

**Fig. 20.9** The transformation of the rural house. 1. Traditional wooden architecture—Zelczyna 2. a combined wooden and brick building—Wielkie Drogi 3. a brick house from 1908—Zelczyna 4. a brick house—Jaśkowice 5. a 'Polish cube house'—Borek Szlachecki 6. ready-made architecture—Zelczyna 7. a house resembling a castle or manor—Facimiech 8. modern architecture, shell finish—Jaśkowice; photo by B. Olczak

**Fig. 20.10** The disharmony in the development layout as seen in Zelczyna. 1. No continuity of single-family housing on the southern side of state road 44, mixed functions on the northern side 2. left side: compact development composing state road 44 frontage; right side: variable setback resulting in spatial chaos enhanced by diverse forms and colours of objects; by B. Olczak based on google.pl/maps

to be a place for fostering neighbourhood relationships and the pride of the owners. A low, wooden openwork fence was another characteristic feature supporting the relationships (Gawryszewska 2013).

The next part of the garden was the agricultural holding area devoted to animal husbandry (horses, poultry, pigs, cows). It had buildings (usually made of wood) for animals, feed storage, farming equipment and so forth. The next part was a kitchen garden, orchard and then farmland (Fig. 20.12).

Today, the appearance of the front garden has changed, partially because of a significant distance between buildings and road centreline (dust and noise). The front garden is now much bigger. It usually contains evergreens and, seldom, annual flowers. High or solid fences compromise the semi-public function of the place. The privacy of the owners is further improved by thujas planted along the fence cut to about 3 m to compose a green barrier from the rest of society. The agricultural

**Fig. 20.11** Disharmonious objects in the landscape. 1. Skawina, high multi-family buildings—blocks of flats among single-family housing, 2–3. Borek Szlachecki—disharmony due to variable heights of buildings 4. Zelczyna—disharmony due to variable forms and heights of buildings 5. Facimiech—the form of the building 6–7. Wielkie Drogi—disharmony due to the form and street furniture 8. Jaśkowice—disharmony due to the size of buildings, lack of vegetation around buildings, is apparent; photo by B. Olczak

**Fig. 20.12** A diagram of a traditional rural garden in the investigated area. 1. a. a residential building, b. a front garden, c. farm buildings, d. a kitchen garden, e. an orchard, f. farmland 2. Borek Szlachecki—a neglected garden, remains of old plantings, numerous volunteer seedlings 3. Zelczyna—apparent negligence, remains of traditional vegetation 4–5. Wielkie Drogi—an attempt to preserve the tradition of the rural garden and farm buildings, old, wooden building materials preserved 6–7. Jaśkowice—a well-maintained rural garden; by B. Olczak, M. Wilkosz-Mamcarczyk

holding area is virtually gone today. The analysed area along the road had merely a few houses with farm buildings, and the only animals kept there were poultry. The farmyard was replaced by a recreational area. Instead of orchards and farmland, the plots have lawns or idle land. Plots are often subdivided and sold to new owners. New access roads to the new buildings are staked out perpendicularly to the state road along the plots (Figs. 20.13 and 20.14).

**Fig. 20.13** A diagram of a modern, increasingly popular type of rural garden. 1. a. a residential building, b. a hedge, c. a lawn 2–4. Zelczyna—a garden with evergreens, so-called low-maintenance garden 5. Facimiech—a typical garden today with boundaries marked with a hedge made of thujas to ensure privacy 6–7. Wielkie Drogi—a garden with park furniture alien to the Polish rural landscape (ancient Greek sculptures, fountains and others) and house surroundings without an extensive garden, just a trimmed lawn; by B. Olczak, M. Wilkosz-Mamcarczyk

**Fig. 20.14** The transformation of the rural garden. 1. A neglected rural garden—Borek Szlachecki 2. a neglected rural garden—Zelczyna 3. a well-maintained rural garden—Zelczyna 4. a rural garden—Jaśkowice 5. a garden drawing on the rural garden—Zelczyna 6. domination of alien species (thujas)—Zelczyna 7. a thuja 'wall'—Facimiech 8. a lawn—Jaśkowice; by B. Olczak, M. Wilkosz-Mamcarczyk

## 20.6 Summary

An analysis of the selected study area of the Municipality of Skawina demonstrated that the landscape of the Polish village is undergoing a metamorphosis. It is embodied in the transformation of former farmland into residential, commercial or industrial developments. Semi-urbanisation entails uncontrolled development of areas that often lack the necessary infrastructure, which results in the built-up areas 'spilling around' and leads to layouts of villages with undefined or chaotic landscapes.

Furthermore, semi-urbanisation generates changes in the rural landscape and eradicates its traditional qualities. Rural residential and farming developments were transformed over time into individual single-family houses. Old, historic cottages are overwhelmed in the study area by large, several-storey high or manor-like residences. Ornamental and kitchen gardens that often used to be the pride of their owners were replaced by lawns with precisely trimmed hedges along their boundaries. Farm animals commonly kept in virtually all rural houses in the past became a nuisance because of the odours or noise they generate.

## 20.7 Conclusions

Suburban villages along state road 44 within the boundary of the Municipality of Skawina exhibit a significant concentration of semi-urbanisation processes, by which there is deterioration of the quality of the space, traditional rural landscapes and the quality of life of many local residents:

- earmarking of excessive areas for residential developments and relevant services;
- dispersion of buildings and significant options regarding the height and roof styles of buildings, finishing materials and colours, which exacerbate spatial chaos;
- abandonment of the traditional forms of vegetation around the home, which had been coherent with buildings and the surrounding landscape in favour of large lawns and hedges, often of alien species of thujas that create green walls in the landscape, sometimes higher than the building they surround.

The intensity of the processes increases towards the city, but it is also a result of poor rural land management over the last 70 years. The lack of consistent policy for managing and functioning of rural areas in Poland and the fact that the study area is located in the suburban zone of Kraków lead to the intensification of rural urbanisation processes and fuel spatial chaos. The intensity of the following is increasing:

- lack of concern about rural developments and gardens that are responsible for the quality of the agrarian cultural landscape;
- dynamic dwindling of natural areas;
- declining quality of space and life of residents; and
- social conflicts between original residents and migrant populations from cities.

To preserve the cultural landscape of villages in Lesser Poland, it is necessary to adapt the traditional models of rural building plot development to modern needs while respecting such components of the cultural landscape as:

1. the rhythm and nature of development responsible for the identity of the village;
2. the minimum surface area of a plot intended for single-family housing and adherence to the setback line;
3. conceiving of the shape and form of building while drawing on and taking account of historical buildings;

4. care about the height of building and roof structure (precise regulations in the zoning plan needed); and
5. construction materials and façade and park furniture colours drawing on the traditions of the region, in particular regarding fencing.

To implement the guidelines mentioned above, it is necessary to introduce precise local zoning plan regulations and educate the residents, preferably already while they are in primary school. Familiarity with the history of the place, the culture and traditions of the region may contribute to the transfer of positive regional models into the landscape. An equally effective solution can be to prepare model examples of garden arrangements so that solutions used in practice will be consistent with the landscape in terms of form and species mix (restriction of the share of alien species).

Technological advancement and changes in the former structure of the village are unstoppable (Uruszczak 2013), but sometimes minor changes can partially restore the past environment.

## References

Act on the protection and care of historical monuments of 23 July 2003, Polish Journal of Laws No. 162, item 1568 as amended

Amended Zoning Conditions and Directions for the City of Kraków adopted with the resolution No. XII/87/03 of the Council of the City of Kraków of 16 April 2003, source bip.krakow.pl. Accessed 8 Feb 2019

Bajwoluk T (2015) Zmiany w strukturze tkanki miejskiej małych miast—studia wybranych przypadków w Obszarze Metropolitalnym Krakowa [Changes in the urban fabric structure of towns. Selected case studies in the Kraków Metropolitan Area]. Acta Universitatis Lodziensis, Folia Geographica Socio-Oeconomica 19:79–93

Facimiech Redevelopment Plan. source bip.malopolska.pl/umigskawina/Download/get/id,549945.html. Accessed 5 Aug 2019

Gawryszewska B (2013) Ogród jako miejsce w krajobrazie zamieszkiwanym [The Garden as a Place in the Residential Landscape]. Wydawnictwo WIEŚ JUTRA, Warsaw

Golachowski S (1969) Urbanizacja wsi w województwie opolskim [Rural urbanisation in the Opole Voivodeship]. Studia nad miastami i wsiami Śląskimi, PWN, Opole-Wrocław, pp 180–193

Górniak A, Kusek R (ed) (2018) Miasta województwa małopolskiego—zmiany, wyzwania i perspektywy rozwoju [Towns and cities of the Małopolskie Voivodeship. Changes, challenges, and perspectives for growth], Małopolska Regional Development Observatory, Department of Regional Policy, Kraków

http://www.borekszlachecki.pl/ Accessed 22 July 2019

Hudzik I (2005) Strefa podmiejska—zarys problemu [The suburban zone. Problem Outlined]. Zarządzanie Publiczne 1/2005 Research Papers of the Public Affairs Institute, Jagiellonian University, pp 147–158

Kowicki M (2010) Patologie/wyzwania architektoniczno-planistyczne we wsi małopolskiej. Studium na tle tendencji krajowych i europejskich [Architecture and planning pathologies and challenges in villages in Lesser Poland. A study with Polish and European trends in the background]. Cracow University of Technology Publishing House

Krzyk P (2010) Strukturalne i krajobrazowe aspekty przemian wiejskich jednostek osadniczych a planowanie przestrzenne [Structural and landscape aspects of transformations of rural settlements in light of spatial planning]. Problemy Rozwoju Miast. 2010(2):25–34

Kucharczyk R (2004) Niektóre karty z historii Parafii Pobiedrskiej [Excerpts from the history of the parish in Pobiedr]. source: archiwum.brzeznica.pl. Accessed 6 Aug 2019

Municipal Revitalisation Programme for Skawina 2016–2020, source: rpo.malopolska.pl. Accessed 22 Mar 2019

Resolution No. IXN/309/06 of the Town Council in Skawina of 15 May 2006 on the adoption of the local zoning plan for the Municipality of Skawina within its administrative boundary. source gminaskawina.pl. Accessed 24 May 2019

Resolution No. IXN/310/06 of the Town Council in Skawina of 15 May 2006 on the adoption of the local zoning plan for the Town of Skawina within its administrative boundary. source gminas kawina.pl. Accessed 24 May 2019

Resolution No. XIIN/456/13 of the Town Council in Skawina of 12 December 2013 on the adoption of the local zoning plan for the Town of Skawina within its administrative boundary. source gmi naskawina.pl. Accessed 22 Mar 2019

Resolution No. XVIIN/217/16 of the Town Council in Skawina of 23 March 2016 on the adoption of the local zoning plan for the Municipality of Skawina within its administrative boundary. source gminaskawina.pl. Accessed 22 Mar 2019

Turkawska Magdalena (ed) (2014) Monografia Skawiny—650 lat tradycji [Monograph of Skawina. 650 Years of Traditions] Skawina Administration, Wydawnictwo Promo

Szymańska D (2009) Geografia osadnictwa [The Geography of Settlements], Wydawnictwo Naukowe PWN, Warsaw

Uruszczak M (2013) Problematyka zachowania tradycyjnej zabudowy wiejskiej, świadectwa tożsamości regionalnej na przykładzie Małopolski. Wybrane zagadnienia i problemy [Problems of preserving traditional rural architecture. Witnesses to the regional identity in Lesser Poland. Selected issues] . Infrastruktura i Ekologia Terenów Wiejskich, No. 3/I/2013, Polish Academy of Science, Kraków Branch: 125–133

Wielkie Drogi Redevelopment Plan 2016–2020 by Local Democracy Foundation, Lesser Poland Local Government and Administration Institute in Kraków (2016). Source: https://bip.malopo lska.pl/e,pobierz,get.html?id=1430421. Accessed 6 Aug 2019

Wójcik M (2015) (2015) Koncepcje badań funkcjonalnych wsi i ich związek z planowaniem rozwoju [Concepts of functional rural studies and the relationship with development planning]. Studia Obszarów Wiejskich 37:43–45

Zelczyna Redevelopment Plan 2010–2017 www.gminaskawina.pl/components/download/send. php?pos_id=3407. Accessed 24 May 2019

Zimnicka A, Czernik L (2007) Kształtowanie przestrzeni wsi podmiejskiej na przykładzie obszaru oddziaływania miasta Szczecin [Shaping the Space of a Suburban Village. The impact of the City of Szczecin], Wydawnictwo Hogben, Szczecin

# Chapter 21
# Assessment of Land-Use and Land-Cover Changes in a Rural Cultural Landscape: A Case Study of a Polish Municipality

**Tomasz Noszczyk, Katarzyna Cegielska, and Anita Kukulska-Kozieł**

**Abstract** The determination and assessment of changes in the rural cultural landscape may provide insight into the complexity of the landscape and environmental transformation dynamics. Land-use and cover changes (LUCC) are caused by both natural and anthropogenic factors. Polish rural areas feature numerous endangered valuable cultural landscapes that require special protection. The comprehension of changes in the cultural landscape over time is vital for the correct programming of actions aimed at preserving such landscapes. The present study analyses, vector land cover maps and official cadastral land, use data for a rural municipality of Wiśniowa (in central Lesser Poland) in order to investigate potential changes in its cultural landscape. The study area features high-value environmental and landscape qualities typical of a tourist and leisure destination. It furthermore has numerous and diverse landscape mosaics that constitute its unique rural cultural landscape. LUCC from a period of 18 years (2000–2018) were assessed using the cadastral data and maps from Corine Land Cover. The study can be a point of reference for the development and implementation of sustainable spatial policies or programmes for managing the rural cultural landscape.

**Keywords** Landscape transformation · Rural areas · Corine Land cover · Cadastral data · Spatial analyses · GIS

T. Noszczyk · K. Cegielska · A. Kukulska-Kozieł (✉)
Department of Land management and Landscape Architecture, Faculty of Environmental Engineering and Land Surveying, University of Agriculture in Krakow, 253c Balicka Street, 30-149 Krakow, Poland
e-mail: anita.kukulska-koziel@urk.edu.pl

T. Noszczyk
e-mail: tomasz.noszczyk@urk.edu.pl

K. Cegielska
e-mail: katarzyna.cegielska@urk.edu.pl

© Springer Nature Switzerland AG 2022

J. Hernik et al. (eds.), *Cultural Heritage—Possibilities for Land-Centered Societal Development*, Environmental History 13,
https://doi.org/10.1007/978-3-030-58092-6_21

## 21.1 Introduction

The year 2018, the European Year of Cultural Heritage, saw numerous initiatives and events all over Europe aimed at bringing cultural heritage closer to the public and encouraging people to improve their involvement in the protection of the heritage. This effort was necessary as according to a survey requested by the European Commission—Special Eurobarometer 466 'Cultural Heritage'. 48% of respondents in EU28 are not involved in the field of cultural heritage (Eurobarometer 2017). At the same time, over 80% of them believed cultural heritage was important for them, and 68% of the respondents declared they would like to know more about Europe's cultural heritage. Besides, cultural heritage enriches the life of residents, helps build a stronger, more coherent society, and is important for the economy through jobs and the growth of tourism (Eurobarometer 2017).

The cultural landscape certainly is part of Europe's cultural heritage. Landscapes that have been shaped by the human-landscape interaction for centuries and exhibit features particularly valued by the society are referred to in the literature as cultural landscapes (Agnoletti 2014; Plieninger et al. 2014). Their value lies in their landscape qualities, cultural heritage, and the inspiration they offer (Schulp et al. 2019). Ziter et al. (2017) believe cultural landscapes may include heritage areas shaped by religious, racial, or cultural groups ('ethnographic landscapes') deliberately designed landscapes such as parks or campuses ('designed landscapes'), and landscapes relevant to historic events ('historic sites') such as the National Park Service in the USA.

Cultural landscapes play a vital role in Europe, mainly because of environmental, political, and socio-economic diversity (Gingrich et al. 2015; Schulp et al. 2019). Their predominant feature is the low intensity of land use. Polish rural cultural landscapes also feature significant diversity (Hernik et al. 2015; Prus et al. 2020). They should be protected because they are part of the regional identity, and their distinctive trends are important for cultural heritage (Hernik 2011).

Humanity has shaped the landscape for centuries through continuous land-use and land-cover change (Aneseyee et al. 2020). The present urbanisation (Cegielska et al. 2017) or mining and consumption of non-renewable energy sources lead to the loss of landscape structures and consequently to a threat to cultural landscapes (Schulp et al. 2019). Almeida et al. (2016) even believe the human demand for food, environment protection, and biomass is at variance with the benefits Europeans derive from cultural landscapes. The conflict stems from the fact that people consume the landscape to satisfy their needs through such activities as mining of non-renewable resources, expansion into protected areas, or urbanisation of culturally valuable areas.

According to a UNESCO report, soil erosion is believed to be one of the primary threats related to climate change that affect cultural heritage and landscapes (Cuca and Agapiou 2018). In their work, Cuca and Agapiou (2018) analysed the possible impact of the land-use change (and soil erosion) on a UNESCO archaeological site on Cyprus. Their results suggest that land-use diversification over time inevitably contributes to the soil loss in any area.

It is important to understand how cultural landscapes can change depending on the land-use and land-cover change so that decisions can be made as to where and how to act to preserve cultural landscapes. According to Schulp et al. (2019), linking past and future land-use and land-cover changes with patterns of cultural landscapes can help assess how cultural landscapes can evolve in the future. It can also help identify potential threats to cultural landscapes caused by land-use change. Therefore, analyses based on GIS can provide useful results regarding land-cover and land-use change, becoming a valuable source of information for decision-makers and researchers who study these phenomena.

The present study analyses potential changes in the cultural landscape of the rural municipality of Wiśniowa (in central Lesser Poland). The work involved analysis of vector land-cover maps and official cadastral data on land use from 2000 to 2018.

## 21.2 Materials and Methods

### *21.2.1 Study Area*

The study involved a rural municipality Wiśniowa situated in the central part of the Małopolskie Province, Myślenicki District (Fig. 21.1). The municipality of Wiśniowa takes up 67 km$^2$ and has 7336 residents, 3674 of which are women (as on 2017)

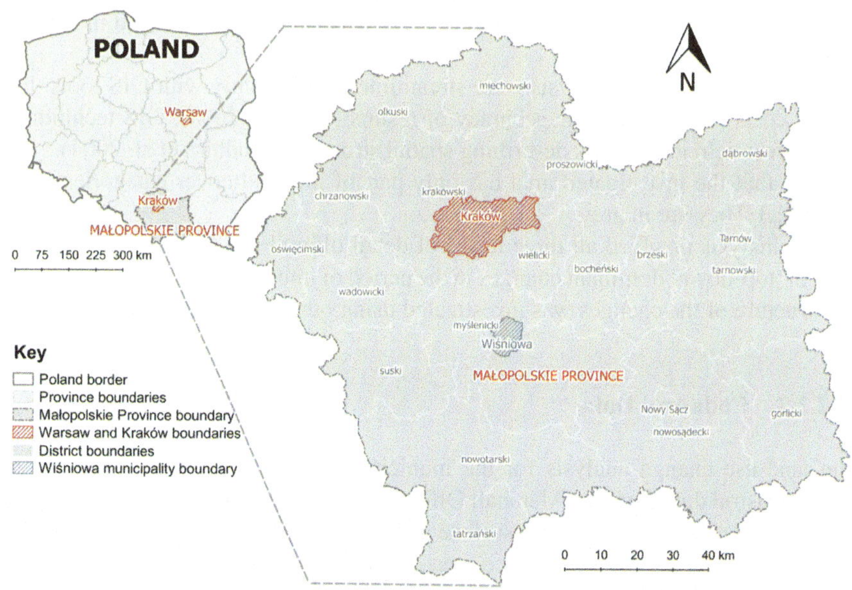


**Fig. 21.1** The study area outlined within Małopolskie Province

(Vademecum 2018). The study area features high-value environmental and landscape qualities typical of a tourist and leisure destination. A large part of the municipality consists of mountainous slopes with multiple-story land cover (Hernik 2008). It furthermore has numerous and diverse landscape mosaics that constitute its unique rural cultural landscape. A remarkable feature of the cultural landscape according to Hernik (2011) are wayside shrines and crosses all over the municipality. The study area has numerous listed monuments. The abundance of the cultural heritage in the municipality of Wiśniowa was discussed by such authors as Hernik (2008, 2011), Piotrowska and Taszakowski (2008), and Dziadowiec-Greganić (2018).

## *21.2.2 Data*

### 21.2.2.1 GIS Data

The input material for GIS analyses was Corine Land Cover (CLC) vector data layers for land cover from 2000, 2006, 2012, and 2018. The spatial diversity of administrative units in the Wiśniowa municipality was juxtaposed using a vector layer of administrative boundaries from the Central Geodetic and Cartographic Documentation Office in Poland.

CLC is the largest database available under the Copernicus project. It provides information on the physical properties of the surface of the Earth (land cover) (Büttner et al. 2004). The project involves 44 classes of land cover grouped into three levels of detail (Table 21.1). The classes represent all land cover categories found in Europe (Burkhard et al. 2009).

Data processing and analysis were streamlined significantly with GIS tools that reduce costs and improve the accuracy of results (Łowicki 2008). GIS techniques are of great help in analyses describing spatial structures (Sudhira et al. 2004).

Note that the investigated area has only part of the land cover categories from Table 21.1 They are in grey.

The analysis involved all three levels of detail offered in CLC. The goal was to present ‘top-down’ dominant changes in the period of interest by using Level 1. Next, the structure of the changes was investigated using other levels.

### 21.2.2.2 Cadastral Data

The land-use change analysis for the municipality of Wiśniowa involved 2002–2018 cadastral data from the Marshall Office of the Małopolskie Province. The data provided four variables: $X_1$—the surface area of agricultural land; $X_2$—the surface area of forest land; $X_3$—the surface area of built-up and urbanised land; $X_4$—the surface area of wasteland. Statistical characteristics of all the variables are shown in Table 21.2.

For variables $X_2$ and $X_3$, the mean is less than the median, which indicates that the empirical distribution is asymmetric. Additionally, the variability of all variables

**Table 21.1** The three-level classification of land cover categories according to Corine Land Cover

| First level of detail | Second level of detail | Third level of detail |
|---|---|---|
| Artificial surfaces | Urban fabric | Continuous urban fabric |
| | | Discontinuous urban fabric |
| | Industrial, commercial and transport units | Industrial or commercial units |
| | | Road and rail networks and associated land |
| | | Port areas |
| | | Airports |
| | Mine, dump and construction sites | Mineral extraction sites |
| | | Dumpsites |
| | | Construction sites |
| | Artificial, non-agricultural vegetated areas | Green urban areas |
| | | Sport and leisure facilities |
| Agricultural areas | Agricultural land | Non-irrigated arable land |
| | | Permanently irrigated land |
| | | Rice fields |
| | Permanent crops | Vineyards |
| | | Fruit trees and berry plantations |
| | | Olive groves |
| | Pastures | Pastures |
| | Heterogeneous agricultural areas | Annual crops associated with permanent crops |
| | | Complex cultivation patterns |
| | | Land principally occupied by agriculture |
| | | Agro-forestry areas |
| Forest and semi-natural areas | Forests | Broad-leaved forest |
| | | Coniferous forest |
| | | Mixed forest |
| | Shrub and/or herbaceous vegetation associations | Natural grassland |
| | | Moors and heathland |
| | | Sclerophyllous vegetation |
| | | Transitional woodland shrub |
| | Open spaces with little or no vegetation | Beaches, dunes and sand plains |
| | | Bare rock |
| | | Sparsely vegetated areas |
| | | Burnt areas |
| | | Glaciers and perpetual snow |
| Wetlands | Inland wetlands | Inland marshes |
| | | Peat bogs |
| | Coastal wetlands | Salt marshes |
| | | Salines |
| | | Intertidal flats |
| Water bodies | Inland waters | Water courses |
| | | Water bodies |
| | Marine waters | Coastal lagoons |
| | | Estuaries |
| | | Sea and ocean |

**Table 21.2** Statistical characteristics of variables

| Characteristics | Variables | | | |
|---|---|---|---|---|
| | $X_1$ | $X_2$ | $X_3$ | $X_4$ |
| Mean [thous. ha] | 3,820.94 | 2,602.65 | 232.41 | 4.00 |
| Median [thous. ha] | 3,814.00 | 2,603.00 | 237.00 | 4.00 |
| Standard deviation [thous. ha] | 17.69 | 1.77 | 16.65 | 0 |
| Coefficient of variation (CV) [%] | 0.46 | 0.07 | 7.16 | 0 |

is low, which is confirmed by the low value of the standard deviation as compared to the mean and the very low value of CV (Table 21.2).

### *21.2.3 Methods*

The analysis, from input material preparation through results compilation, was carried out using GNU-GPL QGIS software. The input layers were pre-processed by adjusting their geographical range to the study area. This was achieved with geoprocessing tools—Clip. Next, the CLC layers were reclassified from numeric codes into text strings separately for each cover type. To this end, original SQL commands were developed based on the logical connective *IF*. This procedure was carried out three times, once for each level of land cover data detail. Objects in the layers were grouped depending on the level of detail assumed for the analysis. The grouping was based on the levels of detail proposed in the CLC project. Each level of detail was assigned a separate display style (QGIS layer style file) to aid the analysis and clarity of resulting maps. Each of them was based on a unique display value related to the attribute table column. The column contained individual names of land cover categories consistent with and unique for the detail level. To facilitate separate calculations for each cadastral district, it was necessary to intersect land cover layers with the cadastral district layer. This was achieved with geoprocessing tools—Intersect. Resulting vector layers were used to calculate the surface of each land cover category with the field calculator. Group Stats Plugin was then used to compute spatial statistics shown in the results.

## 21.3 Results and Discussion

### *21.3.1 Land-Use Changes from 2002 to 2018*

The land use structure of the municipality of Wiśniowa is dominated by agricultural and forest land. In 2018, these took up 56.8% and 38.9% of the surface area of the municipality, respectively. The share of other land-use types was significantly smaller; built-up and urbanised land constituted 3.8% of the municipality, and wasteland was only 0.06% (Table 21.3). The data above confirm that Wiśniowa is a typical agricultural municipality dominated by agricultural land.

According to the analysis of the cadastral data, the most significant changes in the period of interest were observed for agricultural land (42 ha or 0.63% reduction) and built-up and urbanised land, which grew by 39 ha, which is 0.58%. Other types of land use did not change significantly. Forests and forest land grew 0.04%, while wasteland remained constant at 4 ha (Table 21.3). The conclusion is that the land use structure of the municipality of Wiśniowa has not changed significantly for 16 years. Virtually the whole change in the surface area of agricultural land was due to the increase in the surface of built-up and urbanised land and forest land. According to Poławski (2009), this is because of growing urbanisation and afforestation of poor-quality land. Work by Buśko and Szafrańska (2018) are consistent with the trend for diminishing agricultural land in the municipality of Wiśniowa. The Krakowski District has the largest losses of best-quality land out of all districts in the Małopolskie Province. The close distance to the capital of the province, Kraków, positively contributes to the investment attractiveness of the areas around it, resulting in deliberate transformation into residential areas (Buśko and Szafrańska 2018).

### *21.3.2 Land-Cover Changes from 2000 to 2018*

Figure 21.2 shows the spatial distribution of changes (level 1) in the period from 2000 to 2018 in the municipality of Wiśniowa. Significant changes can be identified only for the subperiod of 2006 to 2012 and only for the north-eastern part of the municipality (cadastral districts Poznachowice Dolne and Wiśniowa). The changes were not extensive in the context of the whole municipality but are considered significant due to their dynamics.

Considering the first level of detail with its three land cover categories (agricultural areas, forest and seminatural areas, and artificial surfaces), the most evident change took place for agricultural areas in the village of Wiśniowa. The area shrunk by almost 200 ha over 18 years. This fact is consistent with the trend described by many researchers regarding the prevalent reduction of agricultural land mostly to the benefit of built-up areas (Haack and Rafter 2006). In the case of Wiśniowa, the trend was confirmed completely as 100% of the lost agricultural areas was replaced by an increase in artificial surfaces. Cegielska et al. (2018) presented similar results for land cover changes in Lesser Poland (Poland) and Pest Country (Hungary).

The change consisted in the transformation of land classified as agricultural areas into artificial surfaces. Note that the areas are situated along the voivodeship road No 964, which is the main artery of the Wiśniowa municipality. It is regions near

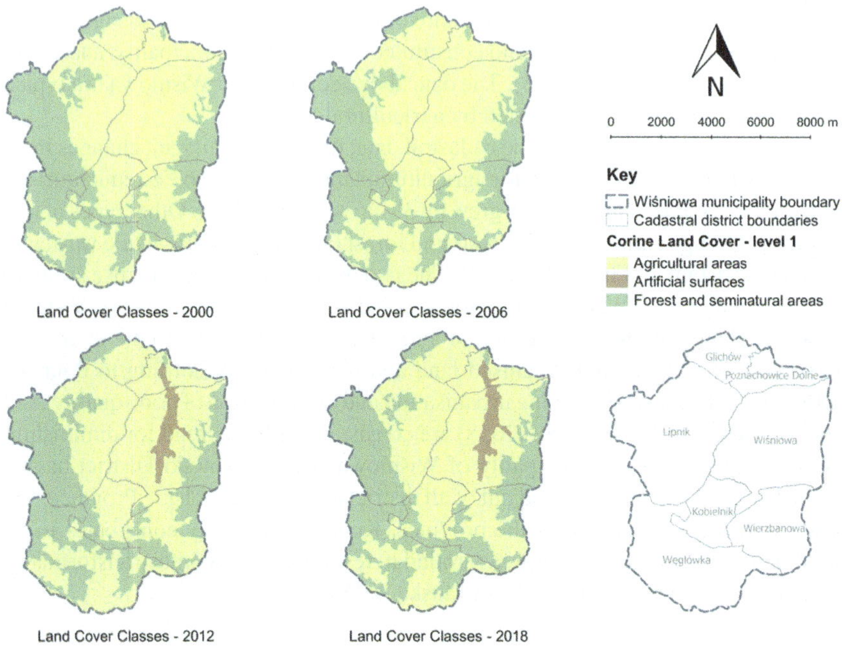


**Fig. 21.2** The spatial distribution of the land-cover change in the municipality of Wiśniowa from 2000 to 2018 (CLC level 1)

transport routes that are developed most dynamically, in particular, in rural areas (Bacior and Prus 2018).

The work of Fiedeń (2019) should be mentioned here. He has demonstrated that the share of agricultural land drops near motorway stretches between junctions to be replaced by forests, while developed land, mostly industrial and commercial, grows significantly in the direct vicinity of motorway junctions (Fiedeń 2019).

The spatial distribution of land-cover categories on CLC level 2 (Fig. 21.3) does not exhibit any changes in significant areas other than those identified on CLC level 1.

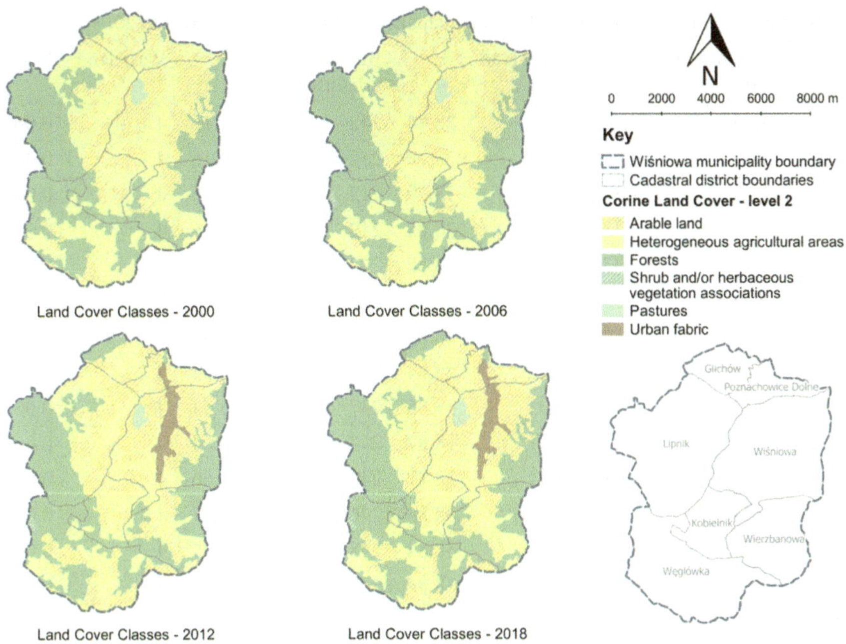


**Fig. 21.3** The spatial distribution of the land-cover change in the municipality of Wiśniowa from 2000 to 2018 (CLC level 2)

It was the introduction of CLC level 3 (Fig. 21.4) that facilitated detection of minor changes in land cover in the north-western part of the Lipnik district between 2012 and 2018. It was a transition from broad-leaved forest to mixed forest. These changes take place mostly on forest land.

The work of Bonilla-Moheno et al. (2012) should be mentioned here. Their results indicated the influence of global economic and demographic factors on land-cover changes, mainly in forest areas. Such changes, mostly deforestation, are affected by urbanisation to a large extent. Densely populated and easily accessible areas located near city centres are much more susceptible to the land-cover change. A similar impact on deforestation was diagnosed in areas with an agrarian structure dominated by holdings focusing on intensive production (Leśniewska-Napierała et al. 2019).

The spatial distribution of land-cover categories does not exhibit many significant land-cover changes even at CLC level 3 (Fig. 21.4). Such a minuscule dynamics of changes can be analysed in-depth only with the values presented in Table 21.4.

**Table 21.3** The land-use structure of the municipality of Wiśniowa from 2002 to 2018

| Year | Agricultural land | % | Forest land | % | Built-up and urbanised land | % | Wasteland | % |
|---|---|---|---|---|---|---|---|---|
| 2002 | 3,846 | 57.40 | 2,602 | 38.84 | 212 | 3.16 | 4 | 0.06 |
| 2003 | 3,846 | 57.40 | 2,599 | 38.79 | 212 | 3.16 | 4 | 0.06 |
| 2004 | 3,844 | 57.37 | 2,599 | 38.79 | 212 | 3.16 | 4 | 0.06 |
| 2005 | 3,841 | 57.33 | 2,603 | 38.85 | 212 | 3.16 | 4 | 0.06 |
| 2006 | 3,841 | 57.33 | 2,603 | 38.85 | 213 | 3.18 | 4 | 0.06 |
| 2007 | 3,839 | 57.30 | 2,603 | 38.85 | 212 | 3.16 | 4 | 0.06 |
| 2008 | 3,827 | 57.12 | 2,602 | 38.84 | 224 | 3.34 | 4 | 0.06 |
| 2009 | 3,814 | 56.93 | 2,602 | 38.84 | 237 | 3.54 | 4 | 0.06 |
| 2010 | 3,815 | 56.94 | 2,603 | 38.85 | 237 | 3.54 | 4 | 0.06 |
| 2011 | 3,808 | 56.84 | 2,604 | 38.87 | 244 | 3.64 | 4 | 0.06 |
| 2012 | 3,804 | 56.78 | 2,604 | 38.87 | 245 | 3.66 | 4 | 0.06 |
| 2013 | 3,804 | 56.78 | 2,602 | 38.84 | 246 | 3.67 | 4 | 0.06 |
| 2014 | 3,803 | 56.76 | 2,602 | 38.84 | 248 | 3.70 | 4 | 0.06 |
| 2015 | 3,805 | 56.79 | 2,602 | 38.84 | 247 | 3.69 | 4 | 0.06 |
| 2016 | 3,808 | 56.84 | 2,605 | 38.88 | 249 | 3.72 | 4 | 0.06 |
| 2017 | 3,807 | 56.82 | 2,605 | 38.88 | 250 | 3.73 | 4 | 0.06 |
| 2018 | 3,804 | 56.78 | 2,605 | 38.88 | 251 | 3.75 | 4 | 0.06 |

Moreover, thanks to the in-depth analysis it was determined that artificial surfaces are the urban fabric in the second level of detail, while for the more accurate third level of detail (Table 21.4), they are the discontinuous urban fabric. A detailed analysis of the loss of agricultural areas demonstrated that in the investigated period, the reduction affected only heterogeneous agricultural areas (–427.56 ha). This change consists of the loss of 463.82 ha of complex cultivation patterns and a negligible increase in the area of land principally occupied by agriculture by 36.25 ha. Surface areas of pastures did not change, but the area of arable land (non-irrigated arable land according to detail level 3) grew by 163.82 ha. Areas of forests and seminatural areas exhibit no significant dynamics. A minor increase in their areas was noticeable by a little over 30 ha in total. The internal dynamics of forest was significant. The loss of nearly 150 ha of broad-leaved forest was replaced with significant growth of the area of mixed forest (+159.96 ha) and a negligible growth in coniferous forest (+0.03 ha). Artificial surfaces have been noted in the municipality of Wiśniowa since 2012: in Poznachowice Dolne and Wiśniowa reaching 226.14 ha. They grew merely by 7.5 ha over the six years.

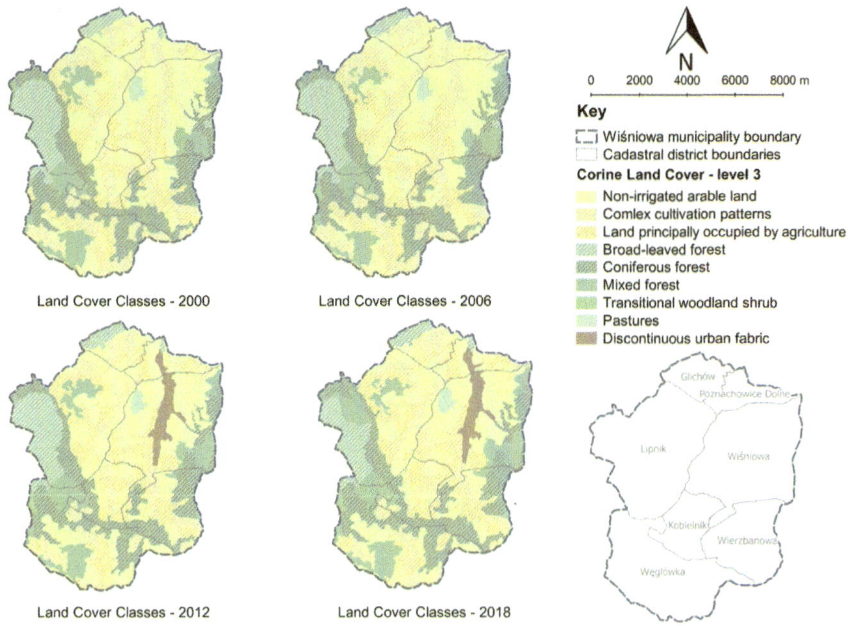


**Fig. 21.4** The spatial distribution of land-cover change in the municipality of Wiśniowa from 2000 to 2018 (CLC level 3)

## 21.4 Conclusions

Europe's cultural landscapes are at particular risk from expanding urbanisation and abandonment of economically questionable land (for example, due to low soil quality class or significant fragmentation). In areas where urbanisation is considered a driver of development, all anthropogenic factors may disturb the rural landscape and affect the agricultural land cover change (Su et al. 2012). The diversity of threats to cultural landscapes and significant dependence on uncertainty regarding requirements and social policy may hinder the development of any means to maintain and strengthen cultural landscapes.

The research has demonstrated that the six-year interval and division of analysis results into cadastral districts facilitated an analysis of spatio-temporal changes. This, in turn, can help build a review of land cover transformations and their impact on cultural landscapes (Cao et al. 2017).

Note, however, that in the case of such insignificant land-use and land-cover changes with a uniform structure as those in the municipality of Wiśniowa from 2000 to 2018, the use of all three levels of detail was not entirely necessary. Similar results and the same conclusions could be reached with just CLC level of detail 3.

**Table 21.4** Surface areas of individual land-cover categories from 2000 to 2018 (level 3)

| | Broad-leaved forest | | | | Coniferous forest | | | | Mixed forest | | | |
|---|---|---|---|---|---|---|---|---|---|---|---|---|
| | 2000 | 2006 | 2012 | 2018 | 2000 | 2006 | 2012 | 2018 | 2000 | 2006 | 2012 | 2018 |
| Poznachowice Dolne | 12.55 | 12.55 | 12.55 | 12.55 | 0.00 | 0.00 | 0.00 | 0.00 | 0.00 | 0.00 | 0.00 | 0.00 |
| Węglówka | 34.70 | 34.70 | 24.09 | 24.09 | 426.76 | 429.30 | 429.26 | 429.26 | 261.88 | 261.88 | 272.52 | 276.21 |
| Wierzbanowa | 0.00 | 0.01 | 0.01 | 0.01 | 358.47 | 358.46 | 362.24 | 362.24 | 19.57 | 19.57 | 19.57 | 19.57 |
| Lipnik | 452.13 | 454.83 | 454.83 | 325.93 | 245.75 | 247.17 | 247.17 | 261.52 | 150.20 | 146.07 | 146.07 | 260.63 |
| Glichów | 86.14 | 86.14 | 86.14 | 86.71 | 2.86 | 2.86 | 2.86 | 2.86 | 0.00 | 0.00 | 0.00 | 0.00 |
| Wiśniowa | 11.32 | 11.32 | 1.06 | 1.06 | 229.94 | 204.99 | 204.99 | 204.99 | 135.02 | 159.97 | 170.22 | 170.22 |
| Kobielnik | 0.00 | 0.00 | 0.00 | 0.00 | 110.49 | 113.42 | 113.42 | 113.42 | 20.83 | 20.83 | 20.83 | 20.83 |
| **Total** | **596.83** | **599.54** | **578.68** | **450.35** | **1,374.27** | **1,356.21** | **1,359.95** | **1,374.30** | **587.50** | **608.31** | **629.21** | **747.46** |
| | Land principally occupied by agriculture | | | | Pastures | | | | Non-irrigated arable land | | | |
| | 2000 | 2006 | 2012 | 2018 | 2000 | 2006 | 2012 | 2018 | 2000 | 2006 | 2012 | 2018 |
| Poznachowice Dolne | 57.36 | 57.36 | 103.49 | 105.09 | 0.00 | 0.00 | 0.00 | 0.00 | 153.42 | 153.42 | 166.15 | 160.17 |
| Węglówka | 260.92 | 247.42 | 159.65 | 154.25 | 0.00 | 0.00 | 0.00 | 0.00 | 102.92 | 102.92 | 326.01 | 306.79 |
| Wierzbanowa | 192.04 | 192.04 | 192.04 | 192.30 | 0.00 | 0.00 | 0.00 | 0.00 | 141.22 | 141.22 | 137.53 | 135.07 |
| Lipnik | 307.65 | 307.65 | 353.06 | 353.06 | 0.08 | 0.08 | 0.08 | 0.08 | 402.75 | 402.75 | 428.20 | 414.16 |
| Glichów | 21.67 | 21.67 | 48.00 | 48.00 | 0.00 | 0.00 | 0.00 | 0.00 | 131.18 | 131.18 | 104.79 | 106.15 |
| Wiśniowa | 262.86 | 262.86 | 275.39 | 275.50 | 47.23 | 47.23 | 47.23 | 47.23 | 689.53 | 689.53 | 679.16 | 675.96 |
| Kobielnik | 169.33 | 166.41 | 166.41 | 179.88 | 0.00 | 0.00 | 0.00 | 0.00 | 133.37 | 133.37 | 133.37 | 119.89 |
| **Total** | **1,271.83** | **1,255.41** | **1,298.04** | **1,308.08** | **47.31** | **47.31** | **47.31** | **47.31** | **1,754.37** | **1,754.37** | **1,975.20** | **1,918.20** |
| | Complex cultivation patterns | | | | Transitional woodland shrub | | | | Discontinuous urban fabric | | | |
| | 2000 | 2006 | 2012 | 2018 | 2000 | 2006 | 2012 | 2018 | 2000 | 2006 | 2012 | 2018 |
| Poznachowice Dolne | 92.68 | 92.68 | 4.10 | 4.09 | 0.00 | 0.00 | 0.00 | 0.00 | 0.00 | 0.00 | 29.71 | 34.10 |
| Węglówka | 313.19 | 310.65 | 175.34 | 196.26 | 0.00 | 13.50 | 13.50 | 13.50 | 0.00 | 0.00 | 0.00 | 0.00 |

(continued)

**Table 21.4** (continued)

| | Broad-leaved forest | | | | Coniferous forest | | | | Mixed forest | | | |
|---|---|---|---|---|---|---|---|---|---|---|---|---|
| | 2000 | 2006 | 2012 | 2018 | 2000 | 2006 | 2012 | 2018 | 2000 | 2006 | 2012 | 2018 |
| Wierzbanowa | 77.16 | 77.16 | 77.06 | 79.26 | 0.00 | 0.00 | 0.00 | 0.00 | 0.00 | 0.00 | 0.00 | 0.00 |
| Lipnik | 193.04 | 193.04 | 122.18 | 136.21 | 0.00 | 0.00 | 0.00 | 0.00 | 0.00 | 0.00 | 0.00 | 0.00 |
| Glichów | 110.93 | 107.81 | 107.88 | 105.94 | 0.00 | 3.11 | 3.11 | 3.11 | 0.00 | 0.00 | 0.00 | 0.00 |
| Wiśniowa | 232.16 | 232.16 | 33.57 | 33.57 | 0.00 | 0.00 | 0.00 | 0.00 | 0.00 | 0.00 | 196.43 | 199.52 |
| Kobielnik | 50.27 | 50.27 | 50.27 | 50.27 | 0.00 | 0.00 | 0.00 | 0.00 | 0.00 | 0.00 | 0.00 | 0.00 |
| **Total** | **1,069.41** | **1,063.76** | **570.39** | **605.60** | **0.00** | **16.61** | **16.61** | **16.61** | **0.00** | **0.00** | **226.14** | **233.62** |

## References


Agnoletti M (2014) Rural landscape, nature conservation and culture: some notes on research trends and management approaches from a (southern) European perspective. Landsc Urban Plan 126:66–73

Almeida M, Loupa-Ramos I, Menezes H, Carvalho-Ribeiro S, Guiomar N, Pinto-Correia T (2016) Urban population looking for rural landscapes: different appreciation patterns identified in Southern Europe. Land Use Policy 53:44–55

Aneseyee AB, Soromessa T, Elias E (2020) The effect of land use/land cover changes on ecosystem services valuation of Winike watershed, Omo Gibe basin Ethiopia. Human Ecol Risk Assessment, 10, 2608-2627. https://doi.org/10.1080/10807039.2019.1675139

Bacior S, Prus B (2018) Infrastructure development and its influence on agricultural land and regional sustainable development. Eco Inform 44:82–93

Bonilla-Moheno M, Aide TM, Clark ML (2012) The influence of socioeconomic, environmental, and demographic factors on municipality-scale land-cover change in Mexico. Reg Environ Change 12(3):543–557

Burkhard B, Kroll F, Müller F, Windhorst W (2009) Landscapes‘ capacities to provide ecosystem services – a concept for land-cover based assessments. Landscape Online 15:1–22

Buśko M, Szafrańska B (2018) Analysis of changes in land use patterns pursuant to the conversion of agricultural land to non-agricultural use in the context of the sustainable development of the Malopolska region. Sustainability 10(1), 136. https://doi.org/10.3390/su10010136

Büttner G, Feranec J, Jaffrain G, Mari L, Maucha G, Soukup T (2004) The corine land cover 2000 project. EARSeL eProceedings 3:331–346

Cao H, Liu J, Fu C, Zhang W, Wang G, Yang G, Luo L (2017) Urban expansion and its impact on the land use pattern in Xishuangbanna since the reform and opening up of China. Remote Sensing 9(2):137. https://doi.org/10.3390/rs9020137

Cegielska K, Noszczyk T, Kukulska A, Szylar M, Hernik J, Dixon-Gough R, Jombach S, Valánszki I, Filepné Kovács K (2018) Land use and land cover changes in post-socialist countries: some observations from Hungary and Poland. Land Use Policy 78:1–18. https://doi.org/10.1016/j.landusepol.2018.06.017

Cegielska KM, Salata T, Gawroński K, Różycka-Czas R (2017) Level of spatial differentiation of anthropogenic impact in Małopolska. J Ecol Eng 18(1):200–209

Cuca B, Agapiou A (2018) Impact of land-use change and soil erosion on cultural landscapes: the case of cultural paths and sites in Paphos district. Appl Geomatics 10(4):515–527

Dziadowiec-Greganić J (2018) Dziedzictwo rzemieślnicze i rękodzielnicze Gminy Wiśniowa w działaniu [Artisan and handicraft heritage of the Municipality of Wiśniowa in action]. Zbiór Wiadomości Do Antropologii Muzealnej 5(2018):325–334. https://doi.org/10.12775/ZWAM.2018.5.20

Eurobarometer (2017) Report – Cultural Heritage. Special Eurobarometer 466, December 2017. https://doi.org/10.2766/576064

Fiedeń L (2019) Changes in land use in the communes crossed by the A4 motorway in Poland. Land Use Policy 85:397–406

Gingrich S, Niedertscheider M, Kastner T, Haberl H, Cosor G, Krausmann F, Kuemmerle T, Mueller D, Reith-Musel A, Jepsen MR, Vadineanu A, Erb K-H (2015) Exploring long-term trends in land use change and aboveground human appropriation of net primary production in nine European countries. Land Use Policy 47:426–438

Haack BN, Rafter A (2006) Urban growth analysis and modeling in the Kathmandu Valley. Nepal, Habitat International 30:1056–1065

Hernik J (2008) Potrzeba uwzględniania walorów krajobrazu kulturowego w zarządzaniu gminą wiejską [Need for considering cultural landscape qualities in rural municipality management]. Prace Komisji Krajobrazu Kulturowego 10:61–68

Hernik (2011) Ochrona wrażliwych krajobrazów kulturowych obszarów wiejskich [Protection of vulnerable rural cultural landscapes]. Zeszyty Naukowe Uniwersytetu Rolniczego w Krakowie, edition: Treatises 474, J 351. Kraków
Hernik J, Czesak B, Noszczyk T, Pazdan M (2015) Cultural landscape and land use in the rural areas of Poland. In: Hernik J, Chen G, Gawroński K (eds) Cultural landscapes in the context of socio-economic changes in Poland and China. Publishing House of the University of Agriculture in Krakow, Krakow, pp 11–21
Leśniewska-Napierała K, Nalej M, Napierała T (2019) The impact of EU grants absorption on land cover changes-the case of Poland. Remote Sens 11(20):2359. https://doi.org/10.3390/rs11202359
Łowicki D (2008) Land use changes in Poland during transformation case study of Wielkopolska region. Landsc Urban Plan 87:279–288
Piotrowska M, Taszakowski J (2008) Rozwój krajobrazów kulturowych na terenach górskich na przykładzie wsi Wiśniowa [Development of cultural landscapes in montane areas. A case study of the village of Wiśniowa]. Infrastruktura i Ekologia Terenów Wiejskich, 3/2008, 63–70
Plieninger T, van der Horst D, Schleyer C, Bieling C (2014) Sustaining ecosystem services in cultural landscapes. Ecol Soc 19(2):59. https://doi.org/10.5751/ES-06159-190259
Poławski Z (2009) Land use changes in Poland during last two centuries. Teledetekcja Środowiska 42:69–82
Prus B, Król K, Gawroński K, Sankowski E, Hernik J (2020) From Classic (Analogue) to Digital Forms of Cultural Heritage Protection in Poland. In: Kremers H. (Ed.), Digital Cultural Heritage. Springer, pp 255–278. https://doi.org/10.1007/978-3-030-15200-0_17
Schulp CJE, Levers C, Kuemmerle T, Tieskens KF, Verburg PH (2019) Mapping and modelling past and future land use change in Europe's cultural landscapes. Land Use Policy 80:332–344
Su S, Xiao R, Zhang Y (2012) Multi-scale analysis of spatially varying relationships between agricultural landscape patterns and urbanization using geographically weighted regression. Appl Geogr 32:360–375
Sudhira HS, Ramachandra TV, Jagadish KS (2004) Urban sprawl: metrics, dynamics and modelling using GIS. Int J Appl Earth Obs Geoinf 5:29–39
Vademecum (2018) Statystyczne Vademecum Samorządowca 2018 [Statistical Handbook for Local Governments 2018]. Statistical Office in Kraków, Kraków 2018. https://krakow.stat.gov.pl/vademecum/vademecum_malopolskie/portrety_gmin/powiat_myslenicki/wisniowa.pdf
Ziter C, Graves RA, Turner MG (2017) How do land-use legacies affect ecosystem services in United States cultural landscapes? Landscape Ecol 32(11):2205–2218

# Chapter 22
# Land Use Change and Landscapes in Rural Areas in China in Forty Years of Reform and Opening Up

**Gaiying Chen, Tomasz Noszczyk, Maria Nawieśniak-Caesar, Maria Pazdan, and Józef Hernik**

**Abstract** Over the past 40 years, great changes have taken place in rural areas in China. The huge changes, including reform of the rural land system, the reform of the purchase and sale system concerning agricultural products, and the rapid development of township enterprises, have had significant impacts on China's economic and social development. From 1979 to 2019, the land system had faced different problems in different historical periods. The reform of the rural land system, the improvement of land use laws, and the protection and management of land rights and farmland have been given attention in Chinese rural land management practice. The reform of and innovation in the rural land system have been remarkable achievements. Rural land circulation, transfer of agricultural land and land with construction, taking the protection of farmers' interests as the basic criterion, have been institutionalized, which is not only conducive to the intensive management of the scale of agricultural activity, but also meets the farmers' social security interests. Land resources' utilization planning had adapted the balance sheet accounting of land resources and the assessment of land ecological conditions, deepening the research on carbon emissions from the perspective of land use. In rural areas, the goal of returning farmland to forests and grassland had been implemented and consolidated the achievements. The scientific management of farmland had been strengthened. Land use has shown evidence of innovations land use patterns, the construction of a land space planning system and furthering of village level planning. Research on land engineering and technology had gradually deepened in fields of comprehensive improvement of agricultural land, technological innovation in reclamation and in methods of land pollution remediation. In the process of urbanization, cultivated land resources and national food security generate seriously questionable and challenging situations, and the comprehensive improvement and reclamation technology of agricultural land has assumed an important role. The constant changes taking

G. Chen
Department of Landscape Architecture, Beijing University of Agriculture, Beijing, China

T. Noszczyk (✉) · M. Nawieśniak-Caesar · M. Pazdan · J. Hernik
Department of Land management and Landscape Architecture, Faculty of Environmental Engineering and Land Surveying, University of Agriculture in Krakow, Krakow, Poland
e-mail: tomasz.noszczyk@urk.edu.pl

© Springer Nature Switzerland AG 2022

J. Hernik et al. (eds.), *Cultural Heritage—Possibilities for Land-Centered Societal Development*, Environmental History 13,
https://doi.org/10.1007/978-3-030-58092-6_22

place in Chinese land use and landscape are caused by many factors. Only rational policy on land management and land use planning can bring tangible benefits to the society. Research about rural landscapes is currently at an early stage in China, but some risks are already noticeable in relation to the country's cultural landscapes. Rapid and dynamic urbanization has produced some negative impacts on rural landscapes in recent years. Planned investments usually interfere with the environment, sometimes extending to the liquidation of valuable green areas, which constitute a cultural heritage. Another problem is the lack of the country's cultural landscape protection in land management. There are not specific rules for determining when planning disturbs the harmony of the landscape.

**Keywords** Land use change · Landscape · Agriculture · China

## 22.1 Introduction

China is the most populous country in the world. Its area includes part of East Asia and Central Asia. The huge territory of the country is characterized by varied terrain, predominantly mountainous and upland. The mountainous areas account for two-thirds of the country's total area, and these large areas create certain difficulties for transportation and agricultural development; however, the mountainous areas can provide forest products, mineral resources, water energy, and tourism resources, which can contribute to the development of the mountainous area economy. China has three great plains, which are distributed on the third step of the terrain in eastern China. The three plains are connected with the north and south, and the soil is fertile. They are the most important agricultural areas in China.

The People's Republic of China was founded in 1949. Before the creation of the Republic of China, the State was semi-feudal, underdeveloped agricultural country, largely dependent on foreign capital (PWN 2019). After 1949, at the beginning of the state that had been founded, the Chinese government successfully completed land reform in areas with more than 90% of the national total agricultural population. From 1953 to 1966, the first five-year plan was implemented and achieved great success. China carried out large-scale socialist construction and advanced the agricultural capital construction and technical transformation. From 1966 to 1976, the country and its people suffered serious setbacks and losses due to the decade of the cultural revolution. Since 1978, the policy of "reform and opening up" has been carried out, and China has experienced profound changes and gradually established a road of socialist modernization with Chinese characteristics. With the state achieving political stability and rapid economic development, the people's living standards have significantly improved.

In an underdeveloped economy, improvement of land use was mainly related to a need to improve national agricultural production, which leads to meeting basic food needs. Population growth is the main driver of growth in food consumption and has exerted great pressure on land resources (Cui et al. 2009). There are significant changes in land use that have occurred over the past 40 years, when market reforms began in this country, and so did active participation in the global economy. Economic development has changed this country into one of the fastest growing countries in the world (Lin and Ho 2003; Kram et al. 2012).

Socio-economic changes that were made after 1949, in particular, land reform and the nationalization of key sectors of the economy, ensured a state-dominant sector in the multi-sector economy of China. In the years 1952–1957 the collectivization of agriculture was carried out. In the late 50 s, only the state sector and collective industries remained. In 1978, the reform brought a new economic development strategy which prioritized agriculture. And thus besides heavy and light industry, agriculture also became a priority sector of the country's economy. From 1979 to 1984, there was a period of liquidation of people's communes and the introduction of individual farms. Their activities were structured for family responsibility for production on land provided periodically for 5–10 years, leased from the state (PWN 2019).

China is also a country that needs to feed 20% of the world's population with only 7% of the world's agricultural land (Zuo et al. 2018). It is a serious research issue for the country because the limited arable land in this country is still declining (Chen 2007). The current agricultural intensification comes at the cost of water resource depletion, soil degradation, and air and water pollution (Zuo et al. 2018). China faces the double pressure of having to increase production on its limited agricultural land resources while at the same time reducing land use pressure on environmental systems (Sun et al. 2018).

The changes that have taken place in Chinese agriculture have led to the economic revival of rural areas. Changes also enabled the gradual transition to a modern commodity economy. The program of modernization of the economy is essential to its policy of opening up to the outside world; it applies both to the development of foreign trade relations and foreign technology imports (mainly from Japan), and foreign investment as well. In China, businesses operate both as solely owned by foreign companies as well as mixed companies (sino-foreign), in which the distribution of the profit is made in accordance with the share capital contribution (PWN 2019).

China has always attached great importance to agriculture, rural development and the people living in rural areas. In the past 40 years, there were many studies concerning land use in China; one of them was a study done in 1996 by Lin and Ho (2003). Also the Ministry of Land and Mineral Resources collects statistical data and regularly updates information on land use. The task of the Ministry was and is fact planning, management, protection and rational use of natural resources in China, including land, mineral and marine resources (MLR 2015). In 2000 the Ministry of Land and Mineral Resources published the first study of land use on a national scale. This study, initiated in 1984 by the State Council, has created a specific database of information about land use in China, studying both factors of quality and quantity (Kram et al. 2012).

The change in the structure of land use is one of the factors that shape cultural landscapes. With the rapid development of an economy, duo to urban sprawl, a great deal of agricultural land has been occupied, and the structure of land use has changed greatly in China, and this is precisely the function of agriculture which has a major impact on the open cultural landscape on the rural areas (Hernik 2011).

## 22.2 Land Use Characteristics in China

Despite the increasing urbanization of China, this is a country with a strong rural roots. Since ancient times, the Chinese economy was built on the foundations of agriculture. Since the economic reform of 1978, China is rapidly changing into a condition heavily urbanized and industrialized from a typical agricultural country. At the same time, the significant changes in land use also occurred, by subjecting the rural and urban transformation. Existing regional disparities, geographical and geophysical between these areas caused a number of problems in land use (Long 2014).

Thorough testing by Lin and Ho (2003) showed that the actual area of China in 1996 was 9.5 million $km^2$ and not 9.6 million $km^2$, as it is commonly assumed. These studies have also shown that fields occupy approx. 130 million hectares, or nearly 40% more than what has been reported to the National Bureau of Statistics. Analyzing the data from the period 1949–1996, it was found that changes in land use were mainly characterized by the expansion of agricultural land and buildings. Restructuring processes of agriculture, rural industrialization and rapid urbanization observed since 1990 in China were the beginning of a new trend of massive loss of agricultural land in relation to the benefit of the market (Lin and Ho 2003).

However research of Lin and Ho (2003) is not the only study that has been carried out in China. Also Cui et al. (2009) examined changes of land use.

The total-surface of China in 2015 was 9.6 million $km^2$. The area of arable land per 1 inhabitant of China was in 1996 only 0.099 ha.; in 2016 it was 0.087. It is approx. 0.265 ha less compared to Central Europe (The World Bank 2019).

The area of agricultural land per capita still continue to decrease, which is caused by an increasing number of population in the country (Lin and Ho 2003) (Table 22.1).

**Table 22.1** Arable land in China and other countries

| Country name | Arable land* (ha. per person) | | | |
|---|---|---|---|---|
| | 1996 | 2000 | 2008 | 2016 |
| China | 0.099 | 0.095 | 0.092 | 0.087 |
| India | 0.164 | 0.152 | 0.132 | 0.118 |
| Central Europe and the Baltics | 0.394 | 0.387 | 0.359 | 0.352 |
| World | 0.233 | 0.221 | 0.200 | 0.187 |

*Source* The World Bank (2019). *Arable land (hectares per person) includes land defined by the FAO as land under temporary crops (double-cropped areas are counted once), temporary meadows for mowing or for pasture, land under market or kitchen gardens, and land temporarily fallow. Land abandoned as a result of shifting cultivation is excluded

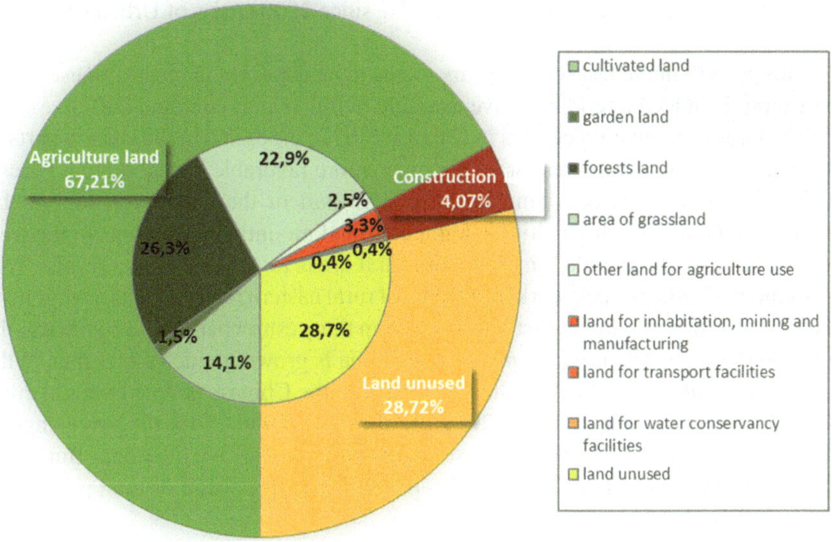


**Fig. 22.1** Land use in China in 2016. *Source* National Bureau of Statistics of China (2019)

In rural areas, the land is divided into three main categories: agriculture land, construction land, and unused land (He et al. 2012). In 2016 agriculture land was 67.21% of total land area, unused land (such as deserts, saltflats from which salt is not harvested, grassland with less than 10% cover, swamps, alpine tundra, glaciers, beaches, etc.; also includes rivers, lakes, streams, etc.) was 28.72% of the total area, and construction lands were 4.07% of total area (Fig. 22.1) (National Bureau of Statistics of China 2019).

Large changes in land use in China are also caused by urbanization, which involves problems of inefficient use of land. One of the biggest problems is the uneven distribution of the population across the country and chaotic growth of cities, which take up too much land. Further problems faced by China include poor supervision of urban land use and the real estate market and the relatively high average area of urban land per 1 inhabitant, which is not fully utilized.

The rapid development of Chinese cities requires the transformation of large areas of land used for agriculture to use for urban areas. Through such persistent uncontrolled expansion over the years, cities can lead to even greater decline in agricultural areas, which in turn may lead to food shortages. The increase in urban land consumption per 1 inhabitant observed in recent years in China should be considered as "waste" land. For example, according to Bertaud (2007) comparing the data on the use of urban land in Tianjin in 1988 and 2000, the population in the city increased by 22%, while the surface area of the city has increased by 63%. Urban land consumption per capita increased by 34%.

The population of Tianjin in 2013 was 10,056,078, increasing at an average annual rate of 5.7% since 2000. The urban extent of Tianjin in 2013 was 236,292 hectares, increasing at an average annual rate of 13.7% since 2000 (Atlas of Urban Expansion 2019).

This growth in the last several years seems to be the norm in most Chinese cities, so Tianjin is indeed a representative example in this regard (Bertaud 2007).

The largest population centers in China are in big cities and in the eastern parts of the country where natural resources are much more favorable for habitation. Forests and agricultural areas occurring in the eastern part of the population support far more than the deserts, high altitude highlands, and mountainous regions to the west of the country. The population shift from rural areas to cities in China results from movement in order to seek work. Residents of rural eastern regions usually migrate to the nearest cities also in the east and those from the western part of the country to the cities in the west. The urban population in China is growing much faster than in the US (Kram et al. 2012). In 1950 the percentage of the Chinese people living in cities was 12% (People's Daily Online 2009). In the years 1990–2016, the percentage of the population living in cities increased from 26.4% to 58.5% (National Bureau of Statistics of China 2019). Figure 22.2 represents the proportion of total population living in urban and rural areas.

The vast majority of arable land in China includes monsoon plains and river basins. According to statistics, almost all the arable land in the country, for a total area of 122 million ha, or 12% of the country is being used. It should also be pointed out that half of the existing arable land in China is artificially irrigated, mostly in the basin of the Yangtze River and the Yellow River (Qiang 2010).

Some of the land is re-created from other lands that had previously been used differently, e.g., as meadows, forests, and wetlands, and existing farmlands are converted to other uses, such as built-up areas, forests and meadows. According to the

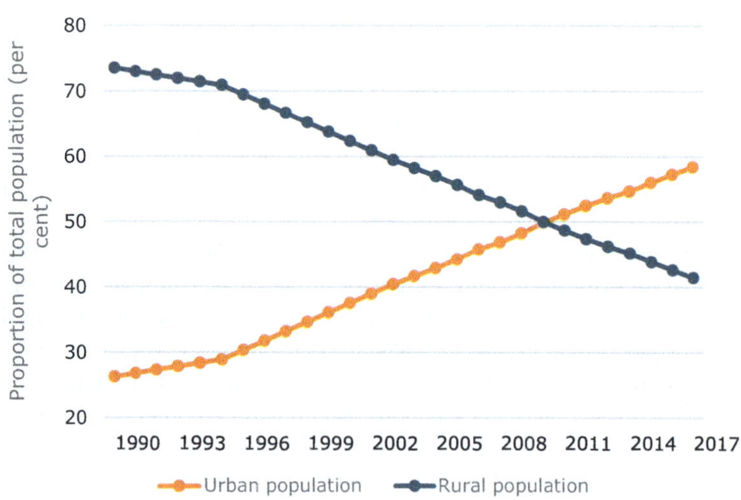


**Fig. 22.2** Urban and rural population in China, 1990–2016. *Source* National Bureau of Statistics of China (2019)

data on the changes in groundwater from the period 1986–2000, approximately 55% of the newly transformed land for cultivation of grasslands, forests 25%, and 20% of wetlands, and other areas are not used for agricultural purposes. Overall, in this period (1986–2000) the increase of the arable land was 2.7 million hectares (1.9%), and the decrease was recorded in meadows, forests, and wasteland as a result of their conversion to arable land. According to Kram et al. (2012) the trend of changes of cultivated land is likely to continue in the coming years. Therefore, there is an urgent need for China to effectively and efficiently use their land resources, not only putting their domestic welfare into consideration but also that of the entire population of the world (Lin and Ho 2003).

## 22.3 Protection of Agricultural Land in China

China has entered a period of rapid urbanization, and the contradiction between ensuring development and protecting resources is increasingly complex. It is an important development trend of land and resources management to transform from extensive utilization of resources to intensive utilization of resources, from quantitative management of resources to comprehensive management of quantity, quality, and ecology, and from pure resource management to tri-une management of resources, assets, and capital. For a long time, the over-expansion of urban and rural construction land makes the limited cultivated land resource bear a huge demand pressure, the contradiction between economic development and cultivated land resource protection is increasingly prominent, and the problems related to cultivated land protection are still the focus of domestic scholars.

Over the past decades, there has been a severe degradation of arable land in China, and this is due to desertification and salinity of the area, as well as through their industrialization. These processes result in a reduced yield and as a result contribute to irreversible degradation of the land. Desertification of generally limited land resources and increased possibility of sandstorms, which in turn cause salt deposition on land. The occurrence of these processes is conducive to the occurrence of erosion, which is greatly intensified (Zhang et al. 2007). Since 1970 more than 40% of arable land in China has degenerated due to soil erosion and desertification progressing (DuBose 2010).

Increased soil pollution in China is due to increased use of fertilizers and pesticides, which clearly increased in the past several years. The use of pesticides in the country far exceeds the average level in the world, and the consumption of fertilizers in China on 1 ha. clearly exceeds the limits of safety. This overuse of fertilizers and pesticides can therefore lead to a decline in agricultural production. In addition, the degradation of agricultural land in China also contributes to industrial pollution. According to some estimates, nearly 1/6 of all arable land is contaminated with heavy metals. The Chinese economy is the world's largest consumer of coal and one of the largest producers of greenhouse gases and sulfur dioxide. Furthermore, the presence of these compounds in the atmosphere contributes to acid rain that have a devastating impact on crops, soil, and water resources (DuBose 2010).

These problems led the State Council to change the law. In 1998, it was stated that "the most economical and rational use of land and good protection of farmland are the basic state policy of China" ("most sparing and rational land utilization and earnest protection of cultivated land constitute China's basic state policy") (Land Administration Law 1998). The government has also introduced a general development plan, which strictly defines the number of farms which may be approved for conversion to non-agricultural purposes. This action was to aimed a significant slowdown in the pace of conversion of land for non-agricultural purposes (DuBose 2010).

A strict agricultural land protection policy (known as the 'land red line' policy) intends to safeguard a '1.8 billion mu' (i.e., 0.12 billion ha) area of quality agricultural land. But maintaining the red line is challenging in the face of urbanization and infrastructure construction that led to the loss of substantial amounts of fertile agricultural lands in the eastern and coastal regions. Unfortunately despite numerous edicts from the central government to protect agricultural land from conversion, agricultural land in China continues to be converted (Bren d'Amour et al. 2016; Sun et al. 2018).

## 22.4 Land Reclamation

In the process of urbanization, the situation of cultivated land resources is becoming serious, and the problem of national food security is becoming acute. The importance of land reclamation is more prominent in urban and rural social and economic development. It has become a basic goal for the transformation of land use patterns, the adjustment of land use structure, and the improvement of land use relationships in China in the new era. The reclamation of rural land results in increasing the effective cultivated land areas, and improving the quality of cultivated land, promoting the large-scale use of land, providing basic guarantees for adhering to the requirements of cultivated land protection, and improving the comprehensive agricultural production capacity.

Against the background of the tensions in relationships between human beings and nature, the arable land was replenished through the reclamation of rural settlements. It is one of the effective ways to realize the dynamic balance of the total cultivated land and relieve the pressure on cultivated land. At the same time, it can improve the rural living environment, improve the intensity of land use of rural residents, promote orderly, rational and scientific land use, and help promote the economic development of urban and rural areas and solve the problems faced by sustainable development.

Regulation, specifically regulation of rural land improvement, and overall planning of urban and township development of rural land in terms of comprehensive improvement are conducive to fully tapping the potential of rural land consolidation and reclamation, increasing the effective cultivated land area, improving the quality of cultivated land, promoting the large-scale use of land, providing a basic guarantee for adhering to the requirements of cultivated land protection, and improving the comprehensive agricultural production capacity.

In 2012, a total of 20,500 land renovation projects and 69.119 billion yuan of renovation funds were supported nationwide, and the total area concerned was 2,504,100 ha., with an additional 544,500 hectares of agricultural land and 465,600 ha. of arable land.

## 22.5 Landscape Changes in the Context of Land Use

Until now, rural landscapes in the area of the Republic of China were dominated by the presence of mainly agricultural land. The traditional Chinese village (Fig. 22.3) is characterized by low buildings with a low density. The buildings are generally made of locally found materials that perfectly fit into the landscape. The whole village is situated in harmony with nature and the landscape (Zhou et al. 2014). However, with the changes in land use and economic reform carried out which resulted in economic growth, we could also readily observe a change in the landscape of the relevant area. It is clear that a large impact on the landscape is an increase in demand and changing preferences in relation to food for the ever increasing population, but

**Fig. 22.3** Traditional village in harmony with nature and landscape. *Source* Photo by Xinmeng

also an important factor in the changes in the landscape is urbanization (Liu et al. 2005).

In some areas of China, landscape appears to become fragmented (broken). What is the reason? It is hard to unequivocally state. The answers can be found in the market economy, which in some cases led to the deepening of the difference between wealthy people and those who are less wealthy. Residents began to form some types of economic classes, which each on their own began to pursue their own goals and interests. The reasons for this problem you can also see in some kind of lack of control in spatial planning. People who are in better financial condition usually build multiple-story buildings and have their own green gardens, while those less wealthy continue to live in low-rise buildings. This situation is liable to adversely affect the harmony of the landscape and even distorts it. Research on the rural landscape in China has not been put on the back burner. Zhou et al. (2005) pointed out that progress of rural landscape research in China recently focused on four aspects:

(1) agriculture landscape research—initially drew attention only on the structure of the rural landscape, but now research focuses more on the processes that take place in this landscape and have a greater significance;
(2) rural culture landscape research—includes preliminary discussions on the types and nature of rural cultural landscape;
(3) rural landscape assessment research—assessing environmental quality, landscape resources, and the global assessment of the countryside;
(4) rural landscape planning research—landscape planning, which has already been used in China, mainly in urban areas.

The constant migration of people from rural to urban areas means a reduction in the population living in rural areas, which in some way "opens" the landscape in areas that are not "eaten" by the growing city. Over time, this trend can provide more opportunities to protect land in rural areas (Kram et al. 2012).

## 22.6 Conclusion

The constant changes taking place in the Chinese land use and landscape are caused by many factors. All of this contributes to the fact that there is a need for effective and efficient management of land, which is a limited resource and especially a non-renewable resource. Only rational policy on land management and land use planning can bring tangible benefits to the society.

Research on rural landscapes in China is currently at an early stage, but some risks in relation to the country's cultural landscapes are already noticeable. Dynamic urbanization that has taken place in recent years has a mainly negative impact on landscapes. Planned investments usually interfere in the environment, sometimes extending to the liquidation of valuable green areas, which constitute an important part of a cultural heritage. Another problem is the lack of control in land management.

There are no very specific rules for determining when planning, and its results disturb the harmony of the landscape.

## References

Atlas of Urban Expansion (2019) http://www.atlasofurbanexpansion.org/cities/view/Tianjin_Tianjin (last visited 04.11.2019)

Bertaud A (2007) Urbanization in China: land use efficiency issues. China land use report. http://alain-bertaud.com

Bren d'Amour C, Reitsma F, Baiocchi G, Barthel S, Guneralp B, Erb K-H, Seto KC (2016) Future urban land expansion and implications for global croplands. Proc Natl Acad Sci 114:8939–8944

Chen J (2007) Rapid urbanization in China: a real challenge to soil protection and food security. CATENA 69:1–15

Cui X, Rounsevell M, Jiang Y, Kang M, Palmer P, Chen W, Dawson T (2009) Simulating land-use change in China from a global perspective. Book Chapter Submitted to "Vulnerability and Resilience of Land Systems in Asia"

Dubose C (2010) China's agricultural land use policy: the growing tension between food security and economic growth. http://www.chinaaviationlaw.com

He C, Huang Z, Wan W (2012) Land use changes and economic growth in China, land lines. https://www.lincolninst.edu/publications/issues/land-lines-october-2012

Hernik J (2011) Ochrona wrażliwych krajobrazów kulturowych obszarów wiejskich, Zeszyty Naukowe UR w Krakowie, ser. Rozprawy, z. 351, s.31, Kraków

Kram M, Bedford C, Durnin M, Luo Y, Rokpelnis K, Roth B, Smith N, Wang Y, Yu G, Yu Q, Zhao X (2012) Protecting China's biodiversity: a guide to land use, land tenure, and land protection tools. In: Smith N (ed) The Nature Conservancy, Beijing

Land Administration law of the People's Republic of China 1998. Adopted at the 16th Meeting of the Standing Committee of the Sixth National People's Congress on June 25, 1986, amended in pursuance of the (Decision on the Amendment of the Land Administration Law of the People's Republic of China) made at the 5th Meeting of the Standing Committee of the Seventh National People's Congress on December 29, 1988 and revised at the 4th Meeting of the Standing Committee of the Ninth National People's Congress on August 29, 1998. http://www.asianlii.org.

Lin G, Ho S (2003) China's land resources and land-use change: insights from the 1996 land survey. Land Use Policy 20:87–107

Liu J, Tian H, Liu M, Zhuang D, Melillo JM, ZHANG Z. (2005) China's changing landscape during the 1990s: Large-scale land transformations estimated with satellite data. Geophys Res Lett 32:L02405

Long H (2014) Land use policy in China: Introduction. Land Use Policy 40:1–5

Land of Ministry and Resources (2009) Report of National land use change investigation. China Land Press, Beijing

MLR (2015) http://www.mlr.gov.cn

National Bureau of Statistic (2011) Communiqué of the National Bureau of Statistics of P.R.C. on major figures of the 2010 population census (No. 1). http://www.stats.gov.cn

National Bureau of Statistics of China (2019) http://www.stats.gov.cn/english/Statisticaldata/AnnualData/ (last visited 04.11.2019)

People's Daily Online (2009) China's urbanization rate expected to reach 48% in 2010. http://english.peopledaily.com.cn

PWN (2019) Encyklopedia. http://encyklopedia.pwn.pl. (last visited 02.11.2019)

Qiang X (2010) China vows to preserve arable land amid urbanization. http://www.chinadaily.com.cn (last visited 14.11.2014)

Sun Z, You L, Müller D (2018) Synthesis of agricultural land system change in China over the past 40 years. J Land Use Sci 13(5):473–479. https://doi.org/10.1080/1747423X.2019.1571120

The World Bank (2019) https://www.worldbank.org/. (last visited 04.11.2019)

Zhang K, Yu Z, Li X, Zhou W, Zhang D (2007) Land use change and land degradation in China from 1991 to 2001. Land Degrad Dev 18(2):209–219

Zhou X, Chen L, Zhang X (2005) Progress of rural landscape research in China. Geogr Geo-inform Sci 2005–02

Zhou Z, Chen Q, Wu S (2014) Study on Rural landscape design method under the background of the population diversification. Int Schol Sci Res Innov 8(4):256–261

Zuo L, Zhang Z, Carlson KM, Macdonald GK, Brauman KA, Liu Y, West PC (2018) Progress towards sustainable intensification in China challenged by land-use change. Nat Sustain 1:304–313

# Chapter 23
# The Rural Area in Historical Cities

**Bohdan Cherkes**

**Abstract** Cities and urban structures could grow thanks to the intensive development of agriculture and the so-called agricultural revolution (Mumford in The city in history. Its origins, its transformations, an its prospects. A Harvest Book Harcourt Inc., San Diego/New York/London, 1989). Historical cities have, therefore, always had a close relationship with agrarian areas and agrarian components of the internal structure of cities. The classification of the primary types of such urbanised agricultural areas and the investigation of the quantitative and qualitative dynamics of their growth based on cartographic sources resulted in the notion of the territorial settlement pyramid (TSP) together with a proposed method for building it. The TSP can be used to identify stages of the territorial development of cities and the relationship between urban and agricultural structures.

**Keywords** Historical cities · Rural areas · Territorial settlement pyramid

## 23.1 European Historical Cities and the Development of Cartography

In the mediaeval period, urban municipalities gradually lost control over the system of dependent villages; the land is increasingly often owned by individuals; and the city is actually bounded by the urban area and areas directly adjacent to it. One could venture that the city-state, perceived as a continuation of the urban traditions of the ancient classical period founded on land property, lost its importance in the Middle Ages. During capitalism, the intensification of agriculture, growth of industry, and

---

B. Cherkes (✉)
Department of Land management and Landscape Architecture, Faculty of Environmental Engineering and Land Surveying, University of Agriculture in Krakow, Balicka 253c, Krakow, Poland

Institute for Architecture and Design, Lviv Polytechnic National University, Bandera Str. 12/328, Lviv, Ukraine

© Springer Nature Switzerland AG 2022

J. Hernik et al. (eds.), *Cultural Heritage—Possibilities for Land-Centered Societal Development*, Environmental History 13,
https://doi.org/10.1007/978-3-030-58092-6_23

increased exchange led to an increase in the non-agrarian population living mostly in cities. Analysing this process, some scholars of the nineteenth century indicated that 'the process of urban–rural separation can also be perceived as the separation of capital from land ownership, as the beginning of a separate functioning independent of land ownership or the growth of capital; the beginning of ownership based solely on work and exchange' (Cherkes 1992, p. 50).

Iconographic sources from the period offer insight into the location of both the city and its rural areas within a single urban entity. The structure of agricultural areas owned by residents of the city is presented in figures and marked separately from extensive farmland by graphic means. The first tome with views of European cities was published in the late fifteenth century (1493) as the *Nuremberg Chronicle* (Reske 2011). The incunabula by Hartmann Schedel features landscapes of 31 cities. The beginning of the practice of regular representation of European cities in iconographic and cartographic sources dates back to the sixteenth century. This period saw the publication of the world-famous *Cosmographia* by a Swiss humanist Sebastian Münster (1545) (Wessel 2004) containing descriptions of cities and the atlas of cities of the world *Civitates Orbis Terrarum* in six volumes by Georg Braun and Frans Hogenberg in Cologne in 1572–1612 (Fig. 23.1; Braun and Hogenberg 2011). Apart from large tomes, albums and even single sheets with landscapes of individual cities were published. They became particularly popular in the seventeenth century. This century saw drawings of almost all historical cities in Europe.

Regardless of the excellency of such images of views and panoramas of fifteenth to seventeenth century cities, it is not possible to base any quantitative measures of urban spatial structure on them. Most of these invaluable works of art reflect the personalities of their authors. In some cases, the artist strove to immortalise the urban solemness of the core, the central part of the city. Sometimes they focused on the beauty of the natural landscape, or the abundance of fruit in gardens and orchards, always omitting some important details because of subjectively notable features.

The cartographic methods for producing city maps were improved at the end of the seventeenth century. Therefore, humanity gained numerous perfectly made, reliable, and clear maps in the eighteenth and nineteenth centuries (Fig. 23.2; Bogen and Thürleman 2009, p. 121). An analysis of individual maps yielded detailed quantitative characteristics of agricultural areas in the urban spatial structure in the investigated time frame. The in-depth analysis involved 103 maps.

## 23.2 Urban and Rural Components of the City and the Methods for Their Measuring

The agrarian component of the spatial structure of the city was considered as a whole comprising four components: large agricultural land, allotment gardens owned by residents of the city, holiday cabin areas, and villas with allotted parcels of land. The surface area of urban zones was calculated using a 2 × 2 mm planimetric grid. As

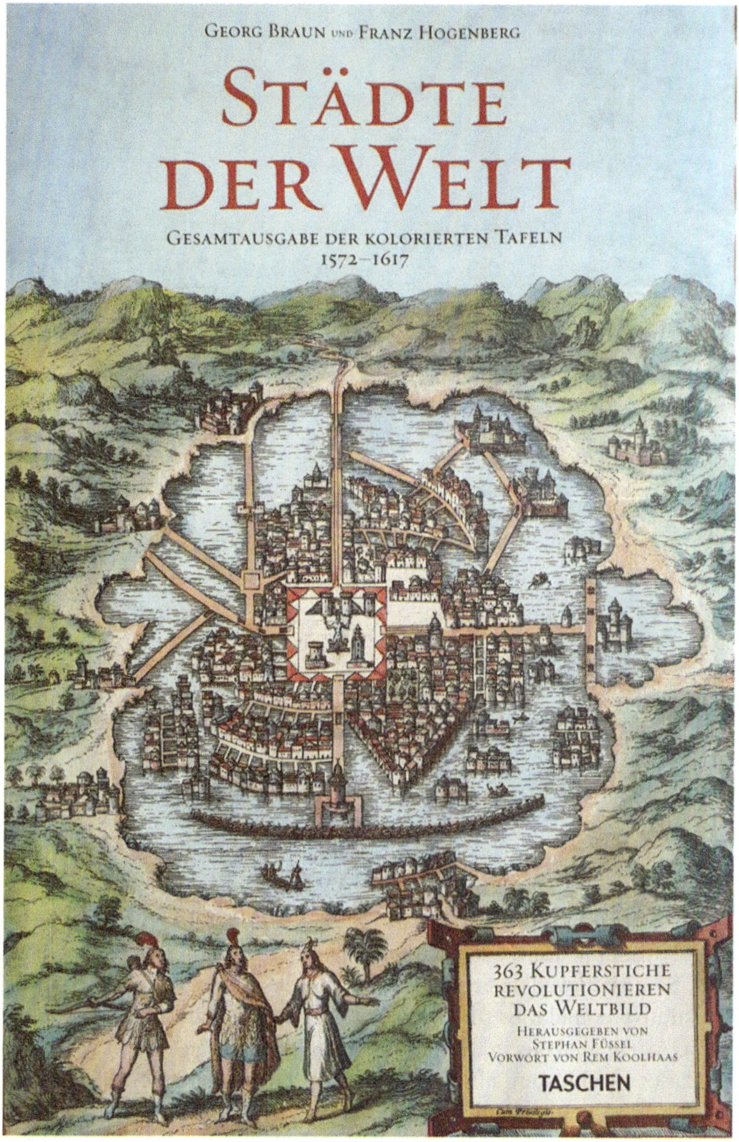


**Fig. 23.1** The title of the first atlas of cities of the world 'Civitates Orbis Terrarium' written and edited by Georg Braun and Franz Hogenberg from 1572 to 1612. *Source* Braun and Hogenberg (2011)

**Fig. 23.2** The map of Rome by Giovanni Battista Falda, published by Giovanni Giacomo de Rossi in 1667. It is a perfect illustration of the presence of agricultural areas in the urban structure of Rome in the second half of the seventeenth century. *Source* Bogen und Thürlemann (2009, pp. 120–128)

historical maps were made according to various scales, the square defined above is assumed the standard unit for calculations. All results are percentage values, where 100% is the total surface area of the city.

The issue of administrative and economic boundaries of the city has been discussed frequently. Some believed that economic boundaries of a locality stretch further than its administrative boundaries (the immediate territorial and economic surroundings of the city). As the analysis in the current paper demonstrates, however, administrative boundaries of cities are always adjusted to their economic impact zones. The boundaries are constantly changing to encompass the largest possible territory of the urban community. This way, the city map produced at any moment in history reflects the best of what people at the time understood as the city and its boundaries.

Analysis of maps of seventeenth to nineteenth century cities revealed that their characteristic feature is their cyclic development. Each stage of the cycle (beginning-homoeostasis-end) has its dynamics between the city and agricultural areas. At first, the agricultural space dominates the urbanised space. During homoeostasis, the structure of agricultural and urbanised areas is balanced by large areas of housing developments with backyards. At the final stage of the cycle, developments displace agriculture so that gardens and holiday cabins of city residents are located outside the administrative boundaries of the city. Then, a next, similar cycle begins.

In the effort to determine quantitative characteristics of the agricultural component of the urban space, authors attempted to encompass the broadest possible geographical extent of European cities. This paper focuses on the most developed cities on the continent as they are the best urban culture models in their respective countries and of their respective period. The investigated cities included Florence and Milan (Fig. 23.3), Rome, Berlin, Paris, Saint Petersburg, and Vienna (Fig. 23.4). The analysed maps of these cities were produced mostly in the eighteenth and the first half of the nineteenth centuries. The map of Saint Petersburg is dated to 1868. The analysed sources belong, therefore, to the period of active development of the capitalist system.

The study shows that all these cities had an agricultural component regardless of their geographical location or social, economic, or national differences. The largest percentage of agricultural areas was identified within the boundaries of Berlin (55.35% of the total area of the city) and Rome (52.78%). Florence also demonstrated a high share of agricultural areas (45.27%). In none of the cities did the agricultural component occupy less than 30% of the surface area. Even in the new capital of the

**Fig. 23.3** The agricultural component in the urban planning structure of Florence (1) and Milan (2) in the mid-eighteenth century: 3—large agricultural land, 4—garden plots, 5—houses with adjacent gardens. *Source* Own work based on *Cartes et Plans de quelques villes en Italie. Library of the National Academy of Sciences of Ukraine in Lviv (Former Ossolineum), Department of Cartography, Inventory Number 7984–8000*

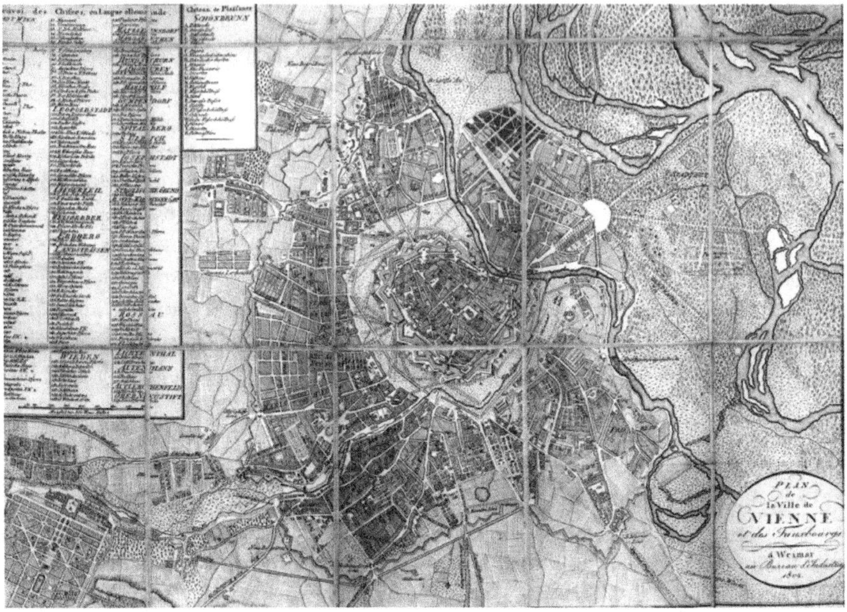


**Fig. 23.4** Map of Vienna with its suburbs from 1802 showing a characteristic combination of urban and agricultural areas in the structure of a large European city in the early nineteenth century. *Source* Schweizer (2010, p. 39)

Russian Empire, Saint Petersburg, where the latest urban-planning ideas of the time were deployed, the share of agricultural areas was 32.41%.

The most characteristic element of the agricultural component were houses with adjacent backyard gardens (Florence, Milan, Berlin), even though it depended on the development cycle of the city and was variable. Saint Petersburg was, for example, already dominated by suburban garden plots, while Rome preferred agrarian villas, most likely in an attempt to stay true to its ancient traditions.

All this further backs the authors' understanding of the cyclical nature of city development (beginning-homoeostasis-end), where the agricultural component is a primary diagnostic. It is only its subcomponents that change, and its total value remains constant above 30% of the area of the city. Therefore, having identified the subcomponents of the agricultural component correctly, one can determine the specific stage of the city's development cycle.

Agricultural areas of cities provided not only food for their residents but aesthetic inspiration as well. When analysing historical gardens, an academic of the Russian Academy of Sciences, Dmitry Likhachov warned against 'overstating its practical importance and juxtaposing it with the aesthetic role of gardens. In the West, gardens, their "green offices", had large numbers of fruit-bearing trees and shrubs as well. Fruit-bearing was a fundamental part of garden aesthetics in every century. The fruit was considered just as beautiful as the flower with its aesthetically pleasing appearance' (Likhachov 1982a, pp. 38–46).

The structure of maps of historical European cities exhibits a great diversity of garden plot organisation arrangement techniques. Apart from a general focus on the species and varieties of cultivated fruit and vegetables, the structure displays attention paid to the configuration of agricultural plots, their sizes, and planting arrangement. Urban gardening made use of various geometric drawings. They suggest that agricultural areas significantly affected the process of development of the urban environment (Cartes et Plans de quelques villes en Italie; Cherkes 1992, p. 57).

Plots surrounded by only low shrubbery made up groups. The groups were merged into larger blocks with paved paths inside. Rather often, the block had a small surface area dominated by a single tree. It was under the tree that plot owners gathered in line with the tradition of holding meetings under trees on central squares. Several blocks made up large clusters in cities with a single parcel of land as the starting point for the architectural and planning arrangement. Its configuration affected the outline of the general layout as well. Plots were of various shapes. Some were rectangular, triangular, trapezoidal, or G-shaped. The actual practice of agricultural plot planning in historical cities was much broader and more varied, however.

An essential factor in shaping the aesthetically pleasing appearance of agricultural land was to determine the direction of planting as it could reinforce or demolish the composition; make parcels stand out or blur in with the background. The variety of leaves and fruit subjected to the strict geometry of the planning drawing created a particular world of urban agriculture elevating it to the level of an aesthetic domain. Such a devoted love for the beauty of farm plants and perfected plant composition and arrangement architecture could be conceived only if the average resident exhibited a high level of artistic culture.

Retaining the Renaissance traditions, the working population of mid-eighteenth century Italian cities was capable of introducing a single element into agricultural plot planning, which was a sublime sense of beauty. The harmonious atmosphere of historical cities nourished the same attitude towards items of everyday use and such objects as land parcels owned by residents of cities.

Still, the architectural organisation of agricultural land in Italian cities was not exceptional. It is present in the axonometric projection of Paris from 1734 known as the Turgot map of Paris (Plan de Paris 1734; Sarazin 2005). It was popular in the development of eighteenth to nineteenth century Russian cities as well. Regions where urban agricultural areas were approached from an artistic angle can be found in all Europe. Regrettably, this widespread embodiment of popular architectural and landscaping creativity based on agriculture has not been studied.

The rural element was a popular component of historical parks. Its detailed description can be found in a letter from Ivan Turgenev to Gustave Flaubert where he depicts his journey 'on avenues of an old rural city full of countryside smells, wild strawberries, sounds of sleepy birds, sunny light and shadow surrounded by two hundred dessiatins of cereal swaying in the wind. Marvellous! One inadvertently halts and immerses oneself in a kind of a state of immobility, solemnity, infinity, and stupidity that merges life, fundamental yearnings, and God' (Likhachov

1982b, p. 40). Landscape parks were a logical combination of the agricultural and environmental elements in a single urban whole.

We now return to the city to consider one of its most challenging problems of the interrelationship between residential developments and farmland areas owned by residents. As was mentioned before, the initial stage of historical city development was dominated by low multi-family residential buildings with adjacent gardens. Low developments were gradually displaced by multi-story buildings, but the point of keeping gardens as close to the houses as possible was still respected. This resulted in districts with typical urban multi-story buildings accompanied by small gardens (plots) in courtyards. Farming there was not cost-efficient. The next stage of the process of compaction of the urban structure eliminated such plots from residential districts. The result was very compact districts with mere patches of greenery left after thriving farming.

Nevertheless, residents of cities still needed agricultural products and areas. Hence spaces outside of city limits where they cultivated fruit and vegetables. This gradually led to whole zones with holiday cabins. The compact urban development induced an increase in suburban agricultural areas owned by residents of cities. There was now a gap between the agricultural and urban components that used to build a single urban whole.

The present historical analysis produced several basic types of urban interrelations between the residential development and areas farmed by residents of cities (Fig. 23.5). This suggests that the agricultural component is interrelated with all four basic types of residential developments. In the manor housing structure (lower level), it takes the form of a backyard of 600–1200 $m^2$. In the terraced housing structure, the surface of the plot is reduced to 100–600 $m^2$. In the low multi-story development structure (three to five stories), the plot surface is much smaller, from 10 to 100 $m^2$. Already for this type of development, some residents seek to compensate for the lack of agricultural area outside of their development. In the case of the three types of development, the home and plot are within the same territorial unit, but for the fourth type, with high multi-story buildings or dense mid-rise buildings, the agricultural component is excluded from the structure. It is located outside of its boundaries (Fig. 23.5), which leads to allotment garden complexes. This way, the relationship between residential buildings and garden plots owned by residents of cities exists on two territorial and urban levels: 1—within one territorial unit; 2—when the territorial gap is created, and the agricultural and urbanised components are separated. In both cases, farming activities by residents of cities should be considered an important factor shaping the environment of not only the historical but also modern city.

Human settlements as the source and entry point for further development exist in specific surroundings. In their attempt to tame them, people face not an abstract space but a tangible landscape, which often sets its own rules and principles. In most cases, it is either natural areas (forest, steppe, coastal region) or agricultural land used by farmers. People come across both types of open space when creating their settlements. The development of any settlement is related to the natural or agricultural component of the landscape. This way, the general structure of the area of each settlement can

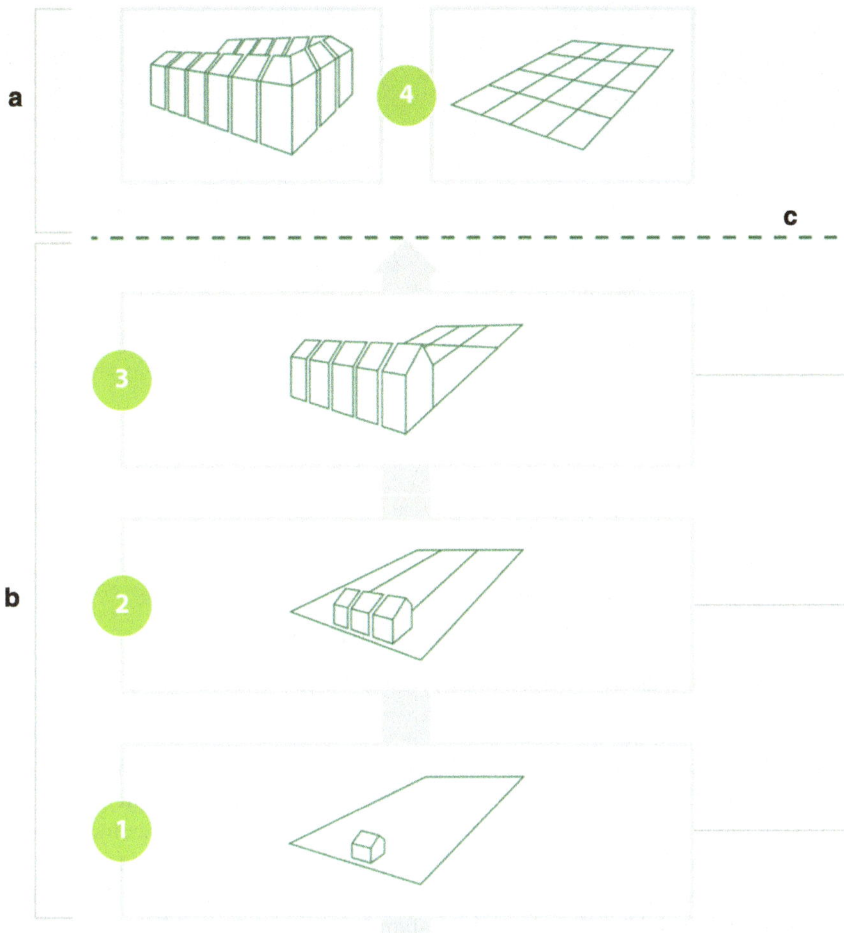


**Fig. 23.5** The basic types of urban interrelations between residential development and areas farmed by residents of cities: **a**—territorial unity, **b**—no territorial unity, **c**—urbanisation line, 1—manor housing, 2—block development with backyards, 3—compact terraced development with small backyards, 4—urbanised areas with allotment gardens outside of the area. *Source* Own work

be divided into two major groups: the area related to the agricultural component and the area without the agricultural component (Fig. 23.6).

Classified according to the development intensity, the areas produce four levels of development: I—green areas; II—manor housing areas; III—low-rise development districts; IV—compact development districts (Fig. 23.6). This division stems from the natural, historical human activities regarding the organisation of the lifestyle and people's surroundings. It reflects the evolution of the activity from the exceptional dominance of the manor housing development in farming settlements to the multi-dimensional method of space formation found in cities. The series was based on

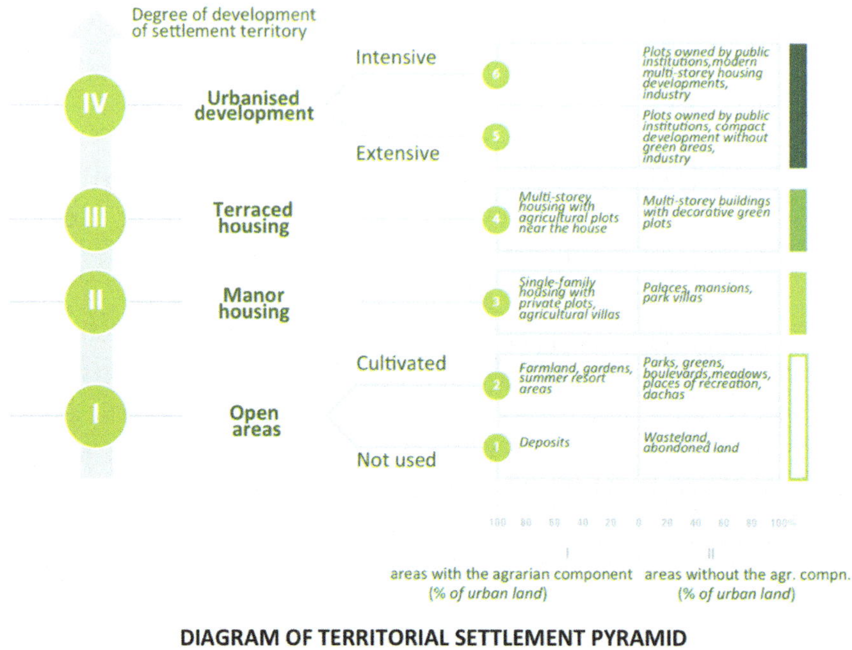


**Fig. 23.6** Diagram of the territorial settlement pyramid. *Source* Own work

historical analysis of the developmental stages of the residential environment as the primary form of the architectural transformation of open spaces in settlements.

One of the main properties of the proposed hierarchy of urban areas focusing on the level of urbanisation is the fact that its upper levels cannot be reduced to the lower levels. It is improbable, for example, for multi-story housing to be demolished and replaced by neighbourhoods of single-family houses or for gardens to eliminate a historical residential district. At the same time, the upper levels of development grow at the expense of the ones below them. Single-family housing takes the place of agricultural land; multi-story multi-family residential buildings replace single-family houses, etc. Urbanised development can use all three lower levels of the hierarchy. It can be prevented only by a high social, cultural, religious, or environmental value of objects that would need to be demolished. This issue resembles the blood group compatibility problem in medicine; one group can be used to help all others and conversely, the fourth one can only use others. In the discussed case, all forms of development can grow at the expense of open areas, while compact (urbanised) developments grow using all others.

Another essential property of the urban area hierarchy is the binary nature of the basic types of territorial development; in other words, the simultaneous development in the park and landscape system and agricultural landscapes. For instance, at the level of open agricultural spaces, farmland and gardens in the agricultural frame of reference correspond to parks and green squares in the park and landscape frame of reference; at the level of single-family housing, single-family houses with agricultural plots correspond to houses with plots of decorative plants in the form of palaces, mansions, or park villas. The same trends can be identified for low-rise terraced housing (Fig. 23.6).

Areas with compact (urbanised) development are the peak of the transformation of the natural habitat. It can be broken down into two stages: a lower one with areas of extensive urbanised development and a higher one with areas of intensive urbanised development. The first stage is visible in densely populated quarters of historical cities, the other one, modern multi-story residential districts. Note that the urbanised level of land development includes industrial premises and public buildings and structures. Can these areas be considered binary? First of all, every urbanised development, and its extensive form, in particular, strives to the maximum negation of the open space within its structure. It is reduced here to an essential existential need. Even if patches of vegetation remain, they are not enough, and a dynamic relationship with nature is replaced by passive contemplation.

The situation is improving in modern multi-story residential districts. Park and landscape areas increase significantly, but there is still no communication with nature. Occasional farming activities by residents of cities around their homes should be considered a desired rather than an actual sign of the binarity. This way, the urbanised development excludes the agricultural component depriving the area of the binarity. It is, nevertheless, compensated for outside of its boundaries as holiday cabins and allotment gardens after it disappears from a specific urban residential unit (quarter, development). Hence, the deprivation of an area of the duality of the agricultural and park component contributes to ecological impoverishment of the urban environment.

## 23.3 Territorial Settlement Pyramid and Its Application in the Determination of the Urban Development of the City

The study resulted in the development of the territorial settlement pyramid (TSP) (Fig. 23.6). Its primary goal is to comprehensively determine the urban environment based on identified properties of the urban area hierarchy of coexisting agricultural and park components (depending on the degree of development and the binarity of the dualism). By putting the sizes of urban areas with the agricultural component on the left-hand side of the zero vertical line and representing sizes of areas without the component right to the line, a specific outline is created that is the territorial settlement pyramid. The sizes of areas used in the pyramid are expressed as the percentage of the total area of the city, which is 100%. The outlines of pyramids reflect a whole array of social, economic, and ecological urban processes. An analysis of the properties of large sets of such pyramids can improve the objectivity of the assessment of the urban environment condition.

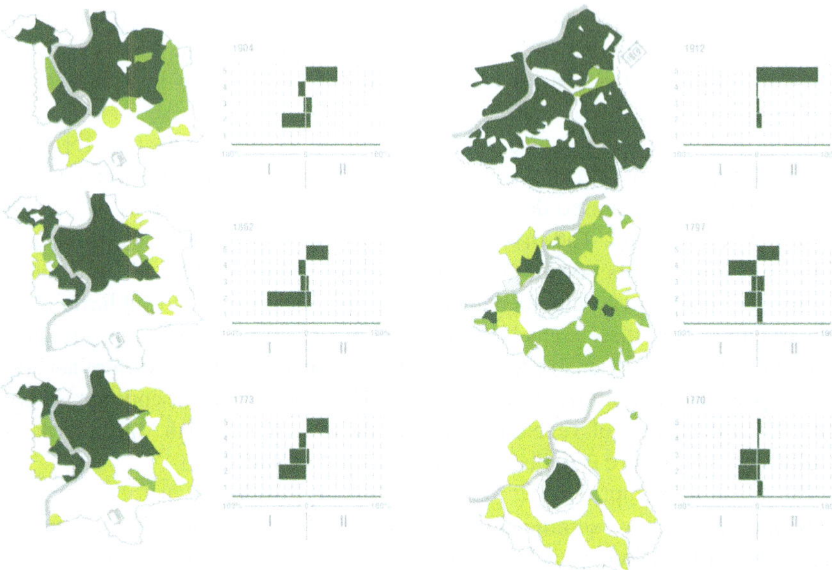

**Fig. 23.7** The evolution of the territorial pyramids of Rome (1773–1904) and Vienna (1700–1912) during the development of their urban structures. *Source* Own work based on Roma (1773, 1862, 1904), Wien 1712, 1797, 1912)

The present study analysed selected historical cities at their various stages of development using the territorial settlement pyramid method. Data regarding surface areas were obtained from city maps produced in the eighteenth and early twentieth centuries. The following cities were investigated: Rome (Fig. 23.7; Roma 1773, 1862, 1904), Vienna (Fig. 23.7; Wien 1712, 1797, 1912), and Cracov (Fig. 23.8; Kraków 1836, 1860, 1912). The primary objective was to determine the nature of the change in the territorial settlement pyramid in time. The work focused particularly on the agricultural component in the cities. Let us consider the results of applying the territorial settlement pyramid to each of the cities.

In the eighteenth century, Rome was a leading European capital. Its maps dated to that period show it within its ancient boundaries, up to the Aurelian Walls built in late third century AD. After the destruction in the 'Dark Ages', Rome's former greatness could not be restored quickly. The Coliseum was turned into gardens, fora, into pastures for goats. The mediaeval city of Rome developed from scratch. The Aurelian Walls contained agricultural areas of city residents that formed an urban whole with urbanised developments in the central part of the city. According to the map from 1773, the area of the agricultural component was 61.34% of the whole territory of Rome. Its size did not change after 89 years, which is demonstrated on the territorial pyramid for 1862 (Fig. 23.7), even though the growth of capitalistic system affected it visibly.

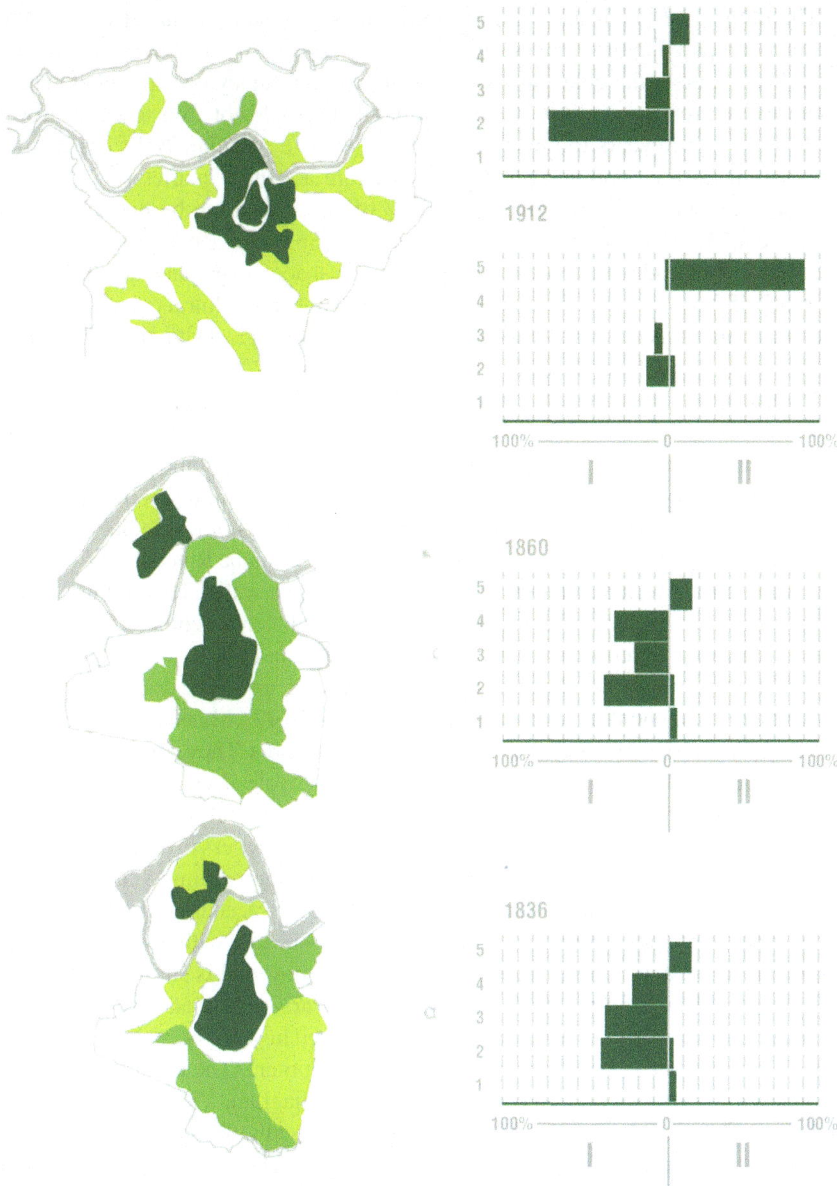


**Fig. 23.8** The evolution of the territorial pyramid of Cracov (1836–1912) during the development of its urban structure. *Source* Own work based on Kraków (1836, 1860, 1912)

The number of small agricultural villas was significantly reduced in the 1862 pyramid compared to the 1773 situation, while the surface of large agricultural plots grew. It was due to the concentration of land and bankruptcy of small landowners. Rome was so great during its best ancient times that it could not grow to that size even in 1904 after 1500 years following its destruction. The Aurelian Walls easily fitted the Termini railway station with its infrastructure and new quarters of urbanised developments and still ‘protected’ over 40% of areas taken up by the agricultural component. All the three pyramids (from 1773, 1862, and 1904) demonstrate an uncompromising advantage of open agricultural spaces over park and landscape areas (47.55% vs 11.64%). To conclude, areas with the agricultural component were of crucial importance for the shaping of the general ecological situation of Rome in the period of the eighteenth to early twentieth century.

Mediaeval Vienna was built in place of a former Roman military camp. The place grew. As the capital of the Habsburg Empire, the city became one of the primary hubs of European culture for a long time. How did its territorial pyramid evolve from the eighteenth century to the early twentieth century? At the beginning of the eighteenth century, Vienna managed to grow outside of the boundaries of its mediaeval cradle. It left undeveloped the esplanade often flooded by the Danube River and Wien River and began an active development within new defensive walls. This initial stage is depicted by the first territorial pyramid (the early eighteenth century) in Fig. 23.7. The fortifications added a large surface of open spaces and single-family houses into the structure of the city that became the primary reserve for future development. The agricultural component of Vienna at the time was 48.74%. Mansions, palaces, and private parks of Vienna’s society occupied a significant area. They covered 16.75% of the total surface of the city. This initial stage of the new cycle of Vienna’s development was, of course, prepared by the whole course of its past evolution.

The 1797 pyramid and diagram show the urbanisation process in the clearly limited space within the new city walls. The upper levels of the pyramid started to grow actively compared to the previous one. It is mainly the terraced housing that expands at the expense of agricultural land and single-family housing with agricultural plots. It remains in the agricultural landscape category for now and is similar to the quarters with small farming plots in the backyards of Paris discussed earlier. Still, the small agricultural plots are developed, and the terraced housing becomes extensively urbanised areas at the next stage of the city’s development. The urban environment becomes wholly urbanised. According to the 1912 diagram, Vienna was unable to build its railway station within the city walls as opposed to Rome. The agricultural component disappeared from the city, and the total number of open park and landscape spacès was about 7.85% of its territory. Even the Esplanada floodplains were drained and developed. It was the time of the cycle ‘threshold’ for the administrative boundaries of the cities at the time.

On the other hand, one could venture that there is no such state for the city as a whole. The territorial pyramid for Vienna in the early eighteenth century, restricted to its boundaries of the first defensive walls and the esplanade, is similar to the 1912 pyramid during the maximum urbanisation level. It would be a mistake to consider it explicitly as an objective and holistic description of the whole urban environment.

From the ecological standpoint, the pyramid will always be balanced by a pyramid with more developed lower agricultural, park, and landscape levels that represent open spaces. At the same time, the area of the agricultural component of the city will not be less than 25% of its total surface. This reasoning is supported by the territorial pyramid for early eighteenth century Vienna.

Cracov has always been a leader of European urban culture. It assimilated the best of the urban heritage of the West and the East. Its territorial pyramids for 1836, 1860, and 1912 present the dynamics of changes in its urban environment during another development cycle (Fig. 23.8).

The other 1912 pyramid shows Cracov in its new administrative boundaries at the beginning of its next evolutionary round. According to the figure, open spaces in the city were created mostly by agricultural land. It was also the main environmental contributor. Agrocoenoses clearly dominated biocoenoses here. There are many similarities between the territorial pyramids of Cracov and Vienna, although the former had more of an agricultural component than the capital of the Austro-Hungarian Empire. What is particularly interesting is the distribution of specific areas within the cities in 1912 as it was the time they both completed their developmental cycles and started new ones. The territorial pyramid of Vienna is shown only at the stage of the end of its evolutionary round when its urbanisation level was the greatest. It apparently repeats the similar pyramid of Cracov for the completed developmental cycle within its old administrative boundaries. A particular conventional nature of the structure and impossibility of objective existence of such a pyramid is conspicuous, however. A city as a whole cannot lose open spaces almost completely. In its current administrative boundaries, the city will always make up for lost space by absorbing new areas and assigning them the function of the open space missing in its internal urban structure. The growth of urban development and open space are always a kind of chain reaction the two opposite phases of which can be seen on the 1912 territorial pyramids of Cracov.

## 23.4 Conclusions

The study has demonstrated that the agricultural component has always occupied the fundamental role in the lives of residents and shaping of the urban structure not only in the Antiquity and Middle Ages but also in the time capitalist cities were born and grew. Modern cities exhibit similar developmental trends. The introduction and application of the territorial settlement pyramid (TSP) facilitate the identification of the stages of territorial development of cities. They demonstrate that an end of a stage of the urban evolution cycle within certain territorial boundaries and the maximum urbanisation of the areas within are always ecologically balanced by a zone where new open space is created outside of the city. When this happens, the administrative boundaries of the city are bound to change, starting a new developmental cycle. At the same time, the area of the agricultural component in the urban landscape usually does not fall below 25% of the total area of the city.

## References


Bogen S, Thürlemann F (2009) Rom. Eine Stadt in Karten von der Antike bis heute. Primus Verlag/Wissenschaftliche Buchgesellschaft WBG, Darmstadt

Braun G, Hogenberg F (2011) Civitates Orbis Terrarum. Städte der Welt. 363 Kupferstiche revolutionieren das Weltbild. Gesamtausgabe der kolorierten Tafeln 1572–1617. Herausgegeben von Stephan Füssel nach dem Original des Historischen Museums Frankfurt. Vorwort von Rem Koolhaas. TASCHEN, Köln

Cartes et Plans de quelques villes en Italie. Library of the National Academy of Sciences of Ukraine in Lviv (Form.Ossolineum), Department of the Cartography, Inventory Number 7984–8000

Cherkes B/Черкес Б. (1992) The Town and its Agrarian Environment/Город и аграрная среда. Svit/Свит, Lwow/Львов (in Russian)

Krakau (1860) Plan von Krakau mit Podgórze und der nächsten Umgebung. Olmütz

Kraków (1836) Plan miasta Krakowa w obrębie okopów. Library of the National Academy of Sciences of Ukraine in Lviv (Form.Ossolineum), Department of the Cartography, Inventory Number 1043. 465x374mm

Kraków (1912) Plan Wielkiego Krakowa. Rysował Kazimierz Stolecki. Muzeum Historyczny miasta Krakowa

Likhachew D/Лихачев Д (1982a) The garden and the culture of Europe/ Сад и культура Европы. In: Decorative Art of the USSR, N3 / Декоративное искусство СССР, N3

Likhachew D/Лихачев Д (1982b) The garden and the culture of Russia / Сад и культура Европы. In: Decorative Art of the USSR, N12 / Декоративное искусство СССР, N12

Mumford L (1989) The city in history. Its origins, its transformations, an its prospects. A Harvest Book Harcourt, Inc. San Diego/New York/London

Plan de Paris (1734) Comencé l'Année 1734–1739. 20 sectional maps each of them 800x500mm. Library of the National Academy of Sciences of Ukraine in Lviv (Form.Ossolineum), Department of the Cartography, Inventory Number 726K

Reske Chr (2011) Schedelsche Weltchronik, publiziert am 11.04.2011; in: Historisches Lexikon Bayerns, URL: http://www.historisches-lexikon-bayerns.de/Lexikon/Schedelsche_Weltchronik (28.07.2019)

Roma (1773) All Emo, e Rmo, Pne. Colmo Il Sig. Cardinale Gio, Carlo Boschi. Emo, e Rmo Principe. Roma il di primo del 1773. Umo Dimo et Obbmo Seruitore Ignazzo Benedelti. Library of the National Academy of Sciences of Ukraine in Lviv (Form.Ossolineum), Department of the Cartography, Inventory Number 2450. 460x683mm

Roma (1862) Pianta della citta di Roma publicata nell'anno MDCCCLXII. Library of the National Academy of Sciences of Ukraine in Lviv (Form.Ossolineum), Department of the Cartography, Inventory Number 4806. 828x597mm

Roma (1904) Pianta di Roma.1904. Library of the National Academy of Sciences of Ukraine in Lviv (Form.Ossolineum), Department of the Cartography, Inventory Number 9721. 510x390mm

Sarazin J-Y (2005) Le Paris des Lumières d'après le plan de Turgot (1734–1739). RMN, Paris

Schweizer H (2010). Historic maps and views of Vienna, h.f.ullmann, Potsdam

Wessel G (2004) Von einem, der daheim blieb, die Welt zu entdecken. Die Cosmographia des Sebastian Münster oder wie man sich vor 500 Jahren die Welt vorstellte. Campus-Verlag, Frankfurt am Main

Wien (1712) Grosser Plan von Wien. Library of the National Academy of Sciences of Ukraine in Lviv (Form.Ossolineum), Department of the Cartography, Inventory Number 2332. 590x470mm

Wien (1797) Grundriss der K.K. Haupt und Residenzstadt Wien mit ihren Vorstädten nach den neuen Hausnummern. Library of the National Academy of Sciences of Ukraine in Lviv (Form.Ossolineum), Department of the Cartography, Inventory Number 2417. 1385x905mm

Wien (1912) Prospect und Grundriss der Kaiserlichen Residenzstadt Wien mit anliegender Gegend und Neuen Linien und Vorstädte. Library of the National Academy of Sciences of Ukraine in Lviv (Form.Ossolineum), Department of the Cartography, Inventory Number 1436. 565x480mm

# Chapter 24
# Sustainable Economic Development and Cultural Landscapes: Some US-Poland Comparisons and Connections, with Global Context

**Edward Sankowski, Betty J. Harris, and Józef Hernik**

**Abstract** This paper discusses some US-Poland comparisons (with global context), emphasizing energy and environmental problems. The paper builds on prior research by Saskia Sassen and others, including Jozef Hernik, and also interprets original site-specific empirical research by us about Poland and the US. This paper proceeds from the following ideas. (1) Urbanization is a major factor in globalization. (2) Urbanization and globalization in their main current forms damage the environment, including rural environments, tend to marginalize and subordinate rural compared to urban development, and are inconsistent with sustainable development. (3) Sustainable development is a defensible concept/value. (4) Economic and environmental policy must address problems of energy policy to align itself with sustainable development. (5) "Cultural landscapes" initiatives that are explicit in Europe have analogues in the US; whether in Europe or the US, cultural landscape initiatives should show themselves to be consistent with sustainable development. (6) Cultural landscape initiatives in Europe and their US analogues must be subjected to critical examination that takes account of both economic development issues and environmental/energy issues, and contributes to reconciliation with sustainable development. In this paper, we analyze in a comparative discussion selected areas in Poland (Małopolska) and the USA (Oklahoma). We use prior research by Józef Hernik about the importance of economic development for protection and improvement of rural or quasi-rural

E. Sankowski (✉)
Department of Philosophy, College of Arts and Sciences, University of Oklahoma, Dale Hall Tower, Norman, OK 73019, USA
e-mail: esankowski@ou.edu

B. J. Harris
Department of Anthropology, College of Arts and Sciences, University of Oklahoma, Dale Hall Tower, Norman, OK 73019, USA
e-mail: bharris@ou.edu

J. Hernik
Department of Land
management and Landscape Architecture, Faculty of Environmental Engineering and Land Surveying, University of Agriculture in Krakow, Balicka 253c, Krakow 30-149, Poland
e-mail: rmhernik@cyf-kr.edu.pl

© Springer Nature Switzerland AG 2022 
J. Hernik et al. (eds.), *Cultural Heritage—Possibilities for Land-Centered Societal Development*, Environmental History 13,
https://doi.org/10.1007/978-3-030-58092-6_24

cultural landscapes not far from Krakow, Malopolska, in Poland, particularly the still quasi-rural setting of the municipality of Miechow. This research by Hernik is supplemented by field work and information gathered by Sankowski, Harris, and Hernik, particularly about Pawhuska, (Osage County), Oklahoma, not far from Tulsa, Oklahoma, in the US. Two other sites in Malopolska and Oklahoma are also discussed much more briefly. Most generally, we aim to put the comparison of US and Poland sites in the context of certain basic conflicts about energy and environments within current globalization.

**Keywords** Sustainable development · Culture · Landscapes · Energy · Environment · Globalization

## 24.1 Introduction

We focus on facts and values concerning sustainable development in planning policy and management decisions about the impact of energy on the environment, including cultural landscapes (especially rural or quasi-rural landscapes) as a major aspect of the environment. We take as one key topic, choices about fossil fuels versus renewable energy sources, and the relevance of globalization for choices about energy and the environment. We take account of the respective roles of government and the market in a "developed" capitalist country (the US) as compared with an increasingly market-oriented post-communist country (post-1989 Transformation) that might be called by some "developing" (Poland), but that is progressing by some reasonable criteria for development. This is an age in which "market" institutions have been globally increasingly in the ascendancy, with not always well understood varying meanings about "markets", depending on historical and cultural contexts.

We claim that global trends result in competing types of energy and environmental advocacy and influence from the endogenous features and exogenous surroundings of each of two locations, Miechow, Poland and Pawhuska, US. Pawhuska represents a fossil-fuel "cultural landscape" with a false image circulated by its advocates (the Tallgrass Prairie Preserve with oil and gas production and fracking nestled within it) and communicated to the rest of the world. That image implicitly conflicts with sustainable development locally and worldwide and is allied with a general trend that its advocates (and loosely affiliated allies) hope to export to many other places, e.g., via fracking for oil and gas. Miechow (at least in some of its rural activism) represents a modest NGO-led rural renewables initiative that offers itself as an alternative to pollution in Krakow, and that is part of an EU where there are even corporate entities such as Enel seeking to extend themselves against opposition into Oklahoma via wind-power strategies. Miechow has the problem of finding an identity for itself that would make it a sustainable and innovative alternative to serving as a merely subordinate area, subsidiary to the urban expansion of Krakow.

One of Józef Hernik's previous research papers is a record of the initial work referred to and extended here on protecting and promoting cultural landscapes in rural areas by economic means. That paper can ideally be better understood if read in the context of his extensive work on cultural landscapes, much of which is cited but not directly examined in this paper. In his paper, Hernik studied pilot projects for the protection and preservation of the cultural landscapes of rural areas in which no activities had so far been undertaken to provide this protection (Hernik 2012). The pilot projects were chosen based on a particular methodological approach (Hernik 2008). To begin with, extensive data on the rural landscapes of Małopolska was methodically researched (broad approach). Based on this analysis, two characteristic communities in Małopolska were chosen for detailed research: Miechów and Wiśniowa. Next, certain aspects of these characteristic landscapes were selected for development as part of pilot projects in the areas concerned (selective approach). In this way, it was possible to test for the most appropriate action to take to conserve and also further to develop the cultural landscapes of these municipalities. (We regard this type of *test* as genuinely empirical, but as presupposing value judgments). In Miechów, the action selected concerned renewable energy, while in Wiśniowa it was agro-tourism. In the studies, it was stated that activities and investments in these areas were based on the assumption that the great potential for development in Małopolska's rural cultural landscapes had not been exploited, partly because of lack of infrastructure (Hernik 2009b; Hernik et al. 2013). The pilot projects, Hernik argued, showed that it is possible to preserve the cultural landscapes while obtaining economic benefits. However, arguably, often the fostering of economically viable cultural landscapes will require active government support, not simply a hands-off "free market" approach (a point not denied by Hernik).

For discussions and illustrations of many issues about (and examples of) cultural landscapes, see: the Council of Europe website, and in particular the European Landscape Convention (2000); Józef Hernik and Jacek M. Pijanowski, eds, Cultural Landscape—Assessment, Protection, Shaping (Hernik and Pijanowski 2007); Józef Hernik, ed., Cultural Landscape—Across Disciplines (Hernik 2009a); Krzysztof Gawroński and Józef Hernik, eds, Spatial Planning and Management as an Instrument of Shaping Cultural Landscapes (Polish and International Experiences) (Gawroński and Hernik 2010); Artur Radecki-Pawlik and Józef Hernik, eds, Cultural Landscapes of River Valleys (Radecki-Pawlik and Hernik 2010); Olaf Kühne, Krzysztof Gawroński and Józef Hernik, eds.,Transformation und Landschaft (Kühne et al. 2015).

Energy policy both before and after the dissolution of the Soviet Union and market "reforms" in China has been particularly important as a factor explaining how contemporary globalizing political economy functions. In energy policy, an increasingly important economic developmental choice is in the relative roles of fossil fuels as contrasted with "renewables". One reason for the importance of this choice is the impact of energy on the environment. Everywhere, including rural areas, one major dimension of the environment can be called its cultural landscapes, a term we use here. And in very many parts of the world, *development* is too easily conceived as if there is a necessary connection with ever-expanding urbanization. We wish to align ourselves with challenges to this idea.

A major source of environmental degradation is ill-conceived energy policy and use, in particular over-reliance or reckless reliance on fossil fuels rather than renewable energy sources. One central concern in this paper is how to change this, to promote development of renewable energy, and reduce reliance on fossil fuels. Issues about such change are particularly relevant to the Miechów example and also to Pawhuska, Osage County, Oklahoma in the USA. In both Miechów and Pawhuska, planning and management issues arise about economic development, cultural landscapes, and fossil fuels in relation to renewables. Also, both areas are rural (or quasi-rural). For useful visually relevant maps and images, see the Google web address listed among references for this paper.

## 24.2 Methods

In this present follow-up multiple-investigator, multiple-author interdisciplinary paper, we do not attempt to repeat Hernik's single-author and other more long-term methodologies (Hernik 2008, 2009b). Rather, we accept these features as a given in his initial research, and build on his results. We ask additional questions and employ somewhat different methods in a comparison and contrast of Hernik's two area examples and two additional areas, each encompassing cultural landscapes, in the USA, as we explain below. However, we concentrate on Miechów (not so much Wisniowa) and Pawhuska (not so much Lawton). We did not select the two areas we did in the USA on the basis of extensive data on rural landscapes in the political units that frame our USA choices, e.g., the central federal government of the nation-state of the USA; Oklahoma, a state in the USA; the counties, such as Osage County, including the Osage Nation, which as a *sovereign* American Indian tribal area in some sense transcends the county governance unit; the Tulsa metropolitan area, which includes Pawhuska (a municipality); Comanche County, or Lawton/Fort Sill, within that county. However, we do have available, and could use in future work, data on the Oklahoma cultural landscapes we designated as objects of interest.

In studying development issues, we assume insights from the work of the economist and philosopher Amartya Sen, particularly his books Development as Freedom (Sen 1999) and The Idea of Justice (Sen 2009). However, we do not here give any detailed exposition of that work; rather we presuppose it. Some key points we take from Sen are that economic development is both an economic and political process, and that development as a concept is value-laden. In critical evaluation and exposition of the values essential to development, one major issue is what the "democratic" role should be of government and markets, respectively, in the development process. Sen maintains in a general way that democratic government requires a development process that includes both activist government and markets. He also maintains that globalization is a major feature of our contemporary situation, both on the explanatory/interpretive side and on the normative ethics/politics side. He has something to say about energy and environmental issues, but not so much in the two books previously mentioned. He does say notable and more detailed things about environmental and energy issues in an article in the USA political periodical The New Republic (Sen 2014). There he argues plausibly that environmental issues involve much more than global warming, that a variety of energy sources is important in order to deal with developmental and environmental issues, and that a normative framework (including an ethics) for reasoning about how to make public choices about these issues is yet to be devised but is very much needed. He acknowledges the transformational importance of the 1987 Brundtland Report.

We chose the areas we did in the USA because they exemplify major issues of political economy that tend to arise both in the USA and in different versions in other nation-states in the world, including Poland. One large feature of the difference between the USA context and the Polish context is this. The USA continues on its course as a primarily capitalist nation-state with dominant established political-economic interests that heavily influence worldwide globalization, whereas Poland has undergone a transformation dating from 1989 from what was arguably a partly autonomous satellite status within the orbit of what was previously the Soviet Union. Much more recently, Russia has evidenced an interest in recovering control or influence over land near its borders, including parts of Ukraine. This probably is partly driven by Russian government concerns about NATO expansion. There are also activities by the Russian government in the Baltic exclave of Kaliningrad which have been concerning to the Poles. Much of this is well-known. It is mentioned here because Russia's actions, and its major role as a fossil-fuel energy supplier to European countries to its west, have sharpened concerns in Poland about Poland's military alliance as part of NATO, and also Poland's interest in greater autonomy about energy sources. One might add to this that there are Polish concerns about costs of oil and gas (although oil prices were slumping as we started the research), and hopes among some decision-makers in Poland that fracking for natural gas may be profitable, and furthermore prudent in pursuing greater political and economic self-sufficiency, as well as cultivating more reliable ties with the US. Then too, there are issues about environmental quality to be considered in Poland, which include addressing the question whether renewable energy sources could become a major part of the national political economy, partially enabling reducing the role of fossil fuels. The regulatory aims of the European Union, also, have included as an implication a push for growth of renewable energy sources as part of the Polish energy economy. Thus we see increasing interest in Poland in biomass as an energy source, as in Miechów, and in energy sources that may be promising for the preservation of agricultural and forested areas that contain exceptionally valuable cultural landscapes as well as serving as economic resources. Some of the collaborators on this present paper are also currently exploring in other related research the potential for development of forested areas in Poland for their potential contribution to the Polish energy economy as well as preservation of certain primarily rural cultural landscapes in that connection.

The question what is the best mixture of government activism (and what levels of government), planning, regulation, and control, on the one hand, and non-governmental institutional measures of varying types, on the other hand, frames many questions of public policy and business activity in the USA. This fact also very much affects political rhetoric. While globalization need not as a matter of logic be constituted by a complete global triumph of specifically US-style free-market capitalism with its supposedly limited role for government, that type of globalization has in empirical fact been particularly influential worldwide after the dissolution of the Soviet Union. For an anthology of papers that study different aspects of globalization, see the edited volume, Globalization (Appadurai 2001). Of particular interest is the essay by Jerome Binde, "Toward an Ethics of the Future" in the Appadurai edited volume (Binde 2001).

With one of the areas we chose, referred to here, Pawhuska and its surroundings, non-governmental factors (such as actions by a global NGO, The Nature Conservancy) have been more prominent than government (in creating and preserving a cultural landscape in the Tallgrass Prairie Preserve). With the other area (much less discussed here), and its sites in and near Lawton, federal governmental factors have been more prominent. This is partly a matter of the US military installation in Lawton, and also a matter of a major federal wildlife refuge nearby. We do not discuss the Lawton area as much in this paper, but mention it as a counterpoint to the private-power message implied by the cultural landscape at Pawhuska. These areas in and near Pawhuska and Lawton were initially chosen because they exemplify the widely supposed thematic conflict (perhaps somewhat exaggerated in the USA) between governmental and non-governmental activism and controls.

We are interested in a subject matter that includes objects of study in the physical and engineering sciences, and also human cultures studied in the social sciences. This is shown in our increased focus on participant observation, fieldwork, and globalization. Hernik's earlier methodology did include extensive personal interactions with inhabitants of rural areas concerned about the future of their rural lives. Hernik was and is interested also in the rural cultural landscapes which he has studied and in which he has intervened. As an environmental engineer, who also has a background as a lawyer, his work shows some resemblances to the methodology of much cultural anthropology in USA-UK-European styles. His earlier approach in Miechów, in particular, we supplemented with a study visit to Miechów in June, 2014, and a study visit by all co-authors to Pawhuska in February, 2015. Our interest in cultural landscapes is likely to continue to require mixed scientific methodologies from physical/biological, social, and planning/managerial sciences, as well as eclectic normative judgment methodologies.

## 24.3 Global Urbanization, Rural Cultural Landscapes, Two Małopolska Examples, and Two Oklahoma Examples

We question uncritical global trends toward overwhelming urbanization that destroys or overwhelms the culturally enabled values of rural life and rural landscapes. A comprehensive and detailed account of what these values can be (which are somewhat culturally variable, but in some ways appear universal) is beyond our aims here. But part of it is an appreciation and love of many aspects of non-human nature, a desire for an economy that allows humans to take up work that directly cultivates and expresses respect for such nature, and that holds out the possibility to promote sustainably the continuation of a way of life in which there is respect and love (even awe) for much of the non-human-created world, and positive attitudes about the human world of transformative work done with a positive relationship toward nature, especially work to accomplish the social provision of food, as well as other central features of our human way of existence as embodied parts of nature. This includes our direct experience (insofar as that is possible) of non-human nature as well as natural aspects of our humanity.

The trends towards greater urbanization are operating in many areas of the world (China, India, Africa, Latin America, etc.), as well as in Poland (Małopolska) and the USA (Oklahoma). Undoubtedly there are major benefits that come from urbanization. But in order for counterbalancing urban–rural, rural (and other less-human-dominated), cultural landscapes to flourish, rural livelihoods are among the vital necessary conditions. The loss of rural residents to city life should not be allowed to depopulate rural areas, or transform too many of them into mere suburbs for cities such as Kraków where commuters engage in their urban working lives and return to an ever-vanishing countryside, where their suburbanizing demand for urban amenities contributes to the loss of the identity of the rural cultural landscape.

We want to compare Małopolska and Miechów in particular to very different cultural landscapes in the USA. The primary one is within Oklahoma, a state itself within the USA, and more specifically Osage County and the town of Pawhuska within Osage County, in Oklahoma. Pawhuska is the main town near or in The Nature Conservancy's Tallgrass Prairie Preserve. The Nature Conservancy is a large private environmental NGO with activities in various parts of the world (Nature Conservancy 2014). A CEO of the Nature Conservancy (a former executive at the major financial services firm Goldman Sachs) has argued that partnerships between corporations and environmental organizations are essential to environmental protection. His contentions about this are worth a separate study and examination (Max 2014; Tercek and Adams 2013).

The Tallgrass Prairie Preserve in Oklahoma is in Osage County. Osage County itself is or overlaps largely with (or is superimposed on) the Osage Indian tribe's reservation. Technically, and some would insist on the normative force of this, the Osage Nation is a sovereign nation under treaties with the USA (Osage Nation 2014). For a discussion of sovereignty as applied to Indian Nations within USA borders, see Hester (2001). The history of Osage County includes the momentous discovery and exploitation of oil and gas as part of the mineral resources of the Osage Nation. At one time, the Osage tribe was considered very rich. The tribe has an elected Minerals Council that oversees its use of oil and gas resources. The tribe has other economic ventures including smoke shops and gambling casinos (so common on Indian reservations in the USA). The tribe retains control (in a sense) of subsurface mineral resources. But leases for exploitation of the oil and gas resources are auctioned off to non-tribal members. A consequence is that on the large expanse of the Tallgrass Prairie Preserve (much of it purchased back by The Nature

Conservancy from non-Indian ranch owners) there are signs with the names of the former ranches in evidence. There on the Preserve landscape a buffalo herd co-exists with plots of land devoted to exploitation of subsurface oil and gas reserves. There is some environmental and nature-oriented tourism in the area, and a small shop, with, among other things, a very modestly produced booklet, probably outdated, attributed to the Oklahoma Geological Survey, a state governmental agency, about topography and geological formations in the area (Suneson 2000). The Preserve also extends into the state of Kansas, but here we focus on the Oklahoma component of the Preserve.

Hernik is surely correct to argue that business or (perhaps more broadly) economic activity (including governmental activism, and NGO spending, as well as for-profit business activity) is necessary to protect or improve cultural landscapes, including rural landscapes such as Miechow and Wisniowa. (This view of Hernik's is somewhat akin to part of Sen's outlook). This argument could be interpreted as an extension of a claim (or practical working hypothesis) about an overall conceptual framework. It is sometimes claimed that there is a conflict or conflicts between economic development on the one hand and environmental protection on the other hand. This claim is often made, for example, in political policy discussions or disagreements in which environmental protection measures taken by government are said to interfere with business or economic activity, and to threaten jobs. This may be true in some social circumstances, e.g., if an undue legal insistence on badly designed environmental impact studies, unfortunately, may prevent the building of infrastructure much needed for economic development in a particular region. Or, e.g., emphasis on coal mining and coal use may damage the environment, but shutting down or severely restricting coal mining activity (e.g., in Silesia in Poland, or for that matter, in West Virginia in the USA) would be very damaging to employment, in the absence of other societal measures. However, in the longer term, arguably there need be no persistent basic conflict between economic development and environmental protection. Thus, here we can recall the remark that the pilot projects showed that it is possible to preserve the cultural landscape while obtaining economic benefits (Hernik 2012). However, we should concede that this may require abandoning the ideal of corporate profit-maximization. Better to adopt an optimizing rather than a maximizing goal for social or corporate policy. (This is easily said, hard to imagine realizing in the current world!). We are here closely associating the preservation of cultural landscapes with environmental protection. Indeed, preservation of cultural landscapes can itself be one type of environmental protection.

## 24.4 Renewable Energy Sources and Fossil Fuels

One important aspect of environmental protection/improvement is widely (but of course not universally) thought to involve greater reliance by society on renewable sources of energy in contrast to current dependence on fossil fuels. For a valuable study of energy in contemporary political economy, Daniel Yergin's work is worth consulting, especially his book The Quest (Yergin 2011). The importance and relevance of renewables is sometimes questioned, of course. However, Yergin writes:

"Even as late as 2010, renewables accounted for only 8% of the U.S. energy supply—about the same share it had in 1980. Remove two items – hydropower (which has been constant for many years) and biomass (primarily ethanol) – and renewables in 2009 constituted less than 1.5% of the total U.S. energy supply. Much the same holds true around the world. (…) Yet today renewables are reenergized to become a growing part of energy supply, embraced as a key solution to the triple challenges of energy supply, security, and climate change" (p. 524).

Yergin is quite sympathetic to what he studies the most, the fossil-fuel industrial complex. However, clearly he acknowledges the need for renewables and its connection with addressing environmental problems.

## 24.5 Renewables and Fossil Fuels in a Variety of Types of Social Systems

The goal of advancing renewables in relation to fossil fuels needs to be pursued in a variety of types of social systems, including Central/Eastern European contexts such as Małopolska (with a tradition of comparatively more regulation, perhaps continuing), and USA contexts such as Osage County (comparatively less state-level regulation, perhaps, in light of current institutions and attitudes, despite the presence of the federal government's Department of the Interior and the Bureau of Indian Affairs), or the mixed civilian and military landscape of Lawton (and Fort Sill), probably both less regulated (civilian Oklahoma context) and at times more highly regulated (federal military), respectively. In fact, both areas, Małopolska (with Miechów and Wiśniowa) and Oklahoma (including Pawhuska/Osage County, and Lawton/Fort Sill/Comanche County) are at present part of different brands of often-avowed *free market* economies evolving in rather different ways, from different histories (towards what?—we do not discover that here!). This may tell us something interesting about the potential future directions for various components of a globalizing political economy, in which there are choices possible about what markets can and should consist in, and the place of markets in an overall institutional mix, and in a governance system. The nature of capitalism and markets, their uses of fossil fuels and renewables, their care for (or carelessness about) the environment and specifically cultural landscapes, their regard or contempt for rural areas, will probably not be uniform throughout the world. This may yield interesting options for political economy and culture.

Thus, the choice of these two parts of the world for comparison and contrast is not arbitrary, but part of an examination of government-and-market societal governance options in a context of globalization, in which varieties of self-described *free-market* capitalism seem to be increasingly the most influential and ascendant organizational forms. And to reiterate, there seems to be a growing conflict between those who advocate one general sort of approach to energy policy (fossil-fuel emphasis) and those who advocate another (renewables). That is so even in the formidably important case of China, which says it is continuing to govern itself under socialism with Chinese characteristics, but where the Chinese Communist Party is an obviously very powerful feature of an overall political economy and cultural system that has also furthered markets as a major factor in development processes. And in China, too, as we all know, energy choices have had environmental impacts that have had some problematic and some promising consequences for valuable cultural landscapes. China would be an important region for extension of the present inquiry about economic development and cultural landscapes (including rural landscapes). The main point of mentioning China here is that like Poland, the type of "capitalism" in China that is emergent there reflects unique national and area characteristics, and is not reducible to the abstraction of *free-market capitalism* that is sponsored by many USA politicians and oligarchs.

## 24.6 Sustainable Development as a Mediating Value Between Economic Development and Environmental Protection/Improvement

The value of sustainable development got its canonical formulation in the Brundtland Report of 1987, produced under the auspices of a UN commission led by Gro Harlem Brundtland. (See www.un-documents.net/our-common-future.pdf). It should be noted that Brundtland herself must have been affected by Norway's political economy and culture. Norway itself has developed its own mode of capitalism, notably including a sovereign wealth fund enhanced largely by fossil-fuel derived wealth. (Very recently, however, Norway has decided to divest itself of investment in certain companies that are dependent on coal). The Brundtland Report diplomatically advises that the world should continue to use fossil fuels, but should also explore expansion of renewables.

## 24.7 Renewables and Fossil Fuels Again

To give some more social context, the state of Oklahoma for years has had an economy with powerful political-economic interests centered on the oil and gas industry. Those interests make their influence felt within the state and even in federal law and policy. However, there is also an important agricultural and ranching sector for economic activity in Oklahoma. Osage County fits into this overall picture. Yet despite the emphasis on fossil fuels, there are also initiatives in Oklahoma driving development of renewable energy sources, including wind power, biomass, biofuels, and solar energy. Nonetheless, renewable energy sources have had to be developed in a business or economic environment often unsympathetic to alternatives to oil and gas. In recent years, there have been controversies over companies that wish to exploit the potential of wind power in the Osage County area. This sort of renewable energy, as developed under these circumstances, somewhat paradoxically, has been resisted significantly within a cultural landscape devoted to landscape preservation and environmental protection. The Osage Nation has expressed both some sympathy for renewables but also concern about sacred sites that some say need to be protected from disturbance from inappropriate wind energy development. The Nature Conservancy has also expressed opposition to what it considers inappropriate wind energy industrial development (Maune 2014; Wertz 2014).

During early 2015, the conflict between more entrenched oil and gas extraction and newly built wind turbines was highlighted, and was quite apparent during a research team visit to Osage County in February, 2015. Oklahoma has been counted in some accounts as the fourth largest state in wind energy activity, an aspect of development that did not start in 2015, but that has increased markedly in Osage County. In the course of these changes, some critics of wind turbines have alleged damage to the cultural landscape. Some of the criticism reveals a visual emphasis in some conceptions of cultural landscapes that has perhaps always been there. Arguably, the visual aspects of a cultural landscape need to be evaluated in conjunction with less exclusively visual and sensory criteria. Arguably, sustainable development, a widely affirmed value, also requires attention to non-visual, non-sensory aspects of cultural landscapes.

It remains to be seen how the conflict in Oklahoma between renewable wind energy and oil and gas reserves will unfold. State government has already made some changes in tax law deemed unfavorable to solar energy, and has tax policies deemed by some observers as increasingly and excessively favorable to oil and gas interests. Besides this more local situation, there has been interest by a large multi-national conglomerate, Enel (based in Italy, but with a presence in 40 countries, including the USA), in wind energy development in Osage County projects. Enel, like many corporations, publicizes an interest in renewable energy, sustainable development, technological innovation, and so forth (www.enel.com/en-GB/group).

It is especially worth mentioning here that there is also perhaps an inherent conflict between a cultural landscape (such as the Tallgrass Prairie Preserve) that is devoted to environmental protection but where there is a fundamental tie with extraction of oil and gas to be sold to a world in which certain fossil fuels have such a major and environmentally problematic impact. It seems to be in the nature of some cultural landscapes such as Osage County that they play a conflict-laden role with regard to economic development and environmental protection, and the messages they convey. A careful description, analysis, and evaluation of the cultural landscape of The Nature Conservancy's Tallgrass Prairie Preserve in Pawhuska, Oklahoma yields the claim that preserving the landscape in certain forms through business or economic activity and technological innovation can appear to contribute to environmental protection, but that the full truth is more complex than this. The Preserve co-exists with and its organization facilitates economic activity that contributes to continuing primary use of fossil fuels as energy sources and consequent environmental damage, both within and outside Osage County itself and its Tallgrass Prairie Preserve. The network of economic and cultural interests in Osage County is somewhat resistant to at least one sort of corporate activity that appears to have the promise of furthering development of a renewable energy source (wind power). It is true that *renewables* can sometimes themselves contribute to environmental damage, some of it more specifically damage to cultural landscapes. This is familiar, for example, with some large hydroelectric dam projects. That was commented on in the Brundtland Report. Nonetheless, many renewables, appropriately developed and used, may well help substantially with environmental protection/improvement and with preservation or improvement of cultural landscapes.

As Miechów is not too far from Kraków, a city of about one million in the metropolitan area, Pawhuska is not too far from Tulsa, a metropolitan area similar to Kraków in the metropolitan area. There seems to be much more pressure towards suburbanization in the Polish case, and comparative underdevelopment, even stagnation in the Pawhuska case. Pawhuska is small and thinly populated, arguably economically underdeveloped, and seems likely to remain so in the near future, despite its proximity to Tulsa, Oklahoma. (Pawhuska is counted in some inventories as part of the Tulsa metropolitan area).

We think it is vital to resist the widespread global trends towards overwhelming urbanization and to plan for the conservation and re-invigoration of rural and quasi-rural areas by preserving and actively constructing cultural landscapes made viable through economic development. However, the cultural landscapes can also be truthful and point to a sustainable and just future.

So both Miechów and Pawhuska are areas in which the issue of rural (or at any rate not entirely urban) cultural survival is posed, and some of the key questions are about energy and the environment. Oklahoma also happens to be, because of its somewhat reactionary but also episodically dynamic oil and gas emphasis (an industry in 2016 in the doldrums), a region with peculiarly problematic cultural landscapes. And the state, once rural, but now tending to urbanize, still holds onto its more rural past, while rural communities often struggle to survive and their advocates nurse hopes for them to flourish.

A point of interest, especially significant for our current research: Miechów and Pawhuska deserve comparison and contrast in part because their histories of disrupted rural life, conflict-ridden jurisdictions, and their vulnerable cultural landscapes may shed light on how the paradigm shift (to borrow the famous phrase from Kuhn) from fossil fuels towards much greater incorporation of renewables might occur (or be stymied). The focus in this paper is both on preserving cultural landscapes and on cultivating constructive future-oriented attitudes about economic development and cultural landscapes. This is aligned with Appadurai's stress on the future. The interest in preservation but with an emphasis on possibilities for future-oriented economic development is also there in Hernik's pioneering work. It can be supplemented in work by cultural anthropology on the future. In this connection, Arjun Appadurai's work in The Future as Cultural Fact is particularly suggestive (Appadurai 2013). The awareness of different attitudes about the future (not reducible to a predictive stance), the ethics of our culturally based attitudes about the future, along with the inclusion of attention to the past, heritage, etc., in our stance towards the future, are all of the utmost importance for the theory and practice of planning and management about economic development and cultural landscapes.

A briefer comparison and contrast can also be sketched between Wiśniowa and Lawton, Oklahoma. Wiśniowa is considered as a landscape in which agro-tourism has the potential to promote economic development and protection of the environment (www.ug-wisniowa.pl). The reference to Lawton picks out a very different cultural space which serves as a contrasting example. We are actually referring to a fairly large urban area by Oklahoma standards, and one dominated by the military functions of the US government's Fort Sill. But not far away is the Wichita Mountains Wildlife Refuge, which is a federal area established years ago for conservation purposes under former President Theodore Roosevelt (U.S. Fish and Wildlife Service (b), Wichita Mountains Wildlife Refuge 2014). Business people and politicians in Lawton have played up energy businesses of both the fossil fuel and emerging renewable varieties. Tourism and the wildlife refuge are also emphasized. Lawton itself (in Comanche County) is not an isolated area. This is not, like Wiśniowa in Małopolska, in the Beskidy Mountains, a relatively isolated area. The refuge is administered by the U.S. Fish and Wildlife Service, who by their own account *steward National Wildlife Refuges for the benefit of wildlife and people* (U.S. Fish & Wildlife Service (a) 2014).

That the Lawton municipal area also has referred significantly in its development planning to emerging renewable energy activities must also be partly attributable to the results of former President Obama's executive orders which furthered focus on renewable energy in US military contexts, and which have been in part an effort to appease Obama's partially environmentalist electoral constituencies. In the era of Donald Trump, of course, the pro-fossil-fuel emphasis has been very different, and would require re-analysis. Obama also faced pressures from other groups to facilitate fossil-fuel industry activity on federal lands. On balance, however, one must recognize both Fort Sill and the Wildlife Refuge as cultural landscapes in which the balance between emphasis on renewable energy and fossil fuels has at times manifested some movement towards the renewables side, despite the locally and even globally powerful oil and gas interests (allied with Trump) influential in the state of Oklahoma.

This strong federal governmental role in Lawton/Fort Sill/Wichita Mountains Wildlife Refuge might be contrasted with the private activism that resulted in the Tallgrass Prairie Preserve and that today embraces advocacy of one style of corporate environmentalism. The interest in renewables in Lawton that results from Fort Sill priorities might of course all too readily be reversed with a change of elected federal administration in Washington, DC. It might also be mentioned that due to its history, in Poland, subsurface mineral rights belong to the government. Thus again, depending on politics, there is profound potential there to alter through central government action the comparative role of renewables versus oil and gas as energy sources, with concomitant business and economic implications.

## 24.8 Institutional Mixtures, Governance Structures, Cultural Landscapes, and Energy Sources

What is obviously salient in this paper is the importance of good decision-making about institutional mixtures and governance structures in any of the quite different cases. The attitude expressed in this paper is that institutions and governance (across a wide variety of histories and regulatory regimes) will need to evolve in complex ways, free from dogmatism about government and markets, for example, and responsive to very particular (and general) local as well as more general (and in some respects particularized) global issues.

In the Polish case, even 30 years since the anti-Soviet, arguably anti-colonialist Transformation of 1989 (in which Poland was empowered by reference to Russia), which has so much changed the country, as well as Poland in a more recent year (2004) becoming a member of the European Union, there are very important and basic questions about re-privatization and land use management. These are fundamental questions of spatial planning. In the USA case, a markedly different vocabulary and politics prevails. However, with globalization, comparisons/contrasts between very different contexts within political economies can generate insight into needed changes in decision-making systems anywhere. Such changes include alterations in the institutional and governance structures that have significant roles in energy policies, so basic for modern political economies and for cultural landscapes.

## 24.9 Conclusions

Land use can be a main element in an environmental studies framework that is innovative and fruitful both for research and for development policy as well as pragmatic real-world interventions. Globalization, historical perspective, and energy issues are particularly important in further work in this type of framework. Comparisons and contrasts of differing land use centered environmental issues (e.g., in Poland and the US) can be productive. Such research can assist in better decision-making by governments at different levels and by non-governmental organizations.

## References

Appadurai A (2001) Globalization. Duke University Press, Durham and London

Appadurai A (2013) The Future as cultural fact. Verso, New York

Binde J (2001) Toward an ethics of the future. In: Appadurai A (ed) Globalization. Duke University Press, Durham and London, pp 90–113

Brundtland GH et al (1987) Report of the World Commission on Environment and Development: Our Common Future. http://www.un-documents.net/our-common-future.pdf. Accessed 14 May 2019

Enel (2019) http://www.enel.com/en-GB/group. Accessed 14 May 2019
European Landscape Convention, Florence (2000) www.coe.int/europeanlandscapeconvention. Accessed 14 May 2019
Gawroński K, Hernik J (eds) (2010) Spatial planning and management as an instrument of shaping cultural landscapes (Polish and International Experiences). ficyna Wydawnicza Branta, Bydgoszcz-Kraków
Geertz C (1973) The interpretation of cultures. Basic Books, New York
Geertz C (2000) Available light-anthropological reflections on philosophical topics. Princeton University Press, Princeton
Google Maps and Google Images. Visual information about Miechów, Pawhuska, and other areas referred to. http://www.google.com. Accessed 14 May 2019
Hernik J (2008) Potrzeba uwzględniania walorów krajobrazu kulturowego w zarządzaniu gminą wiejską. Prace Komisji Krajobrazu Kulturowego 10:61–68
Hernik J (ed) (2009a) Cultural landscapes across disciplines. Oficyna Wydawnicza Branta, Bydgoszcz-Kraków
Hernik J (2009b) Activities for the protection and preservation of cultural landscapes. In Hernik J (ed) Cultural landscapes across disciplines. Oficyna Wydawnicza Branta, Bydgoszcz-Kraków, pp 61–78
Hernik J (2012) Protecting cultural landscapes in rural areas by economic means. Annals Warsaw Univ Life Sci SGGW Horticul Landscape Architect 33:105–112
Hernik J, Pijanowski JM (eds) (2007) Cultural landscape—assessment, protection, shaping. Wydawnictwo AR w Krakowie, Kraków
Hernik J, Gawroński K, Dixon-Gough R (2013) Social and economic conflicts between cultural landscapes and rural communities in the English and Polish systems. Land Use Policy 30:800–813
Hester TL (2001) Political principles and Indian sovereignty. Routledge, New York
IHS Cambridge Energy Research Associates. http://www.ihs.com/index.aspx. Accessed 14 May 2019
Kuhn T (1962) The structure of scientific revolutions. University of Chicago Press, Chicago
Kühne O, Gawroński K, Hernik J (eds) (2015) Transformation und Landschaft. Springer, Wiesbaden
Maune T (2014) Osage County votes "no" on controversial wind farm. http://www.wind-watch.org/news/2014/05/08/osage-county-votes-no-on-controversial-wind-farm. Accessed 14 May 2019
Max DT (2014) Green is good. The New Yorker, May 12, 2014
Radecki-Pawlik A, Hernik J (eds) (2010) Cultural landscapes of River Valleys. Publishing House of the University of Agriculture in Kraków, Kraków
Rittel HWJ, Webber MM (1973) Dilemmas in a general theory of planning. Policy Sci 4:155–169
Sankowski E, Harris B, Hernik J (2016) Some Problems and Possibilities for Sustainable Development. Publishing House of the University of Agriculture, Kraków in Kraków
Sassen S (2009) Chapter in the encyclopedia of life support systems (EOLSS): http://www.eolss.net/sample-chapters/c14/E1-18.pdf. Also see context of the chapter in: Human Settlement Development, Vol. I, ed. Saskia Sassen, EOLSS Publishers/UNESCO, 2009. Accessed 23 May 2016
Sassen S (2012) Cities and the biosphere. http://www.saskiasassen.com/PDFs/publications/cities-and-the-biosphere.pdf
Sassen S (2012) Cities in a world economy, 4th edn. Sage, Los Angeles
Sen A (1999) Development as freedom. Knopf, New York
Sen A (2009) The idea of justice. Harvard University Press, Cambridge
Sen A (2014) Global warming is just one of many environmental threats that demand our attention. The New Republic. http://www.newrepublic.com/article/118969/environmentalists-obsess-about-global-warming-ignore-poor-countries. Accessed 14 May 2019
Sołtys M, Jaszczyński K (2012) 1947 The colors of ruin. The Reconstruction of Warsaw and Poland in the Photographs of Henry N. Cobb. History Meeting House (DSH), Municipal Cultural Institution, Warszawa

Suneson NH (2000) The geology of the Tallgrass Prairie Preserve, Osage County, Oklahoma. Oklahoma Geological Survey Open-File Report. http://www.ogs.ou.edu/pubsscanned/openfile/OF1_2000.pdf. Accessed 14 May 2019

Tercek MR, Adams JS (2013) Nature's fortune: how business and society thrive by investing in nature. Basic Books, New York

The Nature Conservancy. http://www.nature.org/. Accessed 14 May 2019

The Osage Nation. http://www.osagenation-nsn.gov/. Accessed 14 May 2019

U.S. Fish & Wildlife Service (a). National Wildlife Refuge System. http://www.fws.gov/refuges/. Accessed 22 May 2019

U.S. Fish & Wildlife Service (b). Wichita Mountains Wildlife Refuge/Oklahoma. http://www.fws.gov/refuge/wichita_mountains. Accessed 14 May 2019

Van Der Valk A (2009) Multiple cultural landscape: research and planning for living heritage in the Netherlands. In: Hernik

Wertz J (2014) Why Oklahoma's wind energy future could be shaped by Osage County. http://www.wind-watch.org/news/2014/07/24/why-oklahomas-wind-energy-future-could-be-shaped-by-osage-county/. Accessed 22 May 2019

Wiśniowa. http://www.ug-wisniowa.pl. Accessed 14 May 2019

Xiang WN (2013) Working with wicked problems in socio-ecological systems: awareness, acceptance, and adaptation. Landsc Urban Plan 110:1–4

Yergin D (2011) The quest: energy, security, and the remaking of the modern world. Penguin Press, New York

# Summary

This book includes a set of research studies that can be interpreted as contributing to definition of a framework for further interdisciplinary work on cultural heritage, land use, and sustainable (societal) development. The chapters communicate various ideas and extensive data about their own distinctive topics (sometimes quite technical in a social scientific or natural scientific or what could be called a "professional practice" sense). For example, some chapters are intensively about biological subject matter, and sometimes a specifically bio-genetics mode of biology. Other chapters are about topics dealt with in professional practice that has scientific elements, e.g., topics involving spatial planning, agriculture, environmental engineering, veterinary sciences, urban and regional planning, landscape architecture, tourism, real estate, energy issues, etc. (This list is not exhaustive). There are chapters with mixtures of social science, natural science, and professional practice subject matter. There is some esthetics and humanities content in some contributions. Each chapter is innovative in its own right. As a totality, the book also offers an emergent novel perspective. Perhaps, and even likely, there are some precedents for that evolving perspective. That may be a matter for later investigations after the publication of this volume.

The book is diverse in the geographical settings and the cultural boundaries of its subjects. This is quite deliberate, and is intended to encourage comparative studies. Sometimes, too, the studies (taken individually or as sets of chapters) are not exactly comparative, so much as interestingly varied studies of subject matter set in different locales, at varying scale. The chapters themselves are thus individually sometimes studies of dispersed cultures and dispersed terrestrial areas. The dispersed loci can be regarded as invitations to more targeted comparative research.

We would hope that these aforementioned features of the volume may further academic and other contacts, and enable progress in dawning perceptions of possible connections (e.g., similarities and differences) among different places, each place associated with some (perhaps multiple dimensional) cultural heritage. We would hope that this progress will further contacts that gain from both more local and wider, even more global or internationally dispersed foci. Through such work, the

© Springer Nature Switzerland AG 2021

J. Hernik et al. (eds.), *Cultural Heritage—Possibilities for Land-Centered Societal Development*, Environmental History 13,
https://doi.org/10.1007/978-3-030-58092-6

book, we hope, will also productively provoke progress in active cooperation among not only academics, but also governments (at different levels), global-governance organizations such as the UN and its affiliates (UNESCO, FAO, etc.), and independent nonprofit or for-profit non-governmental organizations, or for-profit corporate or other commercial entities (the last type of entities, including unincorporated family farms and small businesses).

In an age of globalization, a complex concept often invoked nowadays, there is also a countervailing tendency to preserve what is more localized or regional and culturally unique. Even some more nation-state defined foci tend this way. We have all recently observed some political, economic, and cultural pushback against globalization (e.g., Brexit), that in some cases rather paradoxically contains elements of variant types of globalization (at least, global contacts) itself, despite its occasionally explicit "anti-globalist" language and practices. We do think, however, that a continuing concern for preservation of unique and more local and regional (or even nation-state) cultural heritage may sometimes be a constructive feature of such pushback. In this book, regions in Central/Eastern Europe are often among the territorial foci of research studies of specific aspects of land-centered cultural heritage. But we also include research here about China, Italy, South Africa, and a concluding self-avowedly comparative chapter about both global and local/regional phenomena in Poland and the US. Our general perspective might have begun by soliciting other territorial research examples. Rather than arbitrariness, this is an indication that variants of the present perspective here could be employed to study other sets of localities, in different sequences, with distinctive rationales, and with other types of outreach to potential research groups with distinctive agendas. It happens, as a matter of contingent fact, that the initial choice of topics here began with what might be called The Krakow Group (a land-centered development research group, many with roots at the University of Agriculture in Krakow) whose attention then extended to other areas in Malopolska, to other parts of Poland, as well as Czechia and Slovakia, to Italy, and to the US and China.

There has been a notable trend in thought and decision-making about societal development, and specifically sustainable development, towards more attention to cultural topics. This can be illustrated in various UN conferences and statements, for example. What is included in referring to "culture" has varied contextually. Not only arts and education, but broader notions of culture, and cultural heritage, have been common.

Overcoming possible tensions between the past as a source of cultural heritage, so often valuable, on the one hand, and a future in which a culture may remember its past, but also in which people will generate new forms of interactions, may well underlie major dimensions of this volume as a whole, as well as some of its chapters.

*Prof. Edward Sankowski, University of Oklahoma, USA*
*Prof. Józef Hernik, University of Agriculture in Krakow, Poland*

The manufacturer's authorised representative in the EU is Springer Nature Customer Service Centre GmbH, Europaplatz 3, 69115 Heidelberg, Germany. If you have any concerns regarding our products, please contact ProductSafety@springernature.com

Printed and bound by CPI Group (UK) Ltd, Croydon, CR0 4YY
28/09/2026
02224065-0004